Physics on Manifolds

MATHEMATICAL PHYSICS STUDIES

Series Editor:

M. FLATO, *Université de Bourgogne, Dijon, France*

VOLUME 15

Physics on Manifolds

Proceedings of the International Colloquium in honour of Yvonne Choquet-Bruhat, Paris, June 3–5, 1992

edited by

M. Flato
Physique Mathématique,
Université de Bourgogne, Dijon, France

R. Kerner
Laboratoire de Gravitation et Cosmologie Relativistes,
UPMC, Paris, France

and

A. Lichnerowicz
Collège de France, Paris, France

SPRINGER SCIENCE+BUSINESS MEDIA, B.V.

Library of Congress Cataloging-in-Publication Data

Physics on manifolds : proceedings of the international colloquium in honour of Yvonne Choquet-Bruhat, Paris, June 3-5, 1992 / edited by M. Flato, R. Kerner, and A. Lichnerowicz.
p. cm. -- (Mathematical physics studies ; v. 15)
English and French.
Includes bibliographical references.
ISBN 978-0-7923-2500-0 ISBN 978-94-011-1938-2 (eBook)
DOI 10.1007/978-94-011-1938-2
1. Manifolds(Mathematics) 2. Mathematical physics. I. Choquet-Bruhat, Yvonne. II. Flato, M. (Moshé), 1937- . III. Kerner, R. IV. Lichnerowicz, André, 1915- . V. Series.
QC20.7.M24P48 1993
530.1'515--dc20 93-31133

Printed on acid-free paper

TABLE DES MATIERES

Tables rondes

Equations différentielles aux dérivées partielles *(présidée par J. VAILLANT)*

Physique Mathématique *(présidée par G. PICHON)*

AVANT-PROPOS

Ce volume contient les actes du Colloque "Analyse, Variétés et Physique", organisé en l'honneur d'Yvonne Choquet-Bruhat, les 3, 4 et 5 juin 1992 à Paris, par ses amis, collaborateurs et anciens élèves. Son titre reflète assez bien les domaines d'activité dans lesquels Yvonne Choquet-Bruhat apporta des contributions essentielles.

Depuis l'avènement de la Relativité Générale, la géométrie des variétés est devenue une partie non triviale de la physique de l'espace-temps. L'analyse fonctionnelle a trouvé en même temps une application essentielle en mécanique quantique, puis en théorie quantique des champs. Son apport devient décisif au moment où l'on considère le comportement *global* des solutions des systèmes différentiels sur les variétés. La Relativité Générale est dans ce sens une théorie exceptionnelle, dans laquelle les solutions d'un système hautement non linéaire d'équations aux dérivées partielles définissent elles-mêmes la variété sur laquelle elles sont supposées exister. C'est pourquoi aucune des solutions des équations d'Einstein ne peut être interprétée physiquement avant que son comportement global, c'est-à-dire sur l'ensemble de la variété hypothétique, ne soit connu.

C'est à ce domaine, se situant au carrefour entre la physique et les mathématiques, que Yvonne Choquet-Bruhat a contribué, de manière spectaculaire dans sa jeunesse, en donnant la preuve de l'existence des solutions des équations d'Einstein sur des variétés différentiables assez générales. Les méthodes qu'elle avait créées ont été élaborées par l'école française, principalement par Jean Leray. Cette première preuve de l'existence et de l'unicité locale pour les équations d'Einstein a été le moteur de la théorie des systèmes hyperboliques généraux de Jean Leray.

Yvonne Choquet-Bruhat a répandu dans la communauté scientifique l'idée de la nécessité de l'étude des problèmes globaux imposés par la physique de systèmes couplés à la gravitation. En ce qui concerne le système d'équations d'Einstein, le but a été atteint très récemment par les travaux de S. Klainerman et de D. Christodoulou, ancien collaborateur d'Yvonne Choquet-Bruhat.

Face au développement de la physique contemporaine, Yvonne Choquet-Bruhat a toujours joué un rôle de pionnier dans l'étude de problèmes globaux, notamment dans la théorie des champs de jauge et la supergravité. Parallèlement, elle a construit une théorie des ondes asymptotiques gravitationnelles et électromagnétiques. Son influence s'est aussi fait sentir par l'élaboration de livres de base, devenus classiques, consacrés à l'analyse fonctionnelle sur les variétés et à ses applications dans différents domaines de la physique.

Yvonne Choquet-Bruhat s'est aussi dévouée à la formation de nombreux jeunes chercheurs français et étrangers, dont certains sont devenus des maîtres.

Sa modestie n'a jamais occulté ses qualités humaines exceptionnelles et son sens profond de l'amitié envers ses collaborateurs, élèves et amis.

Nous dédions ce livre, qui donne la mesure de son rayonnement dans la communauté scientifique, à Yvonne Choquet-Bruhat, en souhaitant qu'il soit digne de sa personne et de son oeuvre.

M. Flato, R. Kerner, A. Lichnerowicz

FOREWORD

This volume contains the proceedings of the Colloquium "Analysis, Manifolds and Physics" organized in honour of Yvonne Choquet-Bruhat by her friends, collaborators and former students, on June 3, 4 and 5, 1992 in Paris. Its title accurately reflects the domains to which Yvonne Choquet-Bruhat has made essential contributions.

Since the rise of General Relativity, the geometry of Manifolds has become a non-trivial part of space-time physics. At the same time, Functional Analysis has been of enormous importance in Quantum Mechanics, and Quantum Field Theory. Its rôle becomes decisive when one considers the *global* behaviour of solutions of differential systems on manifolds. In this sense, General Relativity is an exceptional theory in which the solutions of a highly non-linear system of partial differential equations define by themselves the very manifold on which they are supposed to exist. This is why a solution of Einstein's equations cannot be physically interpreted before its global behaviour is known, taking into account the entire hypothetical underlying manifold.

In her youth, Yvonne Choquet-Bruhat contributed in a spectacular way to this domain stretching between physics and mathematics, when she gave the proof of the existence of solutions to Einstein's equations on differential manifolds of a quite general type. The methods she created have been worked out by the French school of mathematics, principally by Jean Leray. Her first proof of the local existence and uniqueness of solutions of Einstein's equations inspired Jean Leray's theory of general hyperbolic systems.

Yvonne Choquet-Bruhat has promoted in the scientific community the idea of the necessity of studying the global problems posed by the physics of systems coupled with gravity. A complete study of global properties of solutions to Einstein's equations has been achieved recently by S. Klainerman and by her former collaborator D. Christodoulou.

With regard to recent development of contemporary physics, Yvonne Choquet-Bruhat has played a pioneering rôle in the study of global problems, especially in gauge field theory and supergravity. At the same time, she has elaborated a theory of asymptotic gravitational and electromagnetic waves. Her influence has also been felt through new basic books, which have since become classics, concerning functional analysis on manifolds and its applications in various domains of physics. Yvonne Choquet-Bruhat has also devoted herself to the development of many young research students in France and abroad, some of whom have become masters.

Her modesty has never eclipsed her truly exceptional human qualities and her profound sense of friendship towards her collaborators, students and friends.

We dedicate this book, which illustrates the full measure of her influence in the scientific community, to Yvonne Choquet-Bruhat, and we hope that it will be worthy of her and her work.

M. Flato, R. Kerner, A. Lichnerowicz

COMITE D'HONNEUR

Bryce DEWITT (Austin, USA)
Israël GELFAND (Piscataway, USA)
Paul GERMAIN (Paris, France)
Jean LERAY (Paris, France)
André LICHNEROWICZ (Paris, France)
Jacques-Louis LIONS (Paris, France)
Irving E. SEGAL (Cambridge, USA)
Abraham H. TAUB (Berkeley, USA)

COMITE D'ORGANISATION

Daniel BANCEL (Lyon, France)
Moshe FLATO (Dijon, France)
Krzysztof GAWEDZKI (Bures s/Yvette, France)
Richard KERNER (Paris, France)
Lise LAMOUREUX-BROUSSE (Paris, France)
André LICHNEROWICZ (Paris, France)
Charles-Michel MARLE (Paris, France)
Guy PICHON (Villetaneuse, France)
Jean VAILLANT (Paris, France)

LISTE DES CONFERENCIERS

(conférences plénières et tables rondes)

Peter C. AICHELBURG (Vienne, Autriche)
Angelo M. ANILE (Catane, Italie)
Haïm BREZIS (Paris, France)
Francis CAGNAC (Yaoundé, Cameroun)
Esteban CALZETTA (Buenos Aires, Argentine)
Erik A. CHRISTENSEN (Lyngby, Danemark)
Demetrios CHRISTODOULOU (Princeton, USA)
Thibault DAMOUR (Bures-sur-Yvette, France)
Cécile DEWITT-MORETTE (Austin, USA)
Giorgio FERRARESE (Rome, Italie)
Antonio GRECO (Palerme, Italie)
ChaoHao GU (Hefei, Rép. Pop. de Chine)
Nail H. IBRAGIMOV (Moscou, Russie)
Jim ISENBERG (Eugene, USA)
Sergiu KLAINERMAN (Princeton, USA)
Jean LERAY (Paris, France)
Pierre-Louis LIONS (Paris, France)
Waichiro MATSUMOTO (Ohtsu, Japon)
Gérard A. MAUGIN (Paris, France)
Vincent MONCRIEF (New Haven, USA)
Carlos MORENO (Madrid, Espagne)
Olga A. OLEINIK (Moscou, Russie)
Guy PICHON (Villetaneuse, France)
Tommaso RUGGERI (Bologne, Italie)
Irving E. SEGAL (Cambridge, USA)
Nail R. SIBGATULLIN (Moscou, Russie)
Abraham H. TAUB (Berkeley, USA)
Clifford TAUBES (Cambridge, USA)
Jean VAILLANT (Paris, France)
James W. YORK Jr. (Chapel Hill, USA)

Le Colloque a été parrainé et subventionné par les organismes suivants :

- l' Université Pierre et Marie Curie
- le Ministère des Affaires Etrangères
- le Ministère de la Recherche et de l'Espace
- la Fédération des Compagnies d'Assurance
- la Société Mathématique de France
- le Centre National de la Recherche Scientifique
- le Ministère de l'Education Nationale
- le Collège de France
- le G.A.N.

que le Comité d'Organisation remercie chaleureusement , ainsi que pour leur contribution à l'organisation de ce colloque :

- le Centre International de Mathématiques Pures et Appliquées
- les Editions Gauthier-Villars
- les Editions Kluwer
- la Librairie Offilib
- la Société Générale
- l'Université Paris-Nord

La préparation et l'organisation matérielle de ce colloque eût été impossible sans le concours précieux des secrétaires du Laboratoire de Gravitation et Cosmologie Relativistes (Université Pierre et Marie Curie, Unité de Recherche Associée au CNRS) : Mmes F. Allix, D. Goudemand, B. Guibal et C. Trécul, qui par leur disponibilité et leur présence constante auprès des organisateurs ont contribué de manière décisive au bon déroulement du colloque.

LISTE DES PARTICIPANTS

Kossivi ADJAMAGBO (Paris, France)
Lars ANDERSSON (Stockholm, Suède)
Martin AXELRAD (Paris, France)
Alain BACHELOT (Talence, France)
Robin BALEAN (Bonn, Allemagne)
Claude BARRABES (Tours, France)
Robert BARTNIK (Kensington, Australie)
Henri BASART (Reims, France)
Melvyn BERGER (Northampton, USA)
Luc BLANCHET (Meudon, France)
Marco BRUNI (Londres, Royaume-Uni)
Edmond BONAN (Paris, France)
Henri CABANNES (Paris, France)
Emanuele CALLEGARI (Pise, Italie)
Piotr CHRUSCIEL (Varsovie, Pologne)
Bartolomé COLL (Paris, France)
Pierre DAZORD (Villeurbanne, France)
Jacques DELACOUR (Paris, France)
Pierre DOLBEAU (Paris, France)
Jean-Paul DURUISSEAU (Paris, France)
Mouaouia FIRDAOUSS (Orsay, France)
Jean-Pierre FRANCOISE (Paris, France)
Piotr GARBACZEWSKI (Wroclaw, Pologne)
Renée GATIGNOL (Paris, France)
Paul GAUDUCHON (Paris, France)
André GUINIER (Paris, France)
Jean-Pierre GUIRAUD (Paris, France)
Madeleine HUYNH-SERVET (Villetaneuse, France)
Andreadis IOANNIS (Chambéry, France)
Jean-Bertrand KAMMERER (Paris, France)
Hélène KERBRAT (Villeurbanne, France)
Yvan KERBRAT (Villeurbanne, France)
S. KICHENASSAMY (Paris, France)
Joseph KLEIN (Grenoble, France)
Yvette KOSMANN-SCHWARZBACH (Lille, France)
Marie-Hélène KRIVINE (Paris, France)
Simon LABRUNIE (Palaiseau, France)
Michèle LARCHEVEQUE (Paris, France)
Claude LATREMOLIERE (Angers, France)
Eric LEHMAN (Paris, France)
Paulette LIBERMANN (Paris, France)
Jean-Claude LUCQUIAUD (Montgeron, France)
Raymond McLENAGHAN (Waterloo, Canada)
John MADORE (Orsay, France)
Anne MAGNON (Aubière, France)
Béchir MAHJOUB (Paris, France)
Paul MALLIAVIN (Paris, France)
Renato MANFRIN (Pise, Italie)
Danielle MATTHYS (Orvault, France)
Meinhard E. MAYER (Irvine, USA)
Salvatore MIGNEMI (Paris, France)
Irène MORET-BAILLY (Angers, France)
Niall O' MURCHADHA (Cork, Ireland)
Lora NIKOLOVA (Sofia, Bulgarie)
Norbert NOUTCHEGUEME (Yaoundé, Cameroun)
Pawel NUROWSKI (Trieste, Italie)
Abderrahmane OUAQQA (Amiens, France)
Achille PAPAPETROU (Paris, France)
José Antonio PEREIRA DA SILVA (Coïmbra, Portugal)
Antonio PEREZ-RENDON (Salamanca, Espagne)
PHAM TAN HOANG (Montrouge, France)
Jean-François POMMARET (Noisy-le-Grand, France)
Antoine RAUZY (Paris, France)
Cesare REINA (Trieste, Italie)
Alan RENDALL (Garching, Allemagne)
Antonio RIBEIRO GOMES (Coïmbra, Portugal)
Suemi RODRIGUEZ-ROMO (Louvain-la-Neuve, Belgique)
Maurice ROSEAU (Paris, France)
Evariste SANCHEZ-PALENCIA (Paris, France)
Jacqueline SANCHEZ-HUBERT (Caen, France)
Pedro Paulo SCHIRMER (Bonn, Allemagne)
Haydeh SIROUSSE ZIA (Paris, France)
Nick STEWART (Paris, France)
Aline SURIN (Paris, France)
Erik TAFLIN (Courbevoie, France)
Pierre TEYSSANDIER (Paris, France)
Nobuyuki TOSE (Sapporo, Japon)
Yves THIRY (Paris, France)
Fabrizio TINEBRA (Bologne, Italie)
Philip A. TUCKEY (Lancaster, Royaume-Uni)
Liane VALERE BOUCHE (Chambéry, France)
Murthy M.K. VENKATESHA (Pise, Italie)
Mark WILLIAMS (Chapel Hill, USA)

DISCOURS D'OUVERTURE

J.C. LEGRAN

Paris, le 21 Mai 1992

Ne pas vouloir ouvrir un congrès en l'honneur d'un Collègue sans y mettre un peu de son âme, afin d'éviter les inévitables clichés, est une discipline qui m'a paru brusquement périlleuse lorsque Monsieur KERNER m'a demandé d'ouvrir ces journées...

Certes, j'ai tout d'abord accepté très spontanément car je voulais m'associer au plaisir des amis de Madame CHOQUET BRUHAT, de lui dédier ce colloque de très haut niveau et je souhaitais que notre Université, dans laquelle elle fut enseignante depuis 1968, lui dise par ma voix, sa reconnaissance...

Puis, en découvrant une notice biographique sur vous, Madame, j'ai été pris d'un vertige : le médecin que je suis, plus enclin à rechercher les symptômes d'une maladie, à écouter ses patients et soumis à des problèmes quotidiens, était confronté à votre brillante carrière dédiée à la physique mathématique, à l'abstraction.

Appartenant à la même Université, je n'avais pas eu la chance de vous connaître.

Manque de communication? Certainement, accentué par nos vies passionnantes et multiples, nos routes parallèles et aggravé par nos structures cloisonnées.

Incompatibilité entre les mathématiques et la médecine? Je ne le pense pas et les mathématiques viennent aussi au secours des médecins en proposant des modéles permettant de mieux aborder des processus physiologiques complexes. Les médecins se bornent à recevoir les mathématiciens, mais comme clients!

Mesdames, Messieurs, vous l'avez compris je vais profiter honteusement de la grande modestie de Madame CHOQUET BRUHAT pour ne pas vous parler de son oeuvre mais seulement vous dire ce que j'ai perçu à travers les confidences de quelques-uns de ses nombreux amis.

Trois mots résument assez bien les sentiments de notre communauté :

Admiration - respect et amitié.

Admiration tout d'abord car vous avez su, Madame, tirer parti de l'héritage familial, du milieu scientifique, culturel dans lequel vous vous trouviez, de l'exemple de votre père tout en développant votre propre personnalité.

Admiration pour l'oeuvre pédagogique que vous avez entreprise, mais bien entendu, surtout par votre oeuvre scientifique avec un thème difficile entre tous : les problèmes mathématiques posés par la théorie de la relativité générale.

Vous avez su créer des outils mathématiques nouveaux où le sens physique n'apparait qu'en dernier ressort.

Admiration, aussi, pour avoir maintenu les grands équilibres entre votre vie professionnelle, vos enfants, sans vous désintéresser également des problèmes que posent aux femmes la société.

Pour moi, je retiendrai surtout, en élève peu doué, le mystère qui entoure ces grandes découvertes, partageant seulement avec Einstein deux de ses réflexions.

La plus belle chose que nous puissions éprouver c'est le mystère des choses, ou encore que l'imagination est plus importante que le savoir! et nous savons tous le rôle de cette petite étincelle qui transforme le chercheur en découvreur...

A l'admiration de tous vos collègues, initiés ou profanes, j'ai trouvé également un sentiment de respect dû au chef d'Ecole remarqué très tôt par les plus grands Einstein et Pauli.

Mais j'ai trouvé surtout un sentiment profond d'amitié parmi vos nombreux élèves. Quelques semaines après ma prise de fonctions, il y a un an, j'avais rencontré une collègue qui m'avait annoncé cette réunion en des termes qui témoignaient de l'estime et beaucoup d'amité. La preuve aujourd'hui nous est apportée par l'organisation même de ce colloque international qui va maintenant pouvoir s'ouvrir pour laisser enfin parler ceux qui savent et qui savent que vous apprécierez ce geste d'amitié.

RELATIVISTIC DISSIPATIVE FLUIDS

A.M.ANILE, G.ALÍ and V.ROMANO
Dipartimento di Matematica , Universitá di Catania
viale A.Doria 6 - 95125 Catania , Italy

Abstract

A review of relativistic dissipative fluid theories is presented. The basic assumptions under which such theories are formulated are critically revised and the physical ground of validity is assessed.

1 Introduction

Much work about topological and geometrical aspects of many fundamental problems in general relativity has been developed, but only little progress has been obtained with the aim of formulating a consistent theory of the dissipative relativistic phenomena. In astrophysics and cosmology it is customary to assume ideal fluids, avoiding the problem of dealing with non local thermodynamical equilibrium processes. However in several physical situations viscous and thermal effects can play a relevant dynamical role.

In the calculation of the stellar core collapse a very accurate treatment of the dissipative effects of photons and neutrinos is needed in order to determine the critical mass for the formation of black holes or to describe the explosion of supernovae [1,2].

In cosmology the socalled "bulk viscosity inflation" is a fashionable topic [3,4,5,6] while the dissipative inhomogeneous cosmological models seem a viable alternative in order to solve the horizon and flatness problem of the standard FRW cosmologies [7,8].

In the study of phase transitions in the early universe (e.g. quark-gluon phase transition) a description using the formalism of radiation hydrodynamics seems the most viable way to tackle the problem [9,10].

Finally the stability of accretion disks into neutron stars or black holes and the high energy collisions of heavy nuclei require a careful inclusion of viscous processes too.

From the previous review, it is clear that a consistent causal theory of dissipation in relativistic fluids is warranted. In the non relativistic case, the Navier-Stokes-Fourier theory represents a natural and successfull theory for dissipative fluids. So various attempts to generalize it in a relativistic context have been made. The pioneering works of Eckart and Landau-Lifshits, although constitute

M. Flato et al. (eds.), Physics on Manifolds, 1–10.

the most straightforward covariant generalizations of the Navier-Stokes-Fourier theory, are beset by serious inconsistences (acausality and instability), but also the more sophisticated formulations of Israel and Stewart [11] or Liu, Müller and Ruggeri [12] are not free from drawbacks (for a review see [13] or [14]).

An alternative independent approach was followed by Carter, which by using a variational principle has obtained a complete set of constitutive equations for the viscous stresses [15]. In a recent paper D.Priou [16] has performed a comparison between the variational approach of Carter and the the phenomenological one of Israel and Stewart. He showed that the two theories coincide for small perturbations around the equilibrium state, while in general they constitute two members of a more general family of theories, whose formulation would require the introduction of a much more extensive set of phenomenological coefficients.

2 Mathematical framework for relativistic dissipative fluid theories

The basic assumption of fluid theories is that only a finite number of tensor fields is required to describe the state of the fluid. This marks a difference with kinetic theory, where in principle all the moments of the distribution function are needed in order to obtain a macroscopic non equilibrium state. The basic variables which enter in the ideal relativistic fluid theory are the particle number current N^μ and the symmetric stress-energy tensor $T^{\mu\nu}$; they satisfy the local conservation laws

$$\nabla_\mu N^\mu = 0, \tag{1}$$
$$\nabla_\mu T^{\mu\nu} = 0. \tag{2}$$

Out of thermodynamical equilibrium one assumes again that N^μ and $T^{\mu\nu}$ are the only relevant dynamical variables and satisfy eq.s (1) and (2). This is equivalent to neglecting the functional dependence on the infinite set of auxiliary variables which, as indicated by kinetic theory, would appear when thermodynamical equilibrium is violated. In this "minimal" theory eq.s (1) and (2) are supplied by a balance equation

$$\nabla_\mu A^{\mu\alpha\beta} = I^{\alpha\beta}, \tag{3}$$

with $A^{\mu\alpha\beta}$ and $I^{\alpha\beta}$ functions of N^μ and $T^{\mu\nu}$.

As suggested by kinetic theory of gases we suppose that $A^{\mu\alpha\beta}$ and $I^{\alpha\beta}$ are completely symmetric tensors and obey

$$I^{\alpha\beta} = 0, \qquad A^{\alpha\mu}_{\mu} = mc^2 N^\alpha. \tag{4}$$

Moreover, we suppose that there exists a four-vector S^α, called entropy current vector, function of N^μ and $T^{\mu\nu}$ such that the inequality

$$\nabla_\alpha S^\alpha = \sigma(N^\mu, T^{\mu\nu}) \geq 0 \tag{5}$$

holds for every solution of the conservation equations (1)-(3).

The rationale of extended thermodynamics is to study systematically all the consequences of the previous axioms by employing the method of Lagrangian multipliers [11]. It is possible to formulate more general fluid theories describing the state of the fluid by means of a finite set of fields containing $T^{\mu\nu}$ and N^{μ} and imposing balance equations similar to relations (1)-(3) . However, here for sake of simplicity we restrict our treatment to the case where $T^{\mu\nu}$ and N^{μ} are the only dynamical variables.

In account of the formal structure of the relations (1)-(3), the fluids described by this theory are said of divergence-type (see [17]). It is possible to prove that the more general theory of dissipative fluid of divergence-type is determined by specifying a single scalar generating function $\chi = \chi(\zeta, \zeta_\alpha, \zeta_{\alpha\beta})$ and the functional dependence of $I^{\alpha\beta}$ on the new set of dynamical variables ζ, ζ_α, $\zeta_{\alpha\beta}$ with $\zeta_{\alpha\beta}$ obeying $\zeta^\alpha_\alpha = 0$ and $\zeta_{\alpha\beta} = \zeta_{\beta\alpha}$.

The transformation laws between the old dynamical variables and the new ones are given by

$$N^\alpha = \frac{\partial^2\chi}{\partial\zeta\partial\zeta_\alpha} \tag{6}$$

$$T^{\alpha\beta} = \frac{\partial^2\chi}{\partial\zeta_\alpha\partial\zeta_\beta}. \tag{7}$$

Moreover

$$A^{\alpha\beta\gamma} = \frac{\partial^2\chi}{\partial\zeta_\alpha\partial\zeta_{\alpha\beta}}, \tag{8}$$

$$S^\alpha = \frac{\partial\chi}{\partial\zeta_\alpha} - \zeta N^\alpha - \zeta_\beta T^{\alpha\beta} - \zeta_{\beta\gamma}A^{\alpha\beta\gamma} \tag{9}$$

and

$$\sigma = -\zeta_{\alpha\beta}I^{\alpha\beta}. \tag{10}$$

Equations (1)-(3), by using (6)-(10), can be written in a compact form

$$M^{AB\mu}\nabla_\mu\zeta_B = I^A \tag{11}$$

where

$$\zeta_A = (\zeta, \zeta_\alpha, \zeta_{\alpha\beta}), \tag{12}$$

$$I^A = (0, 0^\alpha, I^{\alpha\beta}), \tag{13}$$

$$M^{AB\mu} = \frac{\partial^2\chi^\mu}{\partial\zeta_A\partial\zeta_B}, \tag{14}$$

with

$$\chi^\alpha = \frac{\partial\chi}{\partial\zeta_\alpha}. \tag{15}$$

The matrix $M^{AB\mu}$ is symmetric in the indices A and B and for this reason the system (11) is said symmetric. A symmetric system is said hyperbolic in the direction of the future-directed time-like four-vector w_μ if $M^{AB\mu}$ is negative definite. A symmetric system is causal if $M^{AB\mu}$ is negative definite for all the future-directed time-like four-vectors.

These properties have an immediate physical meaning. Hyperbolicity assures the well posedness of the initial value problem while causality garantees that there is no signal propagating faster than light. Now we consider some examples of divergence-type fluid theory.

Example 1: the Eckart theory.

In this case the dynamical variable N^α and $T^{\alpha\beta}$ and the functions $A^{\alpha\beta\gamma}$, $I^{\alpha\beta}$, S^α and σ are given by

$$N^\alpha = nu^\alpha, \tag{16}$$

$$T^{\alpha\beta} = eu^\alpha u^\beta + (p+\pi)h^{\alpha\beta} + 2u^{(\alpha}u^{\beta)} + \pi^{\alpha\beta}, \tag{17}$$

$$A^{\alpha\beta\gamma} = 2Tu^{(\alpha}h^{\beta\gamma)}, \tag{18}$$

$$I^{\alpha\beta} = -\frac{T}{\eta_1}\pi^{\alpha\beta} - \frac{2T}{3\eta_2}(g^{\alpha\beta} + 4u^\alpha u^\beta)\pi - \frac{2}{\chi}q^{(\alpha}u^{\beta)}, \tag{19}$$

$$S^\alpha = nu^\alpha + \frac{Q^\alpha}{T}, \tag{20}$$

$$\sigma = \frac{\pi^2}{\eta_2 T} + \frac{q^\alpha}{q_\alpha}\chi T^2 + \frac{\pi^{\alpha\beta}\pi_{\alpha\beta}}{2\eta_1 T}. \tag{21}$$

Here u^α is a future-directed time-like four-vector, n is the particle number density, e is the total energy density, p is the hydrostatic pressure, π is the bulk viscosity, q^α is the heat flux (it satisfies $u^\alpha q_\alpha = 0$), $\pi^{\alpha\beta}$ is the shear viscosity (it satisfies $\pi^{\alpha\beta}u_\alpha = 0$, $\pi^\alpha_\alpha = 0$), η_2 and η_1 are the bulk and shear viscosity coefficients, χ is the thermal conductivity, $g^{\alpha\beta}$ is the metric tensor and $h^{\alpha\beta} = g^{\alpha\beta} + u^\alpha u^\beta$ is the projection tensor.

For the stresses we have the following constitutive equations

$$\pi = -\eta_2\theta, \tag{22}$$

$$q^\alpha = -\chi(h^\beta_\alpha T_{,\beta} + Ta_\alpha), \tag{23}$$

$$\pi^{\alpha\beta} = -2\eta_1\sigma^{\alpha\beta}, \tag{24}$$

where T is the absolute temperature, $\theta = \nabla_\alpha u^\alpha$ is the expansion, a_α is the acceleration and $\sigma_{\alpha\beta}$ is the shear.

One can prove that for the Eckart theory the generating function is

$$\chi = \alpha(\zeta,\eta) - \frac{\zeta^{\alpha\beta}\zeta_\alpha\zeta_\beta}{\mu} \tag{25}$$

with $\mu = \zeta^\alpha\zeta_\alpha$ and α an arbitrary smooth function which plays the role of an equation of state.

It's easy to verify that the matrix $M^{AB\mu}w_\mu$ is not negative definite as the term $\frac{\partial^3\chi}{\partial\zeta_\mu\partial\zeta_{\alpha\beta}\partial\zeta_{\gamma\delta}}$ vanishes. Therefore the Eckart theory fails to be causal.

Example 2.

To overcome the drawbacks of the previous example, we now consider the following generating function

$$\chi = \alpha(\zeta,\mu) + \beta(\zeta,\mu)\zeta^{\alpha\beta}\zeta_\alpha\zeta_\beta + \chi_2(\zeta,\zeta_\alpha,\zeta_{\alpha\beta}) \tag{26}$$

with

$$\chi_2 = \frac{\gamma(\zeta,\mu)}{\mu^2}\left(\mu g_{\alpha\beta} - 2\zeta_\alpha\zeta_\beta\right)\left(\mu g_{\gamma\delta} - 2\zeta_\gamma\zeta_\delta\right)\zeta^{\alpha\beta}\zeta^{\gamma\delta}. \tag{27}$$

Geroch and Lindblom [17] showed that in a neighbourhood of $\zeta_{\alpha\beta} = 0$ (which represents the equilibrium states of the fluid) the matrix $M^{AB\mu}w_\mu$ is negative definite provided that $\frac{\partial\gamma}{\partial\mu}$ is sufficientely large and the following two conditions hold

$$\rho + p > 0 \tag{28}$$

$$\left(\frac{\partial\rho}{\partial p}\right)_\zeta > \left(\frac{\partial\rho}{\partial p}\right)_S \geq 1. \tag{29}$$

We note that the latter relation indicates that the adiabatic sound speed must be real and less or equal to the speed of light.

Other more complex examples of causal theories could be given. Among these the one by Liu, Müller and Ruggeri [12,18], which generalizes around the thermal equilibrium the Eckart Theory by means of evolutionary constitutive equations for the stresses

$$\pi = -\left(\frac{1}{3}\eta_2\theta + \beta_0\cdot\pi - \overline{\alpha}_0\nabla_\mu q^\mu\right), \tag{30}$$

$$q^\lambda = -\chi T h^{\lambda\mu}\left(\frac{1}{T}\nabla_\mu T + a_\mu + \overline{\beta}_1\cdot q_\mu - \overline{\alpha_0}\nabla_\mu\pi - \nabla_\nu\pi^\nu_\mu\right), \tag{31}$$

$$\pi^{\mu\nu} = -2\eta_1\left(\sigma^{\mu\nu} + \beta_2\cdot\pi^{\mu\nu} - \overline{\alpha}_1 q^{<\mu;\nu>}\right) \tag{32}$$

with η_1,η_2,$\chi \geq 0$ and β_i and $\overline{\alpha}_j$ phenomenological coefficients which can be determined in some cases by kinetic theory.

3 Difficulties of the previous theory

As just pointed out in section 2, the microphysics of the gas indicates that far from thermodynamic equilibrium a macroscopic state is determined by a set of infinite variables. For example, for a simple gas the distribution function

$$f = f(x^\mu, p^\mu) \tag{33}$$

satisfies the relativistic Boltzmann equation

$$p^\mu\frac{\partial f}{\partial x^\mu} = Q(f,f) \tag{34}$$

where x^μ are the space-time coordinates , p^μ is the four momentum of the gas particles and Q is the collision term.By taking the moments of eq.(34) and by using the properties of the collisional invariants, the following relations for the macroscopic variables can be demonstrated

$$\nabla_\mu N^\mu = 0, \tag{35}$$

$$\nabla_\mu T^{\mu\nu} = 0, \tag{36}$$

$$\nabla_\mu A^{\mu\nu\sigma} = I^{\nu\sigma}, \tag{37}$$

$$\nabla_\mu A^{\mu\nu\sigma\alpha_1\ldots\alpha_k} = I^{\nu\sigma\alpha_1\ldots\alpha_k}. \tag{38}$$

But in the divergence-type fluid theory only the equations (35)-(37) are taken into account, although there is no general physical motivation for neglecting the set of equations (38). As we will see, only in very particolary cases it is possible to evaluate the order of magnitude of the various moments and justify the truncation of the system (35)-(38) to only the eq.s (35)-(37).

This was confirmed by an accurate numerical analysis by W.Weiss [19] who showed that in the classical case the linearized dispersion relation for the first 19.600 moments seems to indicate that the characteristic speeds for the sound modes for a perfect gas diverge monotonically by adding more moments.

An important testing ground for a fluid theory is its ability to describe shock phenomena, that is rapid transition of the fluid state. A carefull study of A.M.Anile and A.Majorana [20] showed in the classical case that Müller-type theories admit a unique, continuous and stable shock structure only for sufficientely low Mach numbers (about 1.67 or 2.07 in according to the model). The same result was obtained by A.Majorana and S.Motta in the relativistic case [21]. Nevertheless the classical Navier-Stokes-Fourier theory and the relativistic Eckart theory yield a very accurate description of shocks [22,23].

In a recent paper Geroch and Lindblom [24] proved that the non existence of a regular shock structure is a general feature of causal theories. However in the non relativistic case D.Jou and D. Pavón [25] by taking into account non local effects and non linear terms in the constitutive equations raised up the critical Mach number above which a regular shock structure ceases to exist up to 4.67. Moreover the procedure indicated by them promises to work well in order to further increase the critical Mach number. Here we point out that in the analysis of D.Jou and D.Pavon not only the fluxes of the usual hydrodynamics theories but also the fluxes of the fluxes and so on are considered and assumed as independent variables. This confirmes again that neglecting the set of eq.s(38) as a general law is totally arbitrary.

4 Simple model: radiation hydrodynamics

In this section we present a physical case where a hydrodynamical theory is adequate: a radiating gas.

When a gas interacts with a radiation field the dissipative processes are almost entirely due to photon transport inasmuch as the mean free path of the gas particles is usually much less than the photons mean free path. For this reason in most applications of astrophysical and cosmological interest, a realistic model is represented by an ideal gas coupled to radiation.

The radiation field is described by a distribution function f which depends on the space-time coordinates and on the photon four momenta,

$$f = f(x^\mu, p^\mu) \tag{39}$$

with $p^\mu p_\mu = 0$.

With respect to the congruence determinated by the observer whose four velocity is u^μ, the following decomposition holds

$$p^\mu = -\nu_0 \left(u^\mu + l^\mu\right), \tag{40}$$

with $l^\mu l_\mu = 1$ and $l^\mu u_\mu = 0$, ν_0 being the frequency measured by the observer whose four velocity is u^μ . Then we can write

$$f = f(x^\mu, \nu_0, l^\mu). \tag{41}$$

For the sake of simplicity we assume a grey medium in order to eliminate the dependence on the frequency in the scattering terms. Moreover we specialize the observer to the comoving one.

Introducing the integrated intensity

$$I = \int_0^\infty f \nu_0^3 d\nu_0, \tag{42}$$

and integrating over all the frequencies the relativistic Boltzmann equation for the photon distribution function, we obtain the relativistic transfer equation [26]

$$\begin{aligned}(u^\mu + l^\mu)\left\{\nabla_\mu I + 4Il^\sigma \nabla_\mu u_\sigma + l^\rho l^\sigma \frac{\partial I}{\partial l^\rho}\nabla_\mu u_\sigma + u^\rho l^\sigma \frac{\partial I}{\partial l^\rho}\nabla_\mu u_\sigma - \frac{\partial I}{\partial l^\rho}\nabla_\mu u^\rho\right\} \\ = \rho_0 \left[\varepsilon_0 - < K >_0\right],\end{aligned} \tag{43}$$

where ρ_0 is the rest frame density, ε_0 and $< K >_0$ are the emissivity and the mean absorption coefficient,

$$< K > = \frac{1}{I}\int_0^\infty k(\nu_0, l^\mu) f \nu_0^3 d\nu_0 \tag{44}$$

with $k(\nu_0, l^\mu)$ absorption coefficient. In the following we suppose the source term

$$B = \frac{\varepsilon}{< K >} \tag{45}$$

to be isotropic. Therefore it assumes a Planckian form.

The moments of the distribution function are defined as

$$M^{\alpha_1 \dots \alpha_k} = \int_0^\infty \int_\Omega f \nu_0^3 l^{\alpha_1} \dots l^{\alpha_k} d\nu_0 d\Omega, \tag{46}$$

with $d\Omega$ element of solid angle over the unit sphere ,

$$d\Omega = \delta(u^\mu l_\mu)\delta(l^\mu l_\mu - 1). \tag{47}$$

The first moments have an immediate macroscopic meaning.

$$J = \int_0^\infty \int_\Omega f \nu_0^3 d\nu_0 d\Omega \tag{48}$$

is the radiation energy density ,

$$H^\mu = \int_0^\infty \int_\Omega l^\mu f \nu_0^3 d\nu_0 d\Omega \tag{49}$$

is the radiative flux and

$$K^{\mu\nu} = \int_0^\infty \int_\Omega l^\mu l^\nu f \nu_0^3 d\nu_0 d\Omega \tag{50}$$

is the radiative stress tensor.

Finally, the energy-momentum tensor of radiation is defined by

$$T^{\mu\nu}_{\text{rad}} = \int_0^\infty \int_\Omega p^\mu p^\nu f \nu_0^3 d\nu_0 d\Omega \tag{51}$$

and satisfies the the balance equation

$$\nabla_\mu T^{\mu\nu}_{\text{rad}} = \rho_0 < K > [u^\nu(B - J) - H^\nu]. \tag{52}$$

By using the decomposition (40), one can write

$$T^{\mu\nu}_{\text{rad}} = J u^\mu u^\nu + u^\mu H^\nu + u^\nu H^\mu + K^{\mu\nu}. \tag{53}$$

For the gas the energy-momentum tensor has the form

$$T^{\mu\nu}_{\text{mat}} = (e + p) u^\mu u^\nu + p g^{\mu\nu}, \tag{54}$$

with e the total energy density and p the hydrostatic pressure of the gas.

Therefore the complete set of equations for a radiating gas is

$$\nabla_\mu (\rho_0 u^\mu) = 0, \tag{55}$$

$$\nabla_\mu (T^{\mu\nu}_{\text{mat}} + T^{\mu\nu}_{\text{rad}}) = 0, \tag{56}$$

$$\nabla_\mu T^{\mu\nu}_{\text{mat}} = -\rho_0 < K > [u^\mu(B - J) - H^\mu], \tag{57}$$

supplied by an equation of state of the type $p = p(\rho, e)$.

Now we note that if we assume J, H^μ and $K^{\mu\nu}$ as independent variables, the previous system is not closed. However in the diffusion limit

$$I = \frac{J}{4\pi}\left(1 + 3\frac{l^\mu H_\mu}{J}\right), \tag{58}$$

with $|H^\mu/J| \ll 1$, and in the streaming limit

$$I = J\delta(l^\mu H_\mu). \tag{59}$$

In both these extreme cases, the radiation intensity depends only on J and H^μ. It seems consequentely reasonable to assume the closure

$$K^{\mu\nu} = K^{\mu\nu}(J, H^\mu), \tag{60}$$

interpolating the expression of the radiation stress tensor obtained in the extreme cases.

From the representation theorems, $K^{\mu\nu}$ has to assume the general form

$$K^{\mu\nu} = J\left(\frac{1-\chi}{2}h^{\mu\nu} + \frac{3\chi - 1}{2}\frac{H^\mu H^\nu}{H^\sigma H_\sigma}\right), \tag{61}$$

with $\chi = \chi(J, H^\sigma H_\sigma)$ subjects to the following conditions: $\chi = 1/3$ for almost isotropic radiation (optically thick medium) and $\chi = 1$ in the streaming limit . In order to obtain the general expression of χ Anile, Sammartino and Pennisi [27,28] (and for the pure radiation case also Kremer and Müller [29]), employed the entropy principle by using the Lagrangian multipliers.

One assumes that a four vector S^α exists such that the relation

$$\nabla_\alpha S^\alpha\left(N^\mu, T^{\mu\nu}_{\text{mat}}, T^{\mu\nu}_{\text{rad}}\right) = g(N^\mu, T^{\mu\nu}_{\text{mat}}, T^{\mu\nu}_{\text{rad}}) \geq 0 \tag{62}$$

holds for every solution of the system (55)-(57).

The following general form of χ was found

$$\chi = \frac{5}{3} - \frac{2}{3}\left(4 - 3f^2\right)^{\frac{1}{2}}, \tag{63}$$

with $0 \leq f = \frac{(H^\sigma H_\sigma)^{\frac{1}{2}}}{J} \leq 1$.

By using the closure (63), the system (56)-(57) fulfils the causality property for $0 < f < 1$ as it is possible to check from the characteristic polynomial .

However a complete analysis of the shock structure of the system (55)-(57) closed by means of the relation (63) is still lacking. Some preliminary numerical results seem to indicate that a regular shock structure would exist. This topic is under current investigation by the authors and it will be the subject of a forthcoming article.

References

1. Bruenn, S.W. (1985), Ap. J., **58** 771
2. Mezzacappa, A. and Matzner, R. (1989), Ap. J., **343** 853.
3. Waga, I., Falcao, R.C. and Chanda, R. (1986), Phys. Rev. D, **33** 1839.
4. Calzetta, E. and Sakellariadou, M. (1992), Phys. Rev. D, **45** 2802.
5. Romano, V. and Pavón, D., submitted to Phys. Rev. D.
6. Hiscock, H. and Salmonson, J. (1991), Phys. Rev. D, **43** 3249.
7. Deng, V. and Mannheim, P.D. (1990), Phys. Rev. D, **42** 371.
8. Deng, V. and Mannheim, P.D. (1991), Phys. Rev. D, **44** 1722.
9. Miller, J. and Pantano, O. (1987), in Bruzzo, U., Cianci, R. and Massa, E. (eds.), General Relativity and Gravitational Physics, World Scientific, Singapore, p. 357.
10. Miller, J. and Pantano, O. (1988), Phys. Rev. D, **40** 1789.
11. Israel, W. and Stewart, M. (1979), Ann. Phys., **118** 341.
12. Liu, I.S., Müller, I. and Ruggeri, T. (1986), Ann. Phys., **169** 191.
13. Israel, W. (1989), in Anile, A.M. and Choquet-Bruhat, Y. (eds.), Relativistic Fluids Dynamics, Lecture Notes in Mathematics 1835, Springer, Berlin, p. 152.
14. Jou, D., Casas-Vazquez, J. and Lebon, G. (1988), Rep. Prog. Phys., **51** 1105.
15. Carter, B. (1989), in Anile, A.M. and Choquet-Bruhat, Y. (eds.), Relativistic Fluids Dynamics, Lecture Notes in Mathematics 1835, Springer, Berlin, p. 1.
16. Proiu, D. (1991), Phys. Rev. D, **43** 1223.
17. Geroch, R. and Lindblom, L. (1990), Ann. Phys.
18. Müller, I. (1967), Z. Phys., **198** 329.
19. Weiss, W. (1990), Zur Hierarchie der Erweiterten Thermodynamik , Dissertation, T.U. Berlin
20. Anile,A.M. and Majorana, A. (1981), Meccanica, **16** 149.
21. Majorana, A. and Motta, S. (1985), J. Non-Equilib. Thermodyn., **10** 29.
22. Cercignani, C. and Majorana, A. (1988), Phys. Fluids, **31** 1064.
23. Majorana, A. and Muscato, O. (1990), Meccanica, **25** 77.
24. Geroch, R. and Lindblom, L. (1990), Phys. Rev. D, **41** 1855.
25. Jou, D. and Pavón, D. (1991), Phys. Rev. A, **15** 6496.
26. Anderson, J.L. and Spiegel, E.A. (1972), Ap. J., **171** 127.
27. Anile, A.M., Pennisi, S. and Sammartino, M. (1991), J. Math. Phys., **32** 544.
28. Anile, A.M., Pennisi, S. and Sammartino, M. (1992), Ann. Inst. Henri Poin., **56** 49.
29. Kremer, G.M. and Müller, I. (1992), J. Math. Phys., **33** 2265.

MATHEMATICAL PROBLEMS RELATED TO LIQUID CRYSTALS, SUPERCONDUCTORS AND SUPERFLUIDS

Haïm BREZIS
Université Pierre et Marie Curie
4 place Jussieu
75252 Paris Cedex 05

Dedicated to Professeur Y. Choquet-Bruhat with esteem and affection.

ABSTRACT. The study of point and line singularities occuring in numerous physical phenomenon, such as liquid crystals, superconductors and superfluids, lead to mathematical problems connected with the singularities of energy-minimizing maps between manifolds.

RÉSUMÉ. L'étude des singularités ponctuelles ou curvilignes qui apparaissent dans de nombreux phénomènes physiques, par exemple les cristaux liquides, les superconducteurs et les superfluides, conduit à des problèmes mathématiques liés à l'analyse des singularités d'applications entre variétés qui minimisent l'énergie.

1. PROBLEMS CONNECTED WITH LIQUID CRYSTALS.

A liquid crystal consists of long molecules with some orientational ordering (see e.g. [12], [10] or [15]). At every point $x \in \Omega$ (the container of the liquid crystal) the optical axis of the molecule has an orientation which we denote $u(x)$ –called the director– so that $|u(x)| = 1$, i.e. $u(x) \in S^2$. Physicists observe that u depends smoothly on x, except at the "defects" (points or sometimes lines) where u is not well defined. Commonly, one associates to every configuration u an energy. For example, the simplest model is

$$E(u) = \int_\Omega |\nabla u|^2. \tag{1}$$

Physicists use more sophisticated models, but they always correspond to <u>integrals</u> over Ω of quadratic expressions in ∇u. On the basis of a mathematical analysis we shall propose a different model for the energy. But, for the moment, let us

M. Flato et al. (eds.), Physics on Manifolds, 11–21.

proceed with E and we shall then discuss its deficiencies. In view of E it is natural to consider the class of all configurations with finite energy :

$$H^1(\Omega; S^2) = \{u \in H^1(\Omega; \mathbb{R}^3); u(x) \in S^2 \text{ a.e.}\}$$

where H^1 is the usual Sobolev space. It is convenient to fix a boundary condition $g : \partial\Omega \to S^2$ (say, smooth) and to consider

$$H_g^1 = \{u \in H^1(\Omega; S^2); u = g \text{ on } \partial\Omega\}.$$

Note that H_g^1 is never empty ; for example, if $\Omega = B^3$ is the unit ball, then $u(x) = g(x/|x|)$ belongs o H_g^1 because $\int_\Omega |\nabla\left(\frac{x}{|x|}\right)|^2 < \infty$.

It is then natural to expect that stable configurations correspond to minimizers of the energy

$$\min_{u \in H_g^1} \int_\Omega |\nabla u|^2. \tag{2}$$

One shows easily that the minimum in (2) is achieved. The corresponding Euler equations is

$$\begin{cases} -\Delta u = u|\nabla u|^2 & \text{on } \Omega, \\ |u(x)| = 1 & \text{on } \Omega, \\ u = g & \text{on } \partial\Omega \end{cases} \tag{3}$$

which is the system of <u>harmonic maps</u> from B^3 to S^2.

We summarize the main properties of the minimizing harmonic maps in the next result :

Theorem 1 (see [24], [9]). *Assume u is a minimizer for (2). Then u has only on finite number of singularities, i.e. u is C^∞ except at isolated points. Moreover if x_0 is such a singularity then*

$$u(x) \simeq \pm R\left(\frac{x - x_0}{|x - x_0|}\right)$$

for some rotation R. As a consequence

$$\deg(u, x_0) = \pm 1$$

where $\deg(u, x_0)$ denotes the Brouwer degree of u restricted to any small sphere around x_0.

<u>Remark 1</u>. The regularity of weak (H^1) solutions of (3) is a very delicate question. Recently, T. Rivière has constructed striking examples of solutions of (3) where the singular set consists of a line or is even a dense in Ω (see [22] and [23]).

One may wonder whether Theorem 1 is optimal, i.e. whether minimizers really have singularities. It is easy to force the occurence of singularities by topological obstructions. For example, if $\deg(g, \partial\Omega) \neq 0$ there is no smooth extension of g inside Ω, i.e. $C_g^1 = C_g^1(\Omega; S^2) = \phi$. However, if $\deg(g, \partial\Omega) = 0$ one may think that since $C_g^1 \neq \phi$ the minimizers in (1) will "prefer" to be smooth. Suprisingly, this is not the case, as was first pointed out by Hardt-Lin [19] (see also [7]) through a very interesting gap phenomenon.

THEOREM 2. *There exist smooth maps $g : \partial\Omega \to S^2$ such that $\deg(g, \partial\Omega) = 0$ and*

$$\min_{u \in H_g^1} \int_\Omega |\nabla u|^2 < \inf_{u \in C_g^1} \int_\Omega |\nabla u|^2. \tag{4}$$

Remark 2. This theorem also shows that the class of smooth maps C_g^1 is not dense in H_g^1. This fact has renewed interest in a question originally raised by Eells-Lemaire [14]. Consider two compact manifolds M and N with $\partial N = \phi$. Is it true or not that $C^1(M, N)$ is dense in $H^1(M, N)$? A remarkable result of F. Bethuel [1] asserts that if dim $M = 1, 2$ the answer is positive for every N. However if dim $N \geq 3$, then there is density iff the homotopy class $\pi_2(N) = 0$. We refer to [1] for the general case of the Sobolev space $W^{1,p}$ over L^p where $1 < p < \infty$ and numerous additional properties.

The gap phenomenon puts us in a rather unconfortable situation : do stable physical configurations "choose" to minimize the energy in H_g^1 or rather in C_g^1 ? At first it would seem more natural to minimize the energy in H_g^1 because it yields the lowest energy among all finite-energy configurations, possibly with singularities.

However in many experiments, singularities move towards each other and eventually they coalesce –sometimes after 2 or 3 days (see e.g. [10]). The phenomenon of "collapse of singularities" suggests that configurations with singularities are not stable and then it is reasonable to search for stable configurations by minimizing the energy in C_g^1 But this is not satisfactory : why would the system choose to end up in a configuration without singularities at a high energy level when it has the possibility of lowering its energy by introducing singularities. We have tried to overcome this paradox by suggesting that the usual energy E given by (1) describes well the energy of a configuration without singularities. However, if a configuration has singularities its energy E is inadequate. The new energy that we propose was first introduced by the author in 1988 (see [8]) and its properties are investigated in a joint work with Bethuel and Coron [2]. The idea is to use a standard tool in functional analysis called relaxation. Namely we define for $u \in H_g^1(\Omega; S^2)$,

$$E_{\text{rel}}(u) = \inf\{\underline{\lim} \int_\Omega |\nabla v_n|^2\},$$

where the infimum is taken over all sequences (v_n) in $C_g^1(\Omega; S^2)$ such that $v_n \to u$ a.e.

It is a remarkable fact that $E_{\text{rel}}(u)$ has a rather simple, explicit form at least in case u has just a finite number of singularities. In order to describe E_{rel} we need some preliminary facts and notation.

LEMMA 1 (see [6]). *The set*

$$R_g = \{u \in H_g^1; u \text{ is smooth except at a finite number of points}\}$$

is dense in H_g^1.

For a given $u \in R_g$ we now introduce $L(u)$, the length of a minimal connection connecting the singularities of u (a notion originally devised in [9]). We first assume that all the singularities of u have degrees ± 1. We denote by p_i (resp. n_i the singularities of degree $+1$ (resp. -1). Note that we have exactly the same number of positive and negative singularities since $\deg(g, \partial\Omega) = 0$. Set

$$L(u) = \min_{\sigma} \sum_i d(p_i, n_{\sigma(i)})$$

where d denotes the geodesic distance in Ω and the minimum is taken over all permutations σ of the integers. In the case where the degrees are not ± 1 we follow the same procedure, except that each point is repeated according to its multiplicity.

One can prove (see [2]) that the map $u \mapsto L(u)$ is locally Lipschitz for the H^1 norm. Hence, in view of Lemma 1, we may now define $L(u)$ by continuity for <u>every</u> $u \in H_g^1$.

On next result provides a striking representation formula for E_{rel} proved in [2].

THEOREM 3. *We have*

$$E_{rel}(u) = \int_\Omega |\nabla u|^2 + 8\pi L(u) \qquad \forall u \in H_g^1.$$

In other words, the relaxed energy involves an additional <u>energy of interaction between the singularities</u>. A configuration with a pair of singularities + 1 and - 1 far apart has a high relaxed energy. In order to minimize the relaxed energy the system may then "try" to bring these singularities close together and possibly they may coalesce.

Finally, the "paradox" described earlier is now elucidated by the following (if we believe that E_{rel} is a physical energy) :

THEOREM 4 (see [2]). *The function* $u \mapsto E_{rel}(u)$ *is lower semi-continuous for the weak* H^1 *topology and consequently*

$$\min_{u \in H_g^1} E_{rel}(u) \quad \text{is achieved.}$$

Moreover

$$\min_{u \in H_g^1} E_{rel}(u) = \inf_{u \in C_g^1} \int_\Omega |\nabla u|^2.$$

The study of the regularity of minimizers for E_{rel} is a very interesting open problem about which little is known. Let us mention that Giaquinta, Modica and Soucek using their theory of Cartesian currents have also derived the relaxed energy as a natural object and they have proved some partial regularity results for the minimizers of E_{rel} ; see [17] and [18]. It is not excluded that minimizers might have singularities of degree zero ; a related phenomenon occurs in [20].

2. PROBLEMS CONNECTED WITH SUPERCONDUCTORS AND SUPERFLUIDS.

It is well known that vortex lines (or filaments) appear in type-II superconductors placed in a magnetic field H ; this is the so-called mixed state, characterized by the coexistence of two phases : the normal phase concentrated in a small region near the vortices and the superconducting state elsewhere (see e.g. [11], [25]). A similar phenomenon occurs in superfluids (see e.g. [13]). When helium II is cooled and rotated in a bucket around the z-axis at constant angular velocity Ω, vortex lines parallel to the z-axis appear. The theory was initiated by L. Onsager [21] and R. Feynman [16] ; the first experimental evidence came in the work of W. Vinen (see e.g. [13]). Mathematically, these phase-transition problems are modelled by introduced an order parameter u which is a <u>complex number</u>, i.e. with two degrees of freedom, and the associated energy is the Ginzburg-Landau energy

$$E_\varepsilon(u) = \frac{1}{2}\int |\nabla u|^2 + \frac{1}{4\varepsilon^2}\int (|u|^2 - 1)^2 \tag{5}$$

where ε is a parameter called the Ginzburg-Landau coherence length ; ε is a constant which depends on the material and its temperature T. In the interesting range of temperature ($T < T_c$, the critical temperature, with T not too close to T_c) ε is extremely small, typically of the order of a some hundreds of Angstroms in superconductors and of the order of a few Angstroms in superfluids. It is therefore of great interest to study the aysmptotics as $\varepsilon \to 0$. In superconductors $|u|^2$ is proportional to the density of superconducting electrons ($|u| \simeq 1$ corresponds to the superconducting state and $|u| \simeq 0$ corresponds to the normal state). In superfluids, $|u|^2$ is proportional to the density of the superfluid.

In order to understand mathematically these line vortices, the simplest geometry corresponds to a cylindrical domain whose axis is denoted by z. Our analysis deals with the study of a <u>2-dimensional</u> cross-section of this cylinder denoted Ω. For simplicity we shall work with a Dirichlet condition. Even through this condition is not physically realistic it is striking to see that it generates "<u>quantized vortices</u>" with propertes very singular to the ones observed in superconductors and superfluids.

The mathematical setting is the following : Ω denotes a smooth bounded domain in $\mathbb{R}^2$ with boundary $\partial\Omega$. We fix a smooth boundary condition $g : \partial\Omega \to \mathbb{C}$ such that $|g| = 1$. This restriction is consistent with the fact that we want to study the limit as $\varepsilon \to 0$ in (5) and, clearly, the "penalty term" $\frac{1}{\varepsilon^2}\int (|u|^2 - 1)^2$ will force u, in the limit, to have $|u| = 1$. The class of testing functions is

$$H_g^1(\Omega;\mathbb{C}) = \{u \in H^1(\Omega;\mathbb{C}); u = g \text{ on } \partial\Omega\}.$$

Let u_ε be a minimizer for

$$\min_{u \in H_g^1(\Omega;\mathbb{C})} E_\varepsilon(u). \tag{6}$$

I will report about recent works with Bethuel and Hélein (see [3], [4], [5]) concerning the limiting behavior of u_ε as $\varepsilon \to 0$. It turns out that

$$d = \deg(g, \partial\Omega), \tag{7}$$

the Brouwer degree of g considered as a map from $\partial\Omega$ into S^1, plays a fundamental role. We have to distinguish two cases :

A) The case $d = 0$,
B) The case $d \neq 0$.

A) The case $d = 0$.

Here the situation is rather simple since no singularity appears. Our main result is

THEOREM 5 (see [4]). *As $\varepsilon \to 0, u_\varepsilon \to u_\star$ is $C^{1,\alpha}(\overline{\Omega})$ $\forall \alpha < 1$, where $u_\star$ is the unique solution of the problem*

$$\min_{u \in H_g^1} \frac{1}{2} \int_\Omega |\nabla u|^2 \tag{6'}$$

and

$$H_g^1 = H_g^1(\Omega; S^1) = \{u \in H_g^1(\Omega;\mathbb{C}); |u| = 1 \text{ a.e. } on\ \Omega\}.$$

Note that since $d = \deg(g, \partial\Omega) = 0$, the class of testing functions H_g^1 is not empty. The Euler equation associated to **(6')** is

$$\begin{cases} -\Delta u = u|\nabla u|^2 & \text{in } \Omega, \\ |u| = 1 & \text{in } \Omega, \\ u = g & \text{on } \partial\Omega. \end{cases} \tag{7'}$$

Even though this system is the same as (3), the fact that, here, $u \in S^1$ (instead of S^2) enables us to solve **(7')** very simply. Indeed, we may write

$$u = e^{i\varphi}$$

where φ is a well-defined smooth singlevalued function from Ω into $\mathbb{R}$. Then (7') is equivalent

$$\Delta\varphi = 0 \text{ in } \Omega. \tag{8}$$

Thus, in order to find $u_\star$ it suffices to solve (8) together with the boundary condition

$$\varphi = \varphi_0 \text{ on } \partial\Omega \tag{9}$$

where φ_0 is such that $e^{i\varphi_0} = g$. Let us emphasize once more that it is possible to find φ_0 as a singlevalued function because we have assumed that $d = \deg(g, \partial\Omega) = 0$.

Remark 3. It is already interesting to point out the contrast between Theorems 2 and 5. In Section 1 singularities could appear even when $\deg(g, \partial\Omega) = 0$. Here, we have no singularity in this case.

B) The case $d \neq 0$.

For simplicity, we may assume $d > 0$ (the case $d < 0$ is similar by passing to complex conjugates). Note that, here, problem (6) –which is formally the limit of (5) as $\varepsilon \to 0$– does not make sense since $H^1_g = \phi$ when $d \neq 0$ (warning : we cannot use $g(x/|x|)$ as in the 3-d problem since, now, $x/|x|$ has infinite energy).

Our first result is the following

Theorem 6 (see [3], [5]). *Assume Ω is starshaped. Then, there exists a sequence $\varepsilon_n \to 0$, d points $a_1, a_2, ..., a_d$ in Ω, and a function $u_\star$ such that*

$$u_{\varepsilon_n} \to u_\star \quad \text{in } C^k_{loc}(\Omega \backslash \cup \{a_i\}) \qquad \forall k.$$

In addition $u_\star$ satisfies

$$\begin{cases} -\Delta u_\star = u_\star |\nabla u_\star|^2 & \text{in } \Omega \backslash \cup \{a_i\} \\ |u_\star| = 1 & \text{in } \Omega \backslash \cup \{a_i\} \\ u_\star = g & \text{on } \partial\Omega \end{cases} \tag{10}$$

and near each singularity a_i we have

$$u_\star(z) \simeq \alpha_i \frac{(z - a_i)}{|z - a_i|}$$

for some complex constants α_i with $|\alpha_i| = 1$. In particular $\deg(u, a_i) = +1\ \forall i$.

Remark 4. It is interesting to point out the analogies and differences between Theorems 1 and 6. In both situations we have a finite number of singularities. In the 3-d problem very little is known about the number of singularities ; in general, it can be much larger that $|\deg(g, \partial\Omega)|$, while in the 2-d problem the number of singularities is exactly equal to $|\deg(g, \partial\Omega)|$. In the 3-d problem the degree of each singularity is $+1$ or -1 ; in general singularities of degrees $+1$ and -1 coexist. In the 2-d problem each singularity has degree + 1 (or -1 if $\deg(g, \partial\Omega) < 0$). This is in complete agreement with the observations of physicists in superconductors and superfluids.

Another striking difference between the results in the 3-d and 2-d cases is the fact that, here, we have a precise information about the location of the singularities (nothing similar is known for the 3-d problem). This is the content of our next result. But we must first introduce some notations.

Given any configuration $b = (b_1, b_2, ..., b_d)$ of d distinct points in Ω, consider

$$\Omega_\delta = \Omega \backslash \bigcup_{i=1}^{d} B(b_i, \rho)$$

where ρ is a small positive paramater. Since we have changed the topology of the domain the class of smooth maps fromΩ_ρ into S^1 such that $u = g$ on $\partial\Omega$ is not empty. More precisely, set

$$\mathcal{E}_\rho = \left\{ u \in H^1(\Omega_\rho; S^1) \middle| \begin{array}{l} \deg(u, \partial B(b_i, \rho)) = +1 \ \forall i \\ \text{and } u = g \text{ on } \partial\Omega \end{array} \right\}$$

Clearly, $\mathcal{E}_\rho \neq \phi$ and we may consider

$$\min_{u \in \mathcal{E}_\rho} \frac{1}{2} \int_{\Omega_\rho} |\nabla u|^2. \tag{11}$$

One can prove (see [5]) that (11) has a unique solution u_ρ which can be related to the solution of a linear problem as in Case A). As $\rho \to 0$ $\int_{\Omega_\rho} |\nabla u_\rho|^2$ blows up like $|\log \rho|$. More precisely we have the expansion (see [5]), as $\rho \to 0$,

$$\frac{1}{2} \int |\nabla u_\rho|^2 = \pi d |\log \rho| + W + 0(\rho) \tag{12}$$

where W is a finite quantity depending only on Ω, g and $b = (b_i)$. Since we assume that Ω and g are fixed we shall consider $W = W(b)$ only as a function of the configuration of the singularities $b = (b_1, b_2, ..., b_2)$. In other words, W is what remains in the energy after the singular "core energy" has been removed. We shall call it the renormalized energy associated to b.

It turns out that there is an explicit formula for computing W. More precisely we have (see [5])

$$W(b) = -\pi \sum_{i \neq j} \log|b_i - b_j| + \frac{1}{2} \int_{\partial\Omega} \Phi(g \wedge g_\tau) - \pi \sum_{i=1}^{d} R(b_i) \tag{13}$$

where Φ is the solution of the linear Neumann problem

$$\Delta\Phi = 2\pi \sum_{i=1}^{d} \delta_{b_i} \text{ in } \Omega \ (\delta = \text{Dirac mass}),$$
$$\frac{\partial\Phi}{\partial\nu} = g \wedge g_\tau \text{ on } \partial\Omega$$

(ν is the unit outward normal to $\partial\Omega$ and τ is the unit tangent vector to ∂G such that (ν, τ) is direct). R denotes the regular part of the "Green's function" Φ, i.e.

$$R(x) = \Phi(x) - \sum_{i=1}^{d} \log|x - b_i|.$$

Note that R is smooth on Ω, so that $R(b_i)$ makes sense.

Using (13) it is not difficult to check that :
(i) $W \to +\infty$ as two of the points b_i coalesce,
(ii) $W \to +\infty$ as one of the points b_i tends to ∂G.
In other words, the singularities b_i repel each other but the boundary has a confinement effect. In particular, W achieves its minimum on G^d and every minimizing configuration consists of d distinct points.

The location of the points (a_i) in Theorem 6 is governed by W through the following :

Theorem 7 (see [5]). *Let $a = (a_i)$ be as in Theorem 6, then*

$$W(a) \leq W(b) \qquad \forall b \in G^d,$$

i.e. a is a minimizing configuration for W.

In general, W can have several minima. Each one of them may arise as the singular set of Theorem 6 associated to different choices of subsequences $\varepsilon_n \to 0$. However, once (a_i) is known, then $u_\star$ given in Theorem 6 is uniquely determined through the following :

Theorem 8 (see [5]). *Let (a_i) and $u_\star$ be as in Theorem 6 then*

$$u_\star(z) = e^{i\varphi(z)} \frac{(z - a_1)}{|z - a_1|} \frac{(z - a_2)}{|z - a_2|} \cdots \frac{(z - a_d)}{|z - a_d|}$$

where φ is the unique solution of the linear Dirichlet problem

$$\Delta\varphi = 0 \qquad \text{in } \Omega,$$
$$\varphi = \varphi_0 \quad \text{on } \partial\Omega,$$

and φ_0 is defined on $\partial\Omega$ by

$$e^{i\varphi_0(z)} = g(z)\frac{|z-a_1|}{(z-a_1)}\,\frac{|z-a_2|}{(z-a_2)}\cdots\frac{|z-a_d|}{(z-a_d)}. \tag{14}$$

Note that the right hand side in (14) is a map from $\partial\Omega$ into S^1 of degree <u>zero</u>, so that φ_0 is a well defined single-valued smooth function.

<u>Remark 5</u>. It is a very interesting open problem to study the pattern formation of the singularities as $d \to +\infty$, for example when Ω is the unit disc and $g = e^{id\theta}$.

REFERENCES.

[1] F. Bethuel, The approximation problem for Sobolev maps between manifolds, Acta Math. **167** (1991), p. 153-206.
[2] F. Bethuel, H. Brezis and J.-M. Coron, Relaxed energies for harmonic maps, in *Variational Problems* (H. Berestycki, J.-M. Coron and I. Ekeland ed.), Birkhauser (1990).
[3] F. Bethuel, H. Brezis and F. Hélein, Limite singulière pour la minimisation de fonctionnelles du type Ginzburg-Landau, C. R. Acad. Sc. **314** (1992), p. 891-894.
[4] F. Bethuel, H. Brezis and F. Hélein, Asymptotics for the minimization of a Ginzburg-Landau functional, Calculus of Variations and PDE (to appear).
[5] F. Bethuel, H. Brezis and F. Hélein, *Ginzburg-Landau vortices* (to appear).
[6] F. Bethuel and X. Zheng, Density of smooth functions between two manifolds in Sobolev spaces, J. Funct. Anal. **80** (1988), p. 60-75.
[7] H. Brezis, Liquid crystals and energy estimates for S^2-valued maps, in *Theory and Applications of Liquid Crystals* (J. Ericksen and D. Kinderlehrer ed.), Springer (1987).
[8] H. Brezis, S^k-valued maps with singularities, in *Topics in the Calculus of Variations* (M. Giaquinta ed.) L.N. vol. 1365, Springer (1989), p. 1-30.
[9] H. Brezis, J.-M. Coron and E. Lieb, Harmonic maps with defects, Comm. Math. Phys. **107** (1986), p. 649-705.
[10] W. Brinkman and P. Cladis, Defects in liquid crystals, Physics Today, May 1982, p. 48-54.
[11] P. De Gennes, *Superconductivity of metals and alloys*, Benjamin (1966).
[12] P. De Gennes, *The Physics of Liquid Crystals*, Clarendon Press, Oxford (1974).
[13] R. Donnelly, *Quantized vortices in helium II*, Cambridge Univ. Press (1991).
[14] J. Eells and L. Lemaire, A report on harmonic maps, Bull. London Math. Soc. **10** (1978), p. 1-68.
[15] J. Ericksen and D. Kinderlehrer ed. *Theory and Applications of Liquid Crystals*, IMA Series vol. 5, Springer (1987).
[16] R. Feynman, Applications of quantum mechanics to liquid helium, in *Progress in Low Temperature Physics*, vol. I (C. Gorter ed.) North-Holland, Chap. 2, (1955).

[17] M. Giaquinta, G. Modica and J. Soucek, Cartesian currents and variational problems for mappings with values into spheres, Ann. Sc. Norm. Sup. Pisa **16** (1989), p. 393-485.
[18] M. Giaquinta, G. Modica and J. Soucek, The Dirichlet energy of mappings with values into the sphere, Manuscripta Math. **65** (1989), p. 489-507.
[19] R. Hardt and F.H. Lin, A remark on H^1 mappings, Manuscripta Math. **56** (1986), p. 1-10.
[20] R. Hardt, F.H. Lin and C.C. Poon, Axially symmetric harmonic maps minimizing a relaxed energy (to appear).
[21] L. Onsager, Discussion on paper by C. Gorter, Nuovo Cimento Suppl. **6** (1949), p. 249-250.
[22] T. Rivière, Applications harmoniques de B^3 dans S^2 ayant une ligne de singularités, C. R. Acad. Sc. **313** (1991), p. 583-587.
[23] T. Rivière, Applications harmoniques de B^3 dans S^2 partout discontinues, C. R. Acad. Sc. **314** (1992), p. 719-723 and detailed paper to appear.
[24] R. Schoen and K. Uhlenbeck, A regularity theory for harmonic maps, J. Diff. Geom. **17** (1982), p. 307-335 and Boundary regularity and the Dirichlet problem for harmonic maps, J. Diff. Geom. **18** (1983), p. 253-268.
[25] D. Tilley and J. Tilley, *Superfluidity and superconductivity*, Adam Hilger (1986).

Microcanonical Action and the Entropy of a Rotating Black Hole

J. David Brown

Departments of Physics and Mathematics
North Carolina State University, Raleigh, NC 27695–8202

James W. York, Jr.

Department of Physics and Astronomy
University of North Carolina, Chapel Hill, NC 27599–3255

The authors have recently proposed a "microcanonical functional integral" representation of the density of quantum states of the gravitational field. The phase of this real–time functional integral is determined by a "microcanonical" or Jacobi action, the extrema of which are classical solutions at fixed total energy, not at fixed total time interval as in Hamilton's action. This approach is fully general but is especially well suited to gravitating systems because for them the total energy can be fixed simply as a boundary condition on the gravitational field. In this paper we describe how to obtain Jacobi's action for general relativity. We evaluate it for a certain complex metric associated with a rotating black hole and discuss the relation of the result to the density of states and to the entropy of the black hole.

1 DEDICATION

We dedicate this paper to Yvonne Choquet–Bruhat in honor of her retirement following a brilliant career. JWY would like to thank her for friendship, support, and their happy collaboration in "analysis, manifolds, and physics" [1].

2 INTRODUCTION

We are concerned here with the description of a stationary and axisymmetric rotating black hole in a closed system in thermodynamic equilibrium with its environment. The equilibrating radiation in the closed system surrounding the hole is the "environment". This "radiation fluid" rotates at a constant angular velocity [2][3]. Unlike the more familiar non–relativistic case, here one must take into account explicitly the spatial finiteness of the system in order to avoid super–luminal velocities resulting from the rotation. Even though we shall ignore the explicit effects of the equilibrating radiation in this work, its presence–in–principle must be kept in mind in order to give the problem a physically and mathematically reasonable formulation.

M. Flato et al. (eds.), Physics on Manifolds, 23–34.

The key conserved quantities in closed systems like ours are the total energy and the total angular momentum. Their "Massieu" conjugates [4] are, respectively, inverse temperature β, and β multiplied by an appropriate angular velocity. An important question is which among these quantities to specify in advance, that is, which "ensemble" picture to employ. We choose here to fix energy and angular momentum as in the microcanonical ensemble. This choice leads us to employ the general relativistic version of Jacobi's action.

Jacobi's form of the action principle [5][6] involves variations at fixed energy, rather than the variations at fixed time used in Hamilton's principle. The fixed time interval in Hamilton's action becomes fixed inverse temperature β in the "periodic imaginary time" formulation, thus transforming Hamilton's action into the appropriate (imaginary) phase for a periodic path in computing the canonical partition function from a "Euclidean" Feynman functional integral [7]. (We are here and in the next paragraph speaking only of energy and inverse temperature in order to simplify the discussion. Similar remarks hold for angular momentum and its conjugate.) In contrast, fixed energy is suitable for the microcanonical ensemble and, correspondingly, Jacobi's action is the phase in an expression for the density of states as a real–time "microcanonical functional integral" (MCFI) [8].

Let us characterize briefly the canonical and microcanonical pictures. Neither picture can hold with perfect precision even in principle when gravity is taken into account because the "infinite" heat bath of the canonical picture and the "perfectly diathermic" walls of the microcanonical picture are both at variance with the known physics of the gravitational field. However, each picture still provides a useful framework for discussion. Furthermore, the long–range, unscreened nature of the gravitational interaction means that simple notions of ordinary statistical thermodynamics like "extensive", "intensive", *etc.*, no longer apply in general. With gravity, statistical mechanics is inherently *global*.

In the canonical picture, with a fixed temperature shared by all constituents of a system, there are no constraints on the energy. This feature simplifies combinatorial (counting) problems and leads to factorization of the partition function for weakly coupled constituents. For gravitating systems in equilibrium, the temperature is not spatially uniform because of red–shift and blue–shift effects. In such cases, the relevant temperature is that determined at the spatial two–boundary B of the system [9]. It can be specified by a boundary condition on the metric [10][11]. It is then used in conjunction with Hamilton's principle, which is the form of the gravity action in which the metric is fixed on the history of the spatial two–boundary [12]. (The metric determines the lapse of proper time along the history of the boundary.) The case of a rotating, charged, stationary, axisymmetric black hole has been treated with (grand) canonical boundary conditions in [13]. See also [10] and [11].

With its constraint on the energy, the microcanonical picture leads to stability properties more robust than in the canonical case. However, the energy constraint can complicate combinatorial problems because the constituents of the system must all share from a common fixed pool of energy. For field theories, with a continuous infinity of degrees of freedom, the energy constraint restricts the entire phase space of the system *unless gravity is taken into account*. For gravitating systems, as a consequence of the equivalence principle, the total energy in a given finite region, including that of matter fields, can be given as an integral of certain geometrically well–defined derivatives of the metric over a finite two–surface bounding the system. In other words, one *can* find and employ a suitable expression for "quasi–local" gravitational energy that does not require an appeal to asymptotic regions. This quasi–local energy has the value, per unit proper time, of the Hamiltonian of the spatially bounded region, as discussed in detail in [12]. Therefore, if we specify as a boundary condition the energy per unit two–surface area, we have constrained the total energy simply by a boundary condition [8][12]. Thus, through the mediation of the gravitational field, the canonical and microcanonical cases are placed on similar footing and differ only in which of the conjugate variables [14], inverse temperature or energy, is specified on the boundary. The corresponding functional integrals, for the partition function or density of states, differ in which action gives the correct phase, Hamilton's or Jacobi's. We regard the MCFI for the density of states as the more fundamental [8].

We shall now outline a recent application of the above reasoning to the case of an axisymmetric stationary black hole. The MCFI, in a steepest descents approximation, shows that the density of states is the exponential of one–fourth of the area of the event horizon, thus confirming in this approximation the Bekenstein–Hawking expression for the entropy of a black hole [8]. Full details of the following are given in [8] and [12]. (An application of the MCFI to a simple harmonic oscillator–without gravity–has been given recently by the authors [15]. No approximations were required in this case and the exact energy spectrum was obtained.)

3 JACOBI'S ACTION FOR GRAVITY

We here analyze the action for pure general relativity. The method, however, can be just as well applied when matter is included and/or for higher–derivative theories of gravity.

Consider a region of spacetime $M = \Sigma \times I$ where Σ are spatial slices and I is an interval of the real line. The two–boundary of space Σ is denoted by B, whose history is ${}^3B = B \times I$. The orthogonal intersection of any Σ with ${}^3B$ is the two–boundary B at some time. Thus a generic B can be regarded as being embedded in Σ with a spacelike unit normal n^μ tangent to Σ at B, and also as being embedded in ${}^3B$ with a timelike unit normal u^μ such that $u_\mu n^\mu = 0$. The subspaces of M that correspond to the endpoints of I are spacelike hypersurfaces $t = t'$ and $t = t''$. The notation $\int_{t'}^{t''} d^3x$ denotes an integral over t'' minus an integral over t'. The spacetime metric is $g_{\mu\nu}$ and its scalar curvature is $\Re$. Then "Hamilton's" action for general relativity, in which the metric is fixed on the boundary, is given by a variant of the Hilbert action, namely [12][16]

$$S[g] = \frac{1}{2\kappa}\int_M d^4x\sqrt{-g}(\Re - 2\Lambda) + \frac{1}{\kappa}\int_{t'}^{t''} d^3x\sqrt{h}K - \frac{1}{\kappa}\int_{{}^3B} d^3x\sqrt{-\gamma}\Theta - S^0 \, , \tag{1}$$

where the metric and extrinsic curvature of the slices Σ are h_{ij} and K_{ij}, while those for ${}^3B$ are γ_{ij} and Θ_{ij}. (Latin letters i, j, ... are used as tensor indices for both Σ and ${}^3B$. No cause for confusion arises from this convention.) In (1), $\kappa = 8\pi G$ and we set Newton's constant $G = 1$ henceforth. The term S^0 in (1) is a functional of the metric γ_{ij} of ${}^3B$. The purpose of this term is to determine the "zero" of energy and momentum. It will not turn out to affect the Jacobi action for general relativity. (S^0 is analyzed in [12] but was not included in [16].)

The canonical form of "Hamilton's" action (1) is [12]

$$S = \int_M d^4x\left[P^{ij}\dot{h}_{ij} - N\mathcal{H} - V^i\mathcal{H}_i\right] - \int_{{}^3B} d^3x\sqrt{\sigma}\left[N\varepsilon - V^i j_i\right] \, , \tag{2}$$

where P^{ij} is the Arnowitt–Deser–Misner momentum conjugate to h_{ij}, and $\mathcal{H}$ and $\mathcal{H}_i$ are the usual Hamiltonian and momentum constraints. The lapse function is denoted by N and the shift vector by V^i. In the surface term of (2), σ denotes the determinant of the two–metric σ_{ij} induced on B as a surface embedded in Σ; likewise, n_i denotes B's unit normal in Σ and k_{ij} denotes the corresponding extrinsic curvature. The energy surface density ε and momentum surface density j_i are given by [12]

$$\begin{aligned} \varepsilon &= \frac{1}{\kappa}k + \frac{1}{\sqrt{\sigma}}\frac{\delta S^0}{\delta N} \, , \\ j_i &= -\frac{2}{\sqrt{h}}\sigma_{ij}n_k P^{jk} - \frac{1}{\sqrt{\sigma}}\frac{\delta S^0}{\delta V^i} \, . \end{aligned} \tag{3}$$

The total quasi-local energy is the integral of ε over B. In obtaining (2), we have assumed that S^0, if present, is a linear functional of N and V^i on ${}^3B$. In the total Hamiltonian extracted directly from (2), the shift vector must satisfy $n_i V^i|_B = 0$. Each of these points is discussed in [12].

In the action (2), one varies h_{ij}, P^{ij}, N, and V^i to obtain

$$\delta S = (\text{terms giving the equations of motion}) + \int_{t'}^{t''} d^3x\, P^{ij}\delta h_{ij} - \int_{^3B} d^3x\sqrt{\sigma}\left[\varepsilon\,\delta N - j_a\delta V^a - (N/2)s^{ab}\delta\sigma_{ab}\right], \tag{4}$$

where indices a, b, ... are used to denote tensors on B, *i.e.*, Σ–tensors that are orthogonal to n^i. The surface stress tensor s^{ab} on B is given in [12] and does not concern us here. We see from (4) that suitable boundary conditions for S are obtained by fixing the metric induced on the boundary elements t', t'', and ${}^3B$ of M. In particular, the lapse function N is fixed on ${}^3B$, where it determines proper time elements Ndt on ${}^3B$ along the unit normal u^μ associated with the foliation of ${}^3B$ by B; hence, we refer to S as "Hamilton's action" in canonical form.

What we define as the microcanonical action S_m is, in essence, Jacobi's action for general relativity. It is obtained from S by a canonical transformation that changes the appropriate boundary conditions on ${}^3B$ from fixed metric components N, V^a, and σ_{ab} to fixed energy surface density ε, momentum surface density j_a, and boundary metric σ_{ab}. Thus, define [8]

$$\begin{aligned} S_m &= S + \int_{^3B} d^3x\sqrt{\sigma}\left[N\varepsilon - V^a j_a\right] \\ &= \int_M d^4x\left[P^{ij}\dot{h}_{ij} - N\mathcal{H} - V^i\mathcal{H}_i\right]. \end{aligned} \tag{5}$$

From (4), it follows that the variation of S_m is

$$\delta S_m = (\text{terms giving the equations of motion}) + \int_{t'}^{t''} d^3x\, P^{ij}\delta h_{ij} + \int_{^3B} d^3x\left[N\,\delta(\sqrt{\sigma}\varepsilon) - V^a\delta(\sqrt{\sigma}j_a) + (N\sqrt{\sigma}/2)s^{ab}\delta\sigma_{ab}\right]. \tag{6}$$

This result shows that solutions of the equations of motion extremize S_m with respect to variations in which ε, j_a, and σ_{ab} are held fixed on the boundary B. Observe that the unspecified subtraction term S^0 does not appear in S_m. Nevertheless, the variation (6) of S_m involves ε, j_a, and s^{ab}, which do depend on S^0 through their definitions. However, all dependences on S^0 in the boundary variation terms of δS_m actually cancel because of the requirement that S^0 be a linear functional of the lapse and shift [8][12]. Thus, neither S_m nor its variation depends on S^0. In other words, as long as energy and momentum are to be fixed, as in S_m, a quantity like S^0 whose only role is to determine their "zero points" is irrelevant.

4 MICROCANONICAL FUNCTIONAL INTEGRAL

In [8] and [15] we showed that for nonrelativistic mechanics the density of states is given by a sum over periodic, real time histories, where each history contributes a phase determined by Jacobi's action. In the case of nonrelativistic mechanics, the energy that is fixed in Jacobi's action is just the value of the Hamiltonian that generates unit time translations. For a self–gravitating system, the Hamiltonian has a "many–fingered" character: space can be pushed into the future in a variety of ways, governed by different choices of lapse function N and shift vector V^i. The value of the Hamiltonian incorporated into (2) depends on this choice. More precisely, the value of the Hamiltonian is determined by the choice of lapse and shift on the boundary B, since the lapse and shift on the domain of Σ *interior* to B are Lagrange multipliers for the (vanishing) Hamiltonian and momentum constraints. Accordingly, the energy surface–density ε and momentum surface–density j_a for a self–gravitating system play a role that is analogous to energy for a nonrelativistic mechanical system.

The above considerations lead us to propose that the density of states for a spatially finite, self–gravitating system is a functional of the energy surface–density ε and momentum surface–density j_a. In addition to these energy–like quantities, the density of states is also a functional of the metric σ_{ab} on the boundary B, which specifies the size and shape of the system. In the absence of matter fields, these make up the complete set of variables and $\nu[\varepsilon, j_a, \sigma_{ab}]$ is interpreted as the density of quantum states of the gravitational field with energy density, momentum density, and boundary metric having the values ε, j_a, and σ_{ab}. The action to be used in the functional integral representation of ν is S_m, which describes the gravitational field with fixed ε, j_a, and σ_{ab}. Note that ε, j_a, and σ_{ab} replace the traditional thermodynamical extensive variables. Our variables are all constructed from the dynamical phase space variables (h_{ij}, P^{ij}) for the system, where the phase space structure is defined using the foliation of M into spacelike hypersurfaces. (We expect this to be a defining feature of extensive variables for general systems of gravitational and matter fields.)

We propose [8] that the density of states of the gravitational field is defined formally by

$$\nu[\varepsilon,j,\sigma] = \sum_M \int \mathcal{D}H \exp(iS_m) \,. \tag{7}$$

(Planck's constant has been set to unity.) The sum over M refers to a sum over manifolds of different topologies. The three–boundary for each M is required to have topology $\partial M = B \times S^1$. If B has two–sphere topology, then the sum over topologies includes $M =$ (ball) $\times S^1$, with $\partial M = \partial$(ball) $\times S^1 = S^2 \times S^1$. Another example is $M =$ (disk) $\times S^2$, with $\partial M = \partial$(disk) $\times S^2 = S^1 \times S^2$. The action S_m that appears in Eq. (7) is the microcanonical action (5) of the previous section applied to the manifolds M with a single boundary component $\partial M = {}^3B = B \times S^1$. The functional integral (7) for ν is a sum over Lorentzian metrics $g_{\mu\nu}$. Note that the microcanonical action may require the addition of terms that depend on the topology of M, such as the Euler number.

In the boundary conditions on $\partial M = B \times S^1$, the two–metric σ_{ab} that is fixed on the hypersurfaces B is typically real and spacelike. Likewise, the energy density ε is real, which requires the unit normal to ∂M to be spacelike. Therefore, the Lorentzian metrics on M must induce a Lorentzian metric on ∂M, where the timelike direction coincides with the periodically identified S^1. Note, however, that there are no nondegenerate Lorentzian metrics on a manifold with topology $M =$ (disk) $\times S^2$ that also induce such a Lorentzian metric on ∂M. This implies that the formal functional integral (7) for the density of states must include degenerate metrics. (For a discussion of the role of degenerate metrics in classical and quantum gravity, see [17].)

Now consider the evaluation of the functional integral (7) for fixed boundary data ε, j_a, σ_{ab} that correspond to a stationary, axisymmetric black hole. That is, start with a real Lorentzian, stationary, axisymmetric, black hole solution of the Einstein equations, and let $T =$ constant be stationary time slices that contain the closed orbits of the axial Killing vector field. Next, choose a topologically spherical two–surface B that contains the orbits of the axial Killing vector field, and is contained in a $T =$ constant hypersurface. From this surface B embedded in a $T =$ constant slice, obtain the data ε, j_a, and σ_{ab}. In the functional integral for $\nu[\varepsilon,j,\sigma]$, fix this data on each $t =$ constant slice of ∂M. Observe that, to the extent that the physical system can be approximated by a single classical configuration, that configuration will be the real stationary black hole that is used to induce the boundary data.

The functional integral (7) can be evaluated semiclassically by searching for four–

metrics $g_{\mu\nu}$ that extremize S_m and satisfy the specified boundary conditions. Observe that the Lorentzian black hole geometry that was used to motivate the choice of boundary conditions is *not* an extremum of S_m, because it has the topology [Wheeler (spatial) wormhole]×[time] and cannot be placed on a manifold M with a single boundary $S^2 \times S^1$. However, there is a related complex four–metric that does extremize S_m, and is described as follows. Let the Lorentzian black hole be given by

$$ds^2 = -\tilde{N}^2 dT^2 + \tilde{h}_{ij}(dx^i + \tilde{V}^i dT)(dx^j + \tilde{V}^j dT) , \tag{8}$$

where $\tilde{N}$, $\tilde{V}^i$, and $\tilde{h}_{ij}$ are T–independent functions of the spatial coordinates x^i. The horizon coincides with $\tilde{N} = 0$. Choose spatial coordinates that are "co–rotating" with the horizon. Then the proper spatial velocity of the spatial coordinate system relative to observers at rest in the $T =$ constant slices vanishes on the horizon, $(\tilde{V}^i/\tilde{N}) = 0$, and the Killing vector field $\partial/\partial T$ coincides with the null generator of the horizon. As shown in [13], the complex metric

$$ds^2 = -(-i\tilde{N})^2 dT^2 + \tilde{h}_{ij}(dx^i - i\tilde{V}^i dT)(dx^j - i\tilde{V}^j dT) , \tag{9}$$

where the coordinate T is real, satisfies the Einstein equations everywhere on a manifold with topology $M =$ (disk) $\times S^2$, with the possible exception of the points $\tilde{N} = 0$ where the foliation $T =$ constant degenerates. The locus of those points $\tilde{N} = 0$ is a two–surface called the "bolt" [18]. Near the bolt, the metric becomes

$$ds^2 \approx \tilde{N}^2 dT^2 + \tilde{h}_{ij} dx^i dx^j , \tag{10}$$

and describes a Euclidean geometry. The sourceless Einstein equations are not satisfied at the bolt if this geometry has a conical singularity in the two–dimensional submanifold that contains the unit normals $\tilde{n}^i$ to the bolt for each of the $T =$ constant hypersurfaces. However, there is no conical singularity if the circumferences of circles surrounding the bolt initially increase as 2π times proper radius. The circumference of such circles is given by $P\tilde{N}$, where P is the period in coordinate time T. Therefore the absence of conical singularities is insured if the condition $P(\tilde{n}^i\partial_i\tilde{N}) = 2\pi$ holds at each point on the bolt, where $\tilde{n}^i$ is the unit normal to the bolt in one of the $T =$ constant surfaces. Because the unit normal is proportional to $\partial_i\tilde{N}$ at the bolt, this condition restricts the period in coordinate time T to be $P = 2\pi/\kappa_H$, where $\kappa_H = [(\partial_i\tilde{N})\tilde{h}^{ij}(\partial_j\tilde{N})]^{1/2}\big|_H$ is the surface gravity at the horizon of the Lorentzian black hole (8).

The lapse function and shift vector for the complex metric (9) are $N = -i\tilde{N}$ and $V^i = -i\tilde{V}^i$. Thus, (9) and the Lorentzian metric (8) differ only by a factor of $-i$

in their lapse functions and shift vectors. In particular, the three–metric $\tilde{h}_{ij}$ and its conjugate momentum $\tilde{P}^{ij}$ coincide for the stationary metrics (8) and (9) [13]. Since the boundary data ε, j_a, and σ_{ab} are constructed from the canonical variables only, the complex metric (9) satisfies the boundary conditions imposed on the functional integral for $\nu[\varepsilon, j, \sigma]$.

The complex metric (9) with the periodic identification given above extremizes the action S_m and satisfies the chosen boundary conditions for the density of states $\nu[\varepsilon, j, \sigma]$. Although this metric is not included in the sum over Lorentzian geometries (7), it can be used for a steepest descents approximation to the functional integral by distorting the contours of integration for the lapse N and shift V^i in the complex plane. Then the density of states is approximated by

$$\nu[\varepsilon, j, \sigma] \approx \exp(iS_m[-i\tilde{N}, -i\tilde{V}, \tilde{h}]) \ , \tag{11}$$

where $S_m[-i\tilde{N}, -i\tilde{V}, \tilde{h}]$ is the microcanonical action evaluated at the complex extremum (9). The density of states can be expressed approximately as

$$\nu[\varepsilon, j, \sigma] \approx \exp(\mathcal{S}[\varepsilon, j, \sigma]) \ , \tag{12}$$

where $\mathcal{S}[\varepsilon, j, \sigma]$ is the entropy of the system. Then the result (11) shows that the entropy is

$$\mathcal{S}[\varepsilon, j, \sigma] \approx iS_m[-i\tilde{N}, -i\tilde{V}, \tilde{h}] \tag{13}$$

for the gravitational field with microcanonical boundary conditions.

In order to evaluate S_m for the metric (9), we start with the microcanonical action written in spacetime covariant form [8] and perform a canonical decomposition under the assumption that the manifold M has the topology of a punctured disk $\times S^2$. That is, the spacelike hypersurfaces Σ have topology $I \times S^2$, and the timelike direction is periodically identified (S^1). The outer boundary of the disk corresponds to the three–boundary element ${}^3B$ of M (denoted ∂M previously) on which the boundary values of ε, j_a, and σ_{ab} are imposed. The inner boundary of the disk, the boundary of the puncture, appears as another boundary element ${}^3H$ for M. (No data are specified at ${}^3H$.) The canonical decomposition results in [8]

$$S_m = \int_M d^4x \left[P^{ij}\dot{h}_{ij} - N\mathcal{H} - V^i\mathcal{H}_i\right] + \int_{{}^3H} d^3x\sqrt{\sigma}\left[n^i(\partial_i N)/\kappa + 2n_iV_jP^{ij}/\sqrt{h}\right] \ , \tag{14}$$

where the expression $a_i = (\partial_i N)/N$ for the acceleration of the timelike unit normal has been used. The boundary term at ${}^3H$ was first given in [13].

Now evaluate the action S_m on the punctured disk $\times S^2$ for the complex metric (9), and take the limit as the puncture disappears to obtain a manifold topology $M = (\text{disk}) \times S^2$. In this limit, the smoothness of the complex geometry is assured by the regularity condition on the period of T. Since the metric satisfies the Einstein equations, the Hamiltonian and momentum constraints vanish, and the terms $P^{ij}\dot{h}_{ij}$ also vanish by stationarity. Moreover, the second boundary term at ${}^3H$ is zero because the shift vector vanishes at the horizon. Thus, only the first of the boundary terms at ${}^3H$ survives. Evaluating this term for the complex metric (9), that is, for the lapse function $N = -i\tilde{N}$, and using the regularity condition $P = 2\pi/\kappa_H$, we find for the microcanonical action

$$S_m[-i\tilde{N}, -i\tilde{V}, \tilde{h}] = -\frac{i}{\kappa}\int_0^P dT \int d^2x\sqrt{\tilde{\sigma}}\tilde{n}^i\partial_i\tilde{N} = -\frac{i}{4}A_H \ , \tag{15}$$

where A_H is the area of the event horizon for the Lorentzian black hole (8).

The result (15) for the microcanonical action evaluated at the extremum (9) leads to the following approximation for the entropy (13):

$$S[\varepsilon, j, \sigma] \approx \frac{1}{4}A_H \ . \tag{16}$$

The generality of the result (16) should be emphasized: The boundary data ε, j_a, and σ_{ab} were chosen from a general stationary, axisymmetric black hole that solves the vacuum Einstein equations within a spatial region with boundary B. Outside the boundary B, the black hole spacetime need not be free of matter or be asymptotically flat. Thus, for example, the black hole can be distorted relative to the standard Kerr family. Furthermore, recall that the quantum–statistical system with this boundary data is classically approximated by the physical black hole solution that matches that boundary data. The result (16) shows that the entropy of the system is approximately 1/4 the area of the event horizon of the physical black hole configuration that classically approximates the contents of the system. It also should be emphasized that the entropy is independent of the term S^0 in (1) as has been shown in the framework of the canonical partition function [9].

Expresson (16) is the principal result we wished to demonstrate here. The physical and mathematical limitations of our analysis and possible ways to overcome them are described in [8], which also discusses canonical and grand canonical boundary conditions for the rotating black hole. A further elaboration of the physical underpinnings and implications of our analysis of relativistic rotating systems is given in [19].

The authors gratefully acknowledge research support received from the National Science Foundation, grant number PHY–8908741.

REFERENCES

[1] Y. Choquet–Bruhat and J.W. York, The Cauchy Problem, pp. 99–172 in *General Relativity and Gravitation*, edited by A. Held (Plenum Press, New York, 1980).

[2] W. Israel and J.M. Stewart, in *General Relativity and Gravitation. II*, edited by A. Held (Plenum Press, New York, 1980).

[3] V. Frolov and K.S. Thorne, Phys. Rev. **D39**, 2125 (1989).

[4] H.B. Callen, *Thermodynamics* (Wiley, New York, 1985).

[5] C. Lanczos, *The Variational Principles of Mechanics* (University of Toronto Press, Toronto, 1970).

[6] J.D. Brown and J.W. York, Phys. Rev. **D40**, 3312 (1989).

[7] R.P. Feynman and A.R. Hibbs, *Quantum Mechanics and Path Integrals* (McGraw–Hill, New York, 1965).

[8] J.D. Brown and J.W. York, Phys. Rev. **D47**, 1420 (1993).

[9] J.W. York, Phys. Rev. **D33**, 2092 (1986).

[10] B.F. Whiting and J.W. York, Phys. Rev. Lett. **61**, 1336 (1988).

[11] H.W. Braden, J.D. Brown, B.F. Whiting, and J.W. York, Phys. Rev. **D42**, 3376 (1990).

[12] J.D. Brown and J.W. York, Phys. Rev. **D47**, 1407 (1993).

[13] J.D. Brown, E.A. Martinez, and J.W. York, Phys. Rev. Lett. **66**, 2281 (1991).

[14] J.D. Brown, G.L. Comer, E.A. Martinez, J. Melmed, B.F. Whiting, and J.W. York, Class. Quantum Grav. **7**, 1433 (1990).

[15] J.D. Brown and J.W. York, "Jacobi's action and the density of states", to appear

in *Festschrift for Dieter Brill*, edited by B.L. Hu and T. Jacobson (Cambridge University Press, Cambridge, 1993).

[16] J.W. York, Found. Phys. **16**, 249 (1986).

[17] G.T. Horowitz, Class. Quantum Grav. **8**, 587 (1991).

[18] G.W. Gibbons and S.W. Hawking, Commun. Math. Phys. **66**, 291 (1979).

[19] J.D. Brown and J.W. York, "Hamiltonian and boundary conditions for rotating general–relativistic equilibrium systems", to be published.

PROBLÈME DE CAUCHY SUR UN CÔNOÏDE CARACTÉRISTIQUE. APPLICATIONS À CERTAINS SYSTÈMES NON LINÉAIRES D'ORIGINE PHYSIQUE

Francis CAGNAC
Marcel DOSSA

Faculté des Sciences
BP 812
YAOUNDE
CAMEROUN

Summary For the general equation of 2nd order, it is proved that the characteristic Cauchy problem on a conoid has a solution in terms of formal series.

For some systems of quasilinear equations of 2nd order, existence and uniqueness of this Cauchy problem are proved in some convenient functional spaces. Applications to physical systems are mentioned.

Introduction

Le problème de Cauchy sur un conoïde a été étudié d'abord pour des équations à coefficients constants par R. d'Adhémar [1] et Marcel Riesz [9].

Pour des équations linéaires du deuxième ordre à coefficients variables, il a été étudié par F. Cagnac sous des hypothèses de classe C^6 [2], et par Friedlander sous des hypothèses C^∞ [7].

Dans [3], F. Cagnac l'a étudié pour des équations quasilinéaires du deuxième ordre. Mais, outre un ordre de différentiation très élevé sur les données, il imposait aux données de Cauchy d'être la trace sur le conoïde caractéristique C de fonctions vérifiant au sommet 0 de C les équations dérivées jusqu'à l'ordre m–2 pour une solution m^{+1} fois dérivable.

Ces conditions au sommet n'étaient pas nécessaires dans le cas des équations linéaires (Friedlander [7]), ni dans le cas de certaines équations quasilinéaires (A. Rendall [8]). Et Dossa [5] a montré qu'elles étaient inutiles.

Ceci a conduit à penser que pour l'équation quasilinéaire du deuxième ordre la plus générale, la donnée sur un conoïde caractéristique C d'une fonction qui est la trace sur C d'une fonction C^∞ (ou C^p) au voisinage du sommet 0 suffit à déterminer au point 0 toutes les dérivées – ou les dérivées jusqu'à l'ordre p – de toute solution du problème de Cauchy.

M. Flato et al. (eds.), Physics on Manifolds, 35–47.

A. Solutions en série formelle du problème de Cauchy sur un conoïde caractéristique

Les conoïdes caractéristiques d'équations du deuxième ordre sont des conoïdes "du deuxième ordre": nous entendons par là que par un changement de coordonnées de classe C^∞ (ou C^p), le conoïde admet une équation

$$x^0 = s \quad , \quad s = \left(\sum_{i=1}^{n} \left(x^i\right)^2 \right)^{\frac{1}{2}}$$

Dans toute la suite nous utiliserons un tel système de coordonnées.

Les indices grecs prendront les valeurs 0, 1, ... n; les indices latins prendront les valeurs 1, ... n.

1. Traces sur un cône du deuxième ordre de fonctions différentiables au sommet du cône Lemme

Lemme. Pour qu'une fonction $\phi(x^i)$ définie sur le cône C d'équation $x^0 = S$, soit la trace sur C d'une fonction u, p fois dérivable au point 0, il faut et il suffit qu'elle admette au voisinage de 0 un développement limité:

$$(1) : \phi\left(x^i\right) = \sum_{m=0}^{p} \left(\frac{1}{m\,!} \phi_{i_1 \dots i_m} x^{i_1} \dots x^{i_m} + \frac{1}{(m-1)\,!} \psi_{i_1 \dots i_{m-1}} s x^{i_1} \dots x^{i_{m-1}} \right) + o\left(s^p\right)$$

($i_k = 1, \dots n$; on utilise la convention d'Einstein)

Ceci résulte immédiatement de l'écriture du développement de Taylor de

$$u\left(x^\alpha\right) = \sum_{m=0}^{p} \frac{1}{m\,!} u_{\alpha_1 \dots \alpha_m} x^{\alpha_1} \dots x^{\alpha_m} + 0\left(|x|^p\right)$$

($\alpha_k = 0, 1, \dots n$)

quand on y substitue $x^0 = s$ et quand on y remplace s^2 par $\delta_{ij} x^i x^j$

En imposant aux $\phi_{i_1 \ldots i_m}$, $\psi_{i_1 \ldots i_{m-1}}$ d'être symétriques par rapport à leurs indices, on obtient

$$(a)\ \phi_{i_1 \ldots i_m} = u_{i_1 \ldots i_m} + \sum_{k=1}^{m/2} u_{\underset{2k}{o \ldots o}(i_1 \ldots i_{m-2k}} \delta_{i_{m-2k+1}\, i_{m-2k}} \ldots \delta_{i_{m-1} i_m)} C_m^{2k}$$

$$(b)\ \psi_{i_1 \ldots i_{m-1}} = u_{o i_1 \ldots i_{m-1}} + \sum_{k=1}^{(m-2)/2} u_{o \ldots o\, i_1 \ldots i_{m-2k-1}} \delta_{i_{m-2k} i_{m-2k+1}} \ldots \delta_{i_{m-2} i_{m-1})} C_m^{2k}$$

(   ) indique la symétrisation par rapport aux indices entre parenthèses.

– Réciproquement la donnée du développement (1) de ϕ ne suffit pas à déterminer les $u_{\alpha_1 \ldots \alpha_r}$: On peut par exemple se donner arbitrarirement les $u_{oo\alpha_3 \ldots \alpha_r}$ et les égalités (2) et (3) déterminent alors les $u_{i_1 \ldots i_r}$. et $u_{oi_1 \ldots i_r}$.

– Cependant pour m = 0 et 1, on a nécessairement:

$$\begin{cases} u_{.} = \phi_{.} & \text{(par u., f., } \psi\text{. on désigne les coefficients sans indice)} \\ u_o = \psi_{.} \\ u_i = \phi_i \end{cases}$$

2) <u>Problème de Cauchy sur un conoïde du 2ème ordre</u>

Soit l'équation quasi-linéaire du 2ème ordre la plus générale

$A^{\lambda\mu}(x^\alpha, u, D_\nu u)\, D_{\lambda\mu} u + f(x^\alpha, u, D_\nu u) = 0$

Soit C le cône d'équation $x^o = s$

Soit ϕ une fonction donnée sur C admettant un développement limité de la forme: (1)

Le problème de Cauchy est: trouver u vérifiant (E) et $u/_C = \phi$.

Si on suppose u p fois dérivable au point 0, les dérivées au point 0 de u, $u_{\alpha_1 \ldots \alpha_m}$, $m \le p$ doivent vérifier les équations (a) et (b).

(4) montre que $u_{.}$, u_o, u_i sont déterminés.

Posons $\phi = a$, $\psi_{.} = a_o$, $\phi_i = a_i$

Si on suppose $A^{\lambda\mu}$ et f p−2 fois dérivables au point $(0, a, a_\nu)$, les développements de Taylor de (E) conduisent à des égalités:

$$\text{(c)}\quad A_{.}^{\lambda\mu}\, u_{\lambda\mu\alpha\ldots\alpha_{m-2}} = P_{\alpha_1\ldots\alpha_{m-2}}\left(D^\theta A^{\lambda\mu}(0,a,a_\nu),\, D^\theta f(0,a,a,a_\nu),\, u_\Lambda\right)$$

où les $P_{\alpha_1 \ldots \alpha_{m-2}}$ sont des polynômes de leurs variables.

. Les θ sont des indices de dérivation par rapport à l'ensemble des variables, $|\theta| \leq m-2$

. On pose $\Lambda = (\alpha_1 \ldots \alpha_k)$, $|\Lambda| = k \leq m-1$

$$A_{.}^{\lambda\mu} = A^{\lambda\mu}(0, a, a_\nu)$$

Les u_Λ, pour $|\Lambda| = m$, sont astreints à vérifier les équations a), b), c).

Si on suppose connus les u_Λ pour $|\Lambda| \leq m-1$, les équations a), b), c) forment un système linéaire S_m ayant autant d'inconnues que d'équations.

Si pour tout m S_m est un système de Cramer, on peut déterminer par récurrence sur m tous les u_Λ.

3) Solution en série formelle du problème de Cauchy sur le cône C dans le cas où le cône est caractéristique en son sommet

Dans ce cas on peut montrer que le système S_m est pour tout m un système de Cramer [4].

On obtient donc le théorème suivant :

Théorème I

Soit l'équations aux dérivées partielles.quasilinéaire sur R^{n+1} :

$$\text{(E)}\quad A^{\lambda\mu}\left(x^\alpha, u, D_\nu u\right)\frac{\partial^2 u}{\partial x^\lambda \partial x^\mu} + f\left(x^\alpha, u, D_\nu u\right) = 0$$

$$\alpha, \lambda, \mu, \nu = 0, 1, \ldots, n.$$

Soit C le cône d'équation $x^o = s$, $s = \left(\sum_{i=1}^{n} (x^i)^2\right)^{1/2}$;

Soit ϕ la trace sur C d'une fonction p fois dérivable au sommet 0 de C, admettant un développement limité:

$$\phi(x^i) = a + \Sigma a_i x^i + a_o s +$$

$$+ \sum_{m=2}^{p} \left(\frac{1}{m!} \phi_{i_1 \ldots i_m} x^{i_1} \ldots x^{i_m} + \frac{1}{(m-1)!} \psi_{i_1 \ldots i_{m-1}} s x^{i_1} \ldots x^{i_{m-1}} \right) + o(s^p)$$

On suppose que les $A^{\lambda\mu}$ et f sont p-2 fois dérivables au point ($x^\alpha = 0$, u = a, $D_\nu u = a_\nu$) et que en ce point $A^{oo} = 1$, $A^{oi} = 0$, $A^{ij} = -\delta^{ij}$. Alors toute solution du problème de Cauchy (E) et $u|_C = \phi$ a toutes ses dérivées au point 0 jusqu'à l'ordre p qui sont déterminées et ne dépendent que des dérivées $D^\theta A^{\lambda\mu}$ et $D^\theta f$ au point (0, a, a_ν) pour $|\theta| \le p-2$, et des $\phi_{i_1 \ldots i_m}$, $\psi_{i_1 \ldots i_{m-1}}$ pour $m \le p$.

4. Application au problème de Cauchy sur un conoïde caractéristique

Soit le problème de Cauchy sur un conoïde caractéristique défini par: $A^{\lambda\mu}$, $\overline{w}_r$, f_r :

$$(E_r) \qquad A^{\lambda\mu}(x^\alpha, w_s) D_{\lambda\mu} w_r + f_r(x^\alpha, w_s, D_\nu w_s) = 0$$

C est le conoïde caractéristique de sommet 0 pour l'opérateur

$$\Sigma A^{\lambda\mu}\left(x^\alpha, \overline{w}_s(x^\alpha)\right) D_{\lambda\mu}$$

et

$$\phi_r = \overline{w}_{r_{|C}}$$

On cherche w_r vérifiant (E_r) et $w_{r_{|C}} = \phi_r$.

On a montré dans [2] que les dérivées premières $\chi_r = D_0 w_{r/C}$ de ce problème sont déterminées de façon unique sur un voisinage C_0 de 0 dans C.

Le théorème I permet de démontrer que toutes les dérivées successives sont déterminées sur C_0 [4]:

<u>Théorème II</u> : Etant donné le problème de Cauchy sur un conoïde caractéristique défini par $A^{\lambda\mu}$, $\overline{w}_r$, f_r, où les $A^{\lambda\mu}$ et $\overline{w}_r$ sont de classe C^p et les f_r de classe C^{p-2}.

Pour $l < \left[\frac{p}{2}\right]$, les $\overset{(l)}{\chi_r} = D_0^l w_{r|C_C}$ sont déterminés de façon unique sur C_0 pour toute solution w_r du problème de Cauchy qui a des dérivées bornées jusqu'à l'ordre l.

Les $\overset{(l)}{\chi_r}$ sont de classe C^{p-2l} sur $C_0 - \{0\}$.

Si l'on désigne par:

$$U_{r,p}(x^\alpha) = \sum_0^p u_{r,\alpha_1 \dots \alpha_m} x^{\alpha_1} \dots x^{\alpha_r}$$

la partie régulière du développement limité à l'ordre p de toute solution du problème de Cauchy p fois différentiable en 0 (cf. Théorème I), on a:

$$\overset{(l)}{\chi_r}(x^i) = D_0^l u_{r,p}(s, x^i) + o(S^{p-2})$$

et les dérivées $\overset{(l)}{\chi_r}$ ont les développements limités dérivés.

B. <u>Solutions du problème de Cauchy sur un conoïde caractéristique pour certains systèmes non linéaires. Applications à certains systèmes d'origine physique</u>.

I. <u>Introduction</u>

La nouveauté des résultats obtenus par Dossa [5], [6] réside dans la petitesse de l'ordre de différentiabilité exigé des données et dans l'absence de toute condition au sommet du conoïde pour les données de Cauchy.

Les espaces fonctionnels utilisés sont, grosso modo, de la forme: $P_m + E_m$ où P_m est l'espace des polynomes de degré $\leq 2(m-1)$ sur R^{n+1}

et E_m est un espace de Sobolev à poids formé de fonctions tendant vers 0 en un certain sens au sommet du conoïde et défini par une norme $\|X\|$ où apparaissent:

1) les dérivées de X sur l'intérieur Y du conoïde C jusqu'à l'ordre m, et les dérivées de X sur le cône C jusqu'à l'ordre 2m-1 (condition nécessaire dans tout problème de Cauchy caractéristique)

2) des poids placés pour contrôler deux phénomènes dûs à la géométrie du cône:

- l'explosion quand $t \to 0$ des constantes de Sobolev associées aux domaines:

$$G_t = Y \cap \{x^o = t\}$$
$$\Sigma_t = C \cap G_t$$

- l'apparition dans les dérivations sur C de facteurs qui explosent au voisinage de 0.

II. Préliminaires techniques

2.1 Cadre géométrique

Soit n un entier ≥ 2. le demi-conoïde de E^{n+1} de sommet 0 et d'équation $x^o = s$ avec $S = \sqrt{\sum_{i=1}^{n} (x^i)^2}$. Soit Y l'intérieur de C.

Pour tout $t \in]o, +\infty[$, on note:

$Y_t = \{ (x^o, ..., x^n) \in Y \mid s \leq x^o < t \}$, $G_t = \{ (x^o, ... x^n) \in Y \mid x^o = t \}$,

$\mathcal{C}_t = Y_t \cap \mathcal{C}$,

$V_t = \{ (x^1, ..., x^n) \in r^n | s, x^1, ..., xn) \in \mathcal{C}_t \}$, $\Sigma_t = \mathcal{C} \cap G_t.$,

Soit U un ouvert de R^{n+1} contenant l'origine 0. Alors il existe un réel $T > 0$ tel que $Y_T \cap U$.

2.2 Notations

On note:

$$\left(x^\alpha\right) = \left(x^0, x^1, \dots x^n\right) \in R^{n+1},\ s = \sqrt{\sum_{i=1}^{n} \left(x^i\right)^2},\ \partial_0^k = \frac{\partial^k}{\left(\partial x^0\right)^k},$$

$$\left(\partial^\beta\right) = \frac{\partial^{|\beta|}}{\left(\partial x^1\right)^{\beta_1} \dots \left(\partial x^n\right)^{\beta_n}} \forall \beta \in N^n \text{ avec } \beta\left(\beta_1, \dots \beta_n \text{ et } |\beta| = \beta_1 + \dots \beta_n.\right)$$

Pour toute fonction v définie sur $\mathcal{C}_t$ (ou V_t) et pour tout $p \in N$, on note:

$$|| V ||_{H^p(\Sigma_t, \mathcal{C})} = \left(\sum_{|\beta| \leq p} \int_{\Sigma_t} | \partial^\beta V |^2 d\sigma\left(\Sigma_t\right) \right)^{\frac{1}{2}}$$

où les $\partial^\beta v$ sont des fonctions, dérivées au sens des distributions, $d\sigma$ (Σ_t) est la mesure induite sur Σ_t par la mesure $dx^1 ... dx^n$. $\forall\ \alpha \in N^{n+1}$, on note:

$$D^\alpha = \frac{\partial^{|\alpha|}}{\left(\partial x^0\right)^{\alpha_0} \dots \left(\partial x^n\right)^{\alpha_n}} \text{ si } \alpha = \left(\alpha_0, \dots \alpha_n\right) \text{ et } |\alpha| = \alpha_0 + \dots \alpha_0 + \dots \alpha_n.$$

Pour toute fonction vectorielle $X = (X^I)$ définie sur Y_T et pour tout $k \in N$, on note:

$$|D^k X|^2 = \sum_I \sum_{|\alpha| \leq k} |D^\alpha X^I|^2,$$

les $D^\alpha X^I$ étant des fonctions, dérivées au sens des distributions des fonctions X^I.

[X] les restrictions de X au conoïde : $[X](x^1, ..., x^n) = X(s, x^1, ..., x^n)$.

2.3 Espaces fonctionnels utilisés

$\forall t \in]o, +\infty[$, on note respectivement:

$C^\infty(Y_t)$ l'espace des restrictions à Y_t des fonctions vectorielles x de classe C^∞ sur R^{n+1} et de type donné (type = dimension de l'espace vectoriel d'arrivée R^l de x).

$C^\infty_\infty(Y_t)$ le sous-espace de $C^\infty(Y_t)$ des fonctions dont les dérivées de tout ordre sont nulles au point 0, sommet du conoïde.

Les espaces de Banach utilisés sont les suivants: $\forall p \in N$, $F^p(Y_t)$ est l'espace de Banach défini par la norme:

$$||x||_{F^p(y_t)} = \underset{\tau \in (0,t)}{\text{Ess sup}}\, \tau^{-\frac{n}{2}} ||X||_{H^p(G_\tau, Y)}$$

avec:

$$||x||^2_{H^p(G_\tau, Y)} = \sum_{k=0}^{p} \int_{G_\tau} |D^k X|^2 dx^1 \ldots dx^n$$

$\forall p \in N^*$, $\widetilde{F}^p(C_t)$ est l'espace de Banach défini par la norme:

$$||x||_{F^p(C_t)} = \sum_I \sum_{r=0}^{2p-1} \underset{\tau \in (0,t)}{\text{Ess sup}}\, \tau^{-2p+r+\frac{3-n}{2}} ||X^I||_{H^q(\Sigma_\tau, C}$$

$\forall\, p \in N^*$, $\widehat{F}^p(C_t)$ est la fermeture dans $\widehat{F}^p(C_t)$ de l'espace des restrictions au conoïde C des fonctions de $C^\infty(V_t)$..

$\forall\, p \in N^*$, $\widetilde{F}^p(V_t)$ est le sous-espace de $F^p(Y_t)$ défini par la norme:

$$||X||_{\widetilde{F}^p(Y)} = ||X||_{F^p(Y)} + \sum_{k=0}^{p-1} ||\partial_0^k X \in||_{\widetilde{F}^{p-k}(C_t)}$$

$\forall\, p \in N^*$, $\widehat{F}^p(V_t)$ est la fermeture de w $C^\infty(V_t)$ dans

III Problème de Cauchy sur un conoïde caractéristique

E On considère le système d'équations

$$(E) \quad A^{\lambda\mu}(x^\alpha, u)\, D_{\lambda\mu} u + f(x^\alpha, u, Du) = 0$$

Hypothèse H_m:

– m entier $< \frac{n}{2} + 1$

– $A^{\lambda\mu}(x^\alpha, u^I)$ est de classe C^{2m-1} sur U x fois W, U ouvert de R^{n+1}, W ouvert de R^N.

– $\forall (x^\alpha, u^I) \in$ U x fois W, $A^{\lambda\mu}(x^\alpha, u^I)$ définit une forme quadratique +, –, –, avec $A^{oo} > 0$ et $A^{ij} x_i x_j$ forme définie négative ;

– $\exists\, (a^I) \in$ W, tel que $A\,(0, a^I) = \eta^{\gamma\mu}$ (métrique de Minkowski)

– $f(x, u^I, D_\nu u^I)$ est de classe C^{2m-3}. Si $\frac{n}{2} + 1 < m < \frac{n}{2} + 2$, les \$ et \$ sont C^{2m-3}.

– $\phi = (\phi^I)$ est définie sur la partie C_T du conoïde C; $\phi^I(0) = a^I$; et C_T est caractéristique pour la donnée de Cauchy $u = \phi$, c'est-à-dire que $\forall\, (x^i) \in V_T$

$$A^{oo}\left(s, x^i; \phi^I\left(x^i\right)\right) - 2 \sum_{k=1}^{n} A^{ok}\left(s, x^i; \phi^I\left(x^i\right)\right) \frac{x^k}{s} + \sum_{k,l=1}^{n} A^{kl}\left(s, x^i; \phi^I\left(x^i\right)\right) \frac{x^k x^l}{s^2} = 0$$

ϕ peut s'écrire:

$$\phi = \bar{\phi}\,|_{C_T} + \phi_1$$

où $\phi_1 \in \widetilde{F}^m(C_T)$ et $\bar{\phi} = \left(\bar{\phi}^I\right)$ est une fonction polynomiale de degré $\leq 2(m-1)$ sur R^{n+1}

<u>Théorème 1 (Existence locale)</u> (cf. [6])

Sous l'hypothèse H_m, $m > \frac{n}{2} + 1$.

1) la donnée f du problème de Cauchy peut se redécomposer en

$$\phi = \bar{u}\,|_{C_T} + \tilde{\phi} \quad ,$$

où $\bar{u}$ est une fonction polynomiale de degré $\leq 2(m-1)$, vérifiant au point 0 le système $(E\ )$ et les équations dérivées jusqu'à l'ordre $2(m-2)$, et $\tilde{\phi}_1 \in \widetilde{F}^m(C_T)$

2) $\exists\ T_0 \in]0,T[$, tel que le problème de Cauchy admet une solution unique

$$u = \left(u^I\right) \in P^{2(m-1)}\left(V_{T_0}\right) + \widetilde{F}^m\left(V_{T_0}\right) \quad ,$$

où $P^{2(m-1)}\left(V_{T_0}\right)$ est l'espace des restrictions à V_{T_0} des fonctions polynomiales sur R^{n+1} de degré $\leq 2(m-1)$.

Cette solution se décompose en

$$u = \bar{u} + u_1 \text{ avec } u_1 \in \widetilde{F}^m\left(V_{T_0}\right),\ u_1(0) = 0,\ Du_1(0) = 0 \quad ,$$

3) Si les $A^{\lambda\mu}$ sont C^{∞} et indépendants de u et si $\phi_1 \in \widehat{F}^m(C_T)$, alors $u_1 \in F^m(V_T)$

4) Si f(x, u, Du) est linéaire en Du, on peut remplacer la condition $m > \frac{n}{2} + 1$ par $m > \frac{n}{2}$.

Théorème 2 (Existence globale dans V_T) (cf. [6])

Mêmes hypothèses et notations que dans le théorème 1.

On pose: $\| \phi \|_{m,T} = | \bar{u} |_m + \| \phi_1 \|_{\widetilde{F}^m(C_T)}$

avec $| \bar{u} |_{Max} | D^{\alpha \bar{u}(0)}$

On suppose en plus que f(x, 0, 0) = 0 et que les dérivées partielles d'ordre 2 par rapport aux variables u et Du sont C^{2m-3}. Alors il existe $d > 0$ tel que si $\| \phi \|_{m,T} \leq d$, la solution $u = \bar{u} + u_1$ est globale: $T_0 = T$.

IV. Applications physiques

Dossa [6] a appliqué les résultats précédents aux problèmes physiques suivants ;

1) Le problème de Cauchy sur un conoïde caractéristique pour les équations de Yang-Mills-Higgs en jauge de Lorentz ;

2) Le problème de Cauchy sur un conoïde caractéristique pour les équations d'Einstein du vide en coordonnées harmoniques ;

3) Le problème de Cauchy sur un conoïde caractéristique pour le système conforme régulier des équations d'Einstein.

Bibliographie

[1] R. d'Adhemar, Rendiconti del Circolo Matematico di Palermo 20 (1905) 142-159.

[2] F. Cagnac, Annali di mat.pura ed appl. (IV) vol CIV (1975), 355-393.

[3] F. Cagnac, Annali di mat. pura ed appl. (IV) vol CXXIX (1980) 13-41.

[4] F. Cagnac, "Problème de Cauchy sur un conoïde caractéristique. Solutions en série formelle et détermination des dérivées successives sur le conoïde", preprint 1992.

[5] M. Dossa, "Problèmes de Cauchy sur un conoïde caractéristique pour des équations quasilinéaires du second ordre", preprint 1992.

[6] M. Dossa, "Espaces de Sobolev non isotropes, à poids et problèmes de Cauchy quasilinéaires du second ordre sur un conoïde caractéristique", preprint 1992.

[7] Friedlander, "The wave equation on a curved space-time" Cambridge Univ. Press, 1975.

[8] A Rendall, Proc.Roy.Soc. London A427 (1990), n°1872, 221-239.

[9] M. Riesz, Acta Mathematica 81 (1949), 107-125.

Recent Progress on the Cauchy Problem in General Relativity

Demetrios Christodoulou

Princeton University

General relativity presents us with a unified description of space, time and gravitation, according to which spacetime is a 4-dimensional manifold M with a metric $g_{\mu\nu}$ of signature $(3,1)$ whose connection represents the gravitational force. The fundamental law of the theory is the Einstein equation:

$$R_{\mu\nu} - \frac{1}{2} g_{\mu\nu} R = 8\pi T_{\mu\nu}$$

where $R_{\mu\nu}$ and R are respectively the Ricci curvature and scalar curvature of $g_{\mu\nu}$ and $T_{\mu\nu}$ is the energy tensor of matter. In the absence of matter the equation reduces to the Einstein vacuum equation for the spacetime manifold:

$$R_{\mu\nu} = 0$$

The central mathematical problem of the theory, as Choquet-Bruhat taught us, is the Cauchy problem. The Cauchy data consist of a 3-dimensional manifold Σ with a positive definite metric $\overline{g}_{ij}$ and a 2-covariant symmetric tensorfield k_{ij}. The problem is to find a 4-dimensional manifold M with a metric $g_{\mu\nu}$ of signature $(3,1)$ satisfying the Einstein vacuum equation and an imbedding of Σ into M such that $\overline{g}_{ij}$ and k_{ij} are respectively the first and second fundamental forms induced on Σ by the imbedding. The Einstein vacuum equation imposes on the Cauchy data the constraint equations:

$$\overline{\nabla}^i k_{ij} - \overline{\nabla}_j trk = 0$$

$$\overline{R} - |k|^2 + (trk)^2 = 0$$

where $\overline{R}$ is the scalar curvature of $\overline{g}_{ij}$. These are respectively the Codazzi and Gauss equations of the imbedding of Σ in M.

The Einstein vacuum equation constitutes at first sight a degenerate differential system. That is, the null space of the symbol σ_ξ at a given covector ξ is nonzero for all covectors ξ. This is due to the fact that the equation is generally covariant; proper account must be taken of the geometric equivalence

M. Flato et al. (eds.), Physics on Manifolds, 49–58.

of metrics related by a diffeomorphism. This is done by considering σ_ξ not on the space of 2-covariant symmetric tensors at a point but rather on the quotient of this space by the equivalence relation induced by the symbol of the diffeomorphisms; two such tensors are equivalent if they differ by $\xi X + X\xi$ for some covector X. The null space of σ_ξ is then found to be nonzero if and only if ξ belongs to the null cone defined by the metric g. The Einstein equations are therefore of hyperbolic character. It is thus reasonable to expect that coordinate systems may be found in which the Einstein vacuum equation reduces to a system of nonlinear wave equations and, indeed, coordinate systems such that each coordinate function satisfies the scalar wave equation are of this type. These "harmonic" coordinate systems played an important role in the proof of the fundamental theorems of Choquet-Bruhat [1] of local existence and uniqueness of solutions of the Cauchy problem.

The Cauchy data is called asymptotically flat if the complement of a compact set in Σ is diffeomorphic to the complement of a ball in $\Re^3$, $\overline{g}_{ij}$ is a complete metric on Σ and the curvature of $\overline{g}_{ij}$ as well as k_{ij} tend to zero at infinity in an appropriate way. By a global solution of the Cauchy problem we mean a solution spacetime $(M, g_{\mu\nu})$ which is geodesically complete. Now the Minkowski spacetime, the spacetime of special relativity, is a trivial global solution of the Einstein vacuum equation arising from asymptotically flat Cauchy data. A basic problem in the theory is whether any asymptotically flat Cauchy data which is in a suitable neighborhood of trivial data gives rise to a global solution of the Einstein vacuum equation. This is the problem of the "stability of Minkowski spacetime". In a recent joint work [2], Sergiu Klainerman and myself have resolved this problem in the affirmative.

Choquet-Bruhat, who early recognized the significance of the problem for the further development of the theory, realized that it could not be resolved by extending the local harmonic coordinate system to a global one, for these coordinate systems become unstable in the large [3]. It is interesting to note that the difficulty would dissappear if the space dimension where greater than 3.

Our resolution uses two main ideas. The first is the relationship between conserved quantities and symmetry and the second is the relationship between symmetry and the causal structure of spacetime. The first idea goes back to Noether. Consider a field theory in a given spacetime whose field equation is derivable from an action S. The energy tensor is defined by:

$$T_{\mu\nu} = \frac{\delta S}{\delta g^{\mu\nu}}$$

and the invariance of S under diffeomorphisms implies that $T_{\mu\nu}$ is divergence-free:

$$\nabla^{\mu} T_{\mu\nu} = 0$$

Now suppose that X is a vectrorfield generating a 1-parameter group of isometries of (M, g) (Killing vectorfield). Then the 1-form

$$P_{\mu} = -T_{\mu\nu} X^{\nu}$$

is divergence-free:

$$\nabla^{\mu} P_{\mu} = 0$$

It follows that the integral

$$\int_{\Sigma} *P$$

on a Cauchy hypersurface Σ is a conserved quantity, that is, its value is the same for all Cauchy hypersurfaces. If the action is invariant under conformal transformations of the metric then the energy tensor is trace free: $trT = 0$ and the above considerations extend to the case where X generates a 1-parameter group of conformal isometries of (M, g) (conformal Killing vectorfield). An important requirement on a physical theory is that the energy tensor should satisfy the positivity condition:

$$T(X_1, X_2) \geq 0$$

for any two future directed timelike vectors X_1, X_2 at a point. If the vectorfield X above is timelike and future directed then the quantity

$$\int_{\Sigma} *P = \int_{\Sigma} T(X, N) d\mu_{\bar{g}}$$

is nonnegative, N being the future directed unit normal to Σ. As its value is the same as that on the Cauchy hypersurface on which the initial data is given, it provides an estimate for the solution in terms of the initial data.

In the case of gravitation the energy tensor defined as above,

$$G_{\mu\nu} = \frac{\delta S}{\delta g^{\mu\nu}}$$

vanishes as this expresses the Euler-Lagrange equation of gravitation, namely the Einstein equation. Thus the above considerations fail at first sight. The way out of this impasse is to consider the Bianchi identities

$$\nabla_{[\alpha}R_{\beta\gamma]\delta\epsilon} = 0$$

(where [ ] stands for cyclic permutation). We define a Weyl field $W_{\alpha\beta\gamma\delta}$, in a given spacetime, to be a 4-covariant tensorfield posessing the algebraic symmetries of the Weyl or conformal curvature tensor. The natural field equation for a Weyl field is the Bianchi equation

$$\nabla_{[\alpha}W_{\beta\gamma]\delta\epsilon} = 0$$

which we write simply as

$$DW = 0$$

In a spacetime which satisfies the the Einstein vacuum equation the curvature is an example of a Weyl field satisfying the Bianchi equation. In a 4-dimensional spacetime the dual $*W$ of a Weyl field W is also a Weyl field and if W satisfies the Bianchi equation so does $*W$. The operator D althogh formally identical to the exterior derivative, is not an exterior differential operator and $D^2 \neq 0$. As a consequence, the Bianchi equation implies an algebraic condition:

$$R_\mu{}^{\alpha\beta\gamma} * W_{\nu\alpha\beta\gamma} - R_\nu{}^{\alpha\beta\gamma} * W_{\mu\alpha\beta\gamma} = 0$$

The Bianchi equation is conformally covariant. If f is a conformal isometry of (M, g) that is $f * g = \Omega^2 g$ for some positive function Ω, and W is a solution of the Bianchi equation then so is $\Omega^{-1} f * W$.

Associated to a Weyl field W is a 4-covariant symmetric tensorfield Q, quadratic in W, called the Bel-Robinson tensor [4]:

$$Q_{\alpha\beta\gamma\delta} = W_{\alpha\rho\gamma\sigma}W_\beta{}^\rho{}_\delta{}^\sigma + *W_{\alpha\rho\gamma\sigma} * W_\beta{}^\rho{}_\delta{}^\sigma$$

It is totally symmetric and trace-free and satisfies the following positivity condition:

$$Q(X_1, X_2, X_3, X_4) \geq 0$$

for any four future directed timelike vectors X_1, X_2, X_3, X_4 at a point, with equality if and only if W vanishes at that point. Furthermore, if W satisfies the Bianchi equation then Q is divergence-free:

$$\nabla^\alpha Q_{\alpha\beta\gamma\delta} = 0$$

It follows that if X_1, X_2, X_3 are three vectorfields each generating a 1-parameter group of conformal isometries of (M, g), then the 1-form

$$P = -Q(\cdot, X_1, X_2, X_3)$$

is divergence-free, consequently the integral

$$\int_\Sigma *P$$

on a Cauchy hypersurface Σ is a conserved quantity which is positive definite in the case that X_1, X_2, X_3 are all timelike and future directed.

Given a Weyl field W and a vectorfield X the Lie derivative $\mathcal{L}_X W$ of W is not in general a Weyl field. However we can define a modified Lie derivative $\hat{\mathcal{L}}_X W$ which is a Weyl field:

$$\begin{aligned}\hat{\mathcal{L}}_X W_{\alpha\beta\gamma\delta} &= \mathcal{L}_X W_{\alpha\beta\gamma\delta} - \frac{1}{8} tr\pi W_{\alpha\beta\gamma\delta} \\ &- \frac{1}{2}(\hat{\pi}_\alpha{}^\mu W_{\mu\beta\gamma\delta} + \hat{\pi}_\beta{}^\mu W_{\alpha\mu\gamma\delta} + \hat{\pi}_\gamma{}^\mu W_{\alpha\beta\mu\delta} + \hat{\pi}_\delta{}^\mu W_{\alpha\beta\gamma\mu})\end{aligned}$$

where $\pi_{\mu\nu} = \mathcal{L}_X g_{\mu\nu}$ and $\hat{\pi}$ is the deformation tensor of X, namely the trace-free part of π. The modified Lie derivative commutes with the Hodge dual:

$$\hat{\mathcal{L}}_X * W = *\hat{\mathcal{L}}_X W$$

As a consequence of the linearity and the conformal covariance of the Bianchi equation, if W is a solution of this equation and X is a vectorfield generating a 1-parameter group of conformal isometries f_t then

$$\hat{\mathcal{L}}_X W = \frac{d}{dt}(\Omega_t^{-1} f_t * W)|_{t=0}$$

is also a solution of the same equation. Therefore the above considerations regarding conserved quantities can be applied to the Weyl field $\hat{\mathcal{L}}_X W$ as well.

Minkowski spacetime has a 15 dimensional conformal group $O(4,2)$ consisting of the Abelian subgroups of translations and inverted translations, the scalings and the Lorenz group $O(3,1)$. The generators of time translations and inverted time translations are the only timelike conformal Killing vectorfields. A general spacetime will not have a nontrivial conformal group.

If the condition that the vectorfields are conformal Killing is dropped, then, although the quantities

$$\int_{\Sigma} *P$$

will not be conserved, their growth shall be determined by the spacetime integrals of expressions which are quadratic in W and linear in $\hat{\pi}$, the deformation tensors of the vectorfields. The idea is to find a subgroup of $O(4,2)$ and an action of this subgroup on the spacetime manifold having the following property: the deformation tensors of the vectorfields generating the action should decay at infinity in such a way that the growth of the corresponding quantities is bounded in terms of the quantities themselves. It turns out that the subgroup consisting of time translations, scalings, inverted time translations and the rotation group $O(3)$, leaving the total energy vector invariant, suffices.

The problem is then how to define the action of these groups on a general spacetime arising from asymptotically flat initial data in such a way as to satisfy the above requirement. Now the group of time translations is the simplest to define and reduces to the choice of a time function t. As our argument is one of continuity starting from the initial Caychy hypersurface, a time function enters the problem naturally. A canonical choice is that of a maximal time function, namely one whose level sets Σ_t are maximal spacelike hypersurfaces. The second fundamental form k_{ij} of Σ_t is the trace-free: $trk = 0$. There is a unique such function t with the property that the total momentum vector relative to Σ_t, that is, the projection to Σ_t of the total energy vector vanishes. To define the action of the other groups we consider the fact that the spacetime is expected to be asymptotically flat. Since these groups act canonically on Minkowski spacetime, there is a canonical action defined at infinity. What is needed is a way to extend this action to the spacetime. This is provided by the causal structure. The causal structure on a manifold M is the fundamental structure defined on M by a metric of signature $(dimM - 1, 1)$. The causal future $J^+(S)$ of a set $S \subset M$ is the set of points q which can be reached by a future directed causal curve initiating at S. Similarly $J^-(S)$ is the set of points q which can be reached, from S, by a past directed causal curve. The specification of $J^+(p)$ and $J^-(p)$ for every $p \in M$ defines the causal structure, which is equivalent to the conformal geometry of M. In our problem we consider the boundaries of the causal pasts of a 1-parameter family of surfaces at "infinity" which are related to each other by time translation. However we cannot quite start from "infinity" since the existence of a global solution is precicely what we wish to establish. Nevertheless, following our continuity argument we can assume that a spacetime slab has been constructed. The role of "infinity" is then played by the final maximal hypersurface Σ_{t*}. We construct a 1-parameter family of surfaces diffeomorphic to S^2 on Σ_{t*} by solving a certain equation of motion for 2-surfaces on a 3-dimensional manifold. We then consider the inner boundaries of the causal pastsof these surfaces in the spacetime slab. These null hypersurfaces C_u are the level sets of what we call the optical function u. Let $S_{t,u}$ be the intersection of the C_u with the Σ_t. Let l and $\underline{l}$ be respectively the outgoing and incoming null normals to $S_{t,u}$ whose component along T, the generator of time translations, is equal to T. Then we have

$$T = \frac{1}{2}(l + \underline{l})$$

and we define the generator of scalings by

$$S = \frac{1}{2}(\underline{u}l + u\underline{l})$$

and the generator of inverted time translations by

$$K = \frac{1}{2}(\underline{u}^2 l + u^2 \underline{l})$$

where

$$\underline{u} = u + 2r, \quad r = \sqrt{\frac{A}{4\pi}}$$

and A is the area of $S_{t,u}$. An action of the rotation group $O(3)$ on Σ_{t*} is defined starting from the standard action on the sphere at infinity in such a way that the group orbits are the level surfaces of u on Σ_{t*}. The action is then extended to the spacetime slab by conjugation: Given an element $O \in O(3)$ and a point $p \in S_{t,u}$, to obtain the point Op we follow the generator of C_u through p toward the future until $p*$ the point of intersection with Σ_{t*}. We then move to $Op*$ and follow the generator of C_u through that point toward the past until the point of intersection with Σ_{t*}. This point, which again lies on $S_{t,u}$, is the sought for point Op. The three rotation vectorfields ${}^{(a)}\Omega$, $a = 1, 2, 3$ generating this action verify:

$$[\, {}^{(a)}\Omega, l] = 0$$

$$g(\,{}^{(a)}\Omega, l) = g(\,{}^{(a)}\Omega, T) = 0$$

and, of course, the commutation relations of the Lie algebra of $O(3)$:

$$[\,{}^{(a)}\Omega,\ {}^{(b)}\Omega] = \epsilon_{abc}\ {}^{(c)}\Omega$$

The group orbits are the surfaces $S_{t,u}$.

By the above construction, the deformation tensors of the generating vectorfields depend entirely on the geometric properties of the hypersurfaces C_u and Σ_t. These properties differ significantly from those in the case of Minkowski spacetime. Consider for example, on a given C_u, the second fundamental form θ of $S_{t,u}$ relative to Σ_t, in particular the ratio

$$f = \frac{|\hat{\theta}|^2}{(tr\theta)^2}$$

where $\hat{\theta}$ is the tace-free part of θ. This ratio, in contrast to the case in Minkowski spacetime, does not tend to zero as $t \to \infty$. In fact, $\lim_{t\to\infty} f$ measures the flux of energy radiated to infinity. Another example is the area of $S_{t,u}$ on a given C_u, which in Minkowski spacetime verifies $r - t = O(1)$ as $t \to \infty$, while in general we find

$$r - t = -2M_0 \log t + O(1)$$

where M_0 is the initial total mass. These differences would dissappear if the space dimension where greater than 3.

The proof of the theorem is by the method of continuity and it involves a bootstrap argument. Using an appropriate version of the local existence theorem we can assume that the spacetime is maximally extended up to a value $t*$ of the time function. This value is defined to be the maximal one such that certain geometric quantities defined by the level sets of the time function and the optical function remain bounded by a small positive number ε_0. These quantities control in particular the isoperimetric constant of the surfaces $S_{t,u}$ on which the Sobolev inequalities depend. It then follows that a certain norm of the deformation tensors of the vectorfields T, S, K and ${}^{(a)}\Omega$ in the spacetime slab bounded by Σ_0 and Σ_{t*} is less that another small positive number ε_1. We then consider the 1-form P, where

$$P = P_0 + P_1 + P_2,$$

$$P_0 = -Q(R)(\cdot, \overline{K}, T, T),$$
$$P_1 = -Q(\hat{\mathcal{L}}_O R)(\cdot, \overline{K}, \overline{K}, T) - Q(\hat{\mathcal{L}}_T R)(\cdot, \overline{K}, \overline{K}, \overline{K}),$$

$$\begin{aligned} P_2 &= -Q(\hat{\mathcal{L}}_O^2 R)(\cdot, \overline{K}, \overline{K}, T) - Q(\hat{\mathcal{L}}_O \hat{\mathcal{L}}_T R)(\cdot, \overline{K}, \overline{K}, \overline{K}) \\ &- Q(\hat{\mathcal{L}}_S \hat{\mathcal{L}}_T R)(\cdot, \overline{K}, \overline{K}, \overline{K}) - Q(\hat{\mathcal{L}}_T^2 R)(\cdot, \overline{K}, \overline{K}, \overline{K}) \end{aligned}$$

and

$$\overline{K} = K + T$$

while O stands for the collection $\{ {}^{(a)}\Omega \, : \, a = 1, 2, 3\}$. We define the quantity $E = \max\{E_1, E_2\}$ where

$$E_1 = \sup_t \int_{\Sigma_t} *P, \quad E_2 = \sup_u \int_{C_u} *P$$

and everything is restricted to the spacetime slab under consideration. The crucial point is the estimate of the error terms which control the growth of $\int_{\Sigma_t} *P$ and $\int_{C_u} *P$. Using the bound for the deformation tensors we are able to estimate the integral in the spacetime slab of these error terms by $c\varepsilon_1 E$ and thus arrive at an inequality of the form

$$E \leq c(D + \varepsilon_1 E)$$

where D stands for initial data. When ε_1 is chosen sufficiently small, which is achieved by choosing ε_0 suitably small, this implies $E \leq cD$. On the other hand we can show that the aforementioned geometric quantities associated to the level sets of u and t are bounded by cE. Thus if D is chosen suitably small this bound does not exceed $\varepsilon_0/2$, which by continuity contradicts the maximality of $t*$, unless of course $t* = \infty$, in which case the theorem is proved.

The estimate of the error terms would fail if it were not for the fact that the worst error terms vanish due to a simple algebraic identity: if A, B, C are any three symmetric trace-free 2-dimensional matrices then $tr(ABC) = 0$. The reason why such matrices appear here can be traced back to the symbol of the Einstein equation. As I mentioned at the beginning the null space of σ_ξ is nontrivial if and only if ξ is a null covector. But we can say more; when ξ is a (nonzero) null covector the null space of σ_ξ is isomorphic to the space of symmetric trace-free 2-dimensional matrices. This is therefore the space of dynamical degrees of freedom of the gravitational field at a point. There is no product in this space because for any two such matrices A, B we have $AB + BA - Itr(AB) = 0$. This implies the identity mentioned above.

In concluding, I would like to emphasize our indebtedness to Choquet-Bruhat, for it is her vision that has lead to progress in this difficult field.

References

[1] Y.Choquet-Bruhat: "Theorem d'existence pour certain systems d,equations aux derivees partielles nonlineaires", Acta Math. **88** (1952), 141-225.

[2] D.Christodoulou-S.Klainerman: "The global nonlinear stability of the Minkowski space, *Annals of Mathematics Studies, Princeton University Press*, 1993.

[3] Y.Choquet-Bruhat: "Un theorem d'instabilite pour certaines equations hyperboliques nonlineaires", C.R.Acad.Sci.Paris **276A** (1973), 281-284.

[4] L.Bel: "Introduction d'un tenseur du quatrieme ordre", C.R.Acad.Sci.Paris **248** (1959), 1094-1096.

On some links between mathematical physics and physics in the context of general relativity

Thibault Damour
Institut des Hautes Etudes Scientifiques, 91440 Bures sur Yvette, France
and DARC - CNRS - Observatoire de Paris-Meudon, 92195 Meudon, France
(Received: 7 December 1992)

1 Introduction

It is an honour and a pleasure to associate myself to the celebration of the very fruitful career of Yvonne Choquet–Bruhat in the field of mathematical physics. Being neither a mathematician nor a mathematical physicist, I am incompetent for writing about the mathematical impact of her work. However, as a physicist I am deeply convinced of the importance of mathematics and mathematical physics in the scientific description of reality. I view physics as a buffer zone between "reality" and mathematics. If we consider the side of physics turned towards reality, there are no definite criteria allowing one to decide when physics succeeds in making a close contact with reality, i.e. in uncovering some definite physical "truth". The link between physics and reality may well be of the nature of a regressio ad infinitum. On the other hand, if we consider the side of physics turned towards mathematics, it is essential to the consistency and meaning of the whole scientific endeavour that this side establish a close contact with mathematical physics (defined as the outer layer of mathematics aimed towards the physics buffer). Any gap at the physics-mathematical-physics interface is an unacceptable break of continuity in the scientific representation of reality. In the present contribution, I will discuss some examples (taken within the context of general relativity and its generalisations) where the work of Yvonne Choquet-Bruhat has played a crucial role in providing tools or models for closing such gaps.

M. Flato et al. (eds.), Physics on Manifolds, 59–65.

2 Exact results and perturbation methods in general relativity.

In 1952, Y. Choquet-Bruhat [1] suceeded for the first time in proving the local existence of solutions of Einstein's equations. The first step in her proof was to reduce Einstein's (vacuum) equations to a quasi-linear diagonal hyperbolic system by the use of the harmonic-coordinates condition ($\Box_g x^\mu = 0$). Suppressing indices, the harmonically relaxed vacuum Einstein's equations have the form

$$\Box_g g = Q(g, \partial g) , \tag{1}$$

where Q is quadratic in the first derivatives of the metric $g(\equiv g^{\mu\nu})$. She then introduced an iterative way of solving Eq. (1). Namely, given the nth iterate g_n, the next one satisfies the following linear equation

$$\Box_{g_n} g_{n+1} = Q(g_n, \partial g_n) . \tag{2}$$

In Ref. [1], she proved that the sequence (g_n) defined by (2) (given some initial conditions) converged (in suitable functional spaces) to a solution of (1).

This result was a landmark in the mathematical history of general relativity. However, in spite of its importance, it left open a gap at the physics-mathematics interface. Indeed, most applications of general relativity do not use the convergent iteration (2) (which cannot be used in practice because one does not know the Green function in a generic curved spacetime), but other iterations based on the systematic use of the flat-spacetime Green function.

An important step towards closing the gap between mathematics and physics has been taken in 1979 by Demetrios Christodoulou and Bernd Schmidt [2]. They considered the case of the reduced Einstein equations with trivial initial conditions and prescribed source (i.e. an additional term $\rho(x)$ on the right-hand side of Eq. (1)). They introduced the following iteration scheme

$$\Box_f \bar{g}_{n+1} = Q(\bar{g}_n, \partial \bar{g}_n) + (\Box_f - \Box_{\bar{g}_n})\bar{g}_n + \rho(x) . \tag{3}$$

In Eq. (3) f denotes the flat (Minkowski) metric. It is clear that the solution of the iteration (3) uses only the knowledge of the Green function on a flat spacetime.

Ref. [2] showed that the sequence $(\bar{g}_n)$ defined by Eq. (3), though generically not convergent, was however asymptotic to an exact solution g of Einstein's equations. The asymptotic nature of a sequence of iterates is sufficient (in most cases) to justify the physical use of this sequence. (Indeed, observational data have a limited accuracy, and, in many applications, it suffices to compare them to some low-order iterate, say $\bar{g}_2$ or $\bar{g}_3$).

However, this is not the end of the story. The result of Ref. [2] realizes only a near contact with the actual needs of physics because it considered only the reduced Einstein system, with prescribed sources and very special (trivial) Cauchy data. Recently, progress has been made in establishing a close contact between exact mathematical results and the perturbation methods actually used by physicists working in general relativity [3]. One of the main results of Ref. [3] is a proof that any formal perturbation series,

$$g^{\text{pert}}(\lambda) = g_0 + \lambda h_1 + \lambda^2 h_2 + \cdots + \lambda^n h_n + \cdots , \tag{4}$$

which satisfies (order by order) Einstein's (vacuum) equations when formally inserted and expanded, is asymptotic (when $\lambda \to 0$) to some one-parameter family of exact solutions $g^{\text{exact}}(\lambda)$. The proof of this result relies heavily on exact mathematical results due notably to Y. Choquet-Bruhat [4] and A. Rendall [5]. The result of Ref. [3] is valid locally, and admits also some physically interesting semi-global generalizations (on domains including future null infinity). (The latter generalization was made possible by some mathematical results about the regularization of the conformal Einstein vacuum equations [6], [7]). Finally, let us note that the results of Ref. [3] are directly applicable to actual perturbation calculations used in general relativity, notably in the study of gravitational radiation [8].

To end this section one can also mention another type of link between exact mathematical concepts and approximate physical studies. As introduced by Whitney in 1936 [9], the concept of differentiable manifold is based on the simultaneous use of a complementary set of local coordinate charts to describe any global differentiable structure living on a continuum. Yet, most physical studies of curved spacetimes try, as much as possible, to describe physical events in spacetime by using only one coordinate system. However, it has been recently realized that there are cases where the mathematical idea of using several different complementary coordinate systems is physically necessary [10], [11], or at least extremely well adapted to present high-precision technologies. The latter case arises in the relativistic description of the solar-system gravitational field where it is indeed advantageous to introduce as many local coordinate systems as there are separate bodies in gravitational interaction [12].

3 Mathematics and physics in the context of Einstein's "last unified theory"

The present section will give an indirect proof of the scientific flair of Yvonne Choquet-Bruhat. [It serves also the purpose of associating Stanley Deser to the celebration of her career]. Indeed, while in the fifties many scientists did work to unravel the physical and mathematical status of Einstein's last unified theory [13] (notably Schrödinger, Bose, Lichnerowicz, Papapetrou, Tonnelat), Y. Choquet-Bruhat decided to stay away from this theory. A recent work done in collaboration with Stanley Deser and Jim McCarthy [14] shows that she was well inspired in doing so. Indeed, in spite of the superficial similarity between this theory and general relativity, the results of Ref. [14] show that it suffers from serious problems of physical consistency.

Let us recall that the ("first order") Lagrangian density defining the theory is simply (see e.g. [15])

$$\mathcal{L}^{\text{1st order}}(g,\Gamma) = \sqrt{g}\, g^{\mu\nu} R_{\mu\nu}(\Gamma) \ , \tag{5}$$

where $g_{\mu\nu}$ is an asymmetric tensor ($g_{\mu\nu} \neq g_{\nu\mu}$) and $\Gamma^\lambda_{\mu\nu}$ an independent, arbitrary connection ($\Gamma^\lambda_{\mu\nu} \neq \Gamma^\lambda_{\nu\mu}$). Eq. (5) is formally similar to the first-order (Palatini) Lagrangian density defining general relativity. However, the theory now contains new fields (besides the familiar massless spin 2 one of general relativity), associated with the antisymmetric pieces of $g_{\mu\nu}$ and $\Gamma^\lambda_{\mu\nu}$. To exhibit the dynamics of these new fields let us define

$$\begin{aligned} G_{\mu\nu} &\equiv \tfrac{1}{2}(g_{\mu\nu} + g_{\nu\mu}) \ , \\ B_{\mu\nu} &\equiv \tfrac{1}{2}(g_{\mu\nu} - g_{\nu\mu}) \ , \\ \Gamma_\mu &\equiv \tfrac{1}{2}(\Gamma^\alpha_{\mu\alpha} - \Gamma^\alpha_{\alpha\mu}) \ , \end{aligned} \tag{6}$$

and let us expand the action in powers of the antisymmetric $B_{\mu\nu}$ field. At the quadratic level one finds that the <u>second-order</u> Lagrangian density defining the theory reads (modulo a total derivative)

$$\mathcal{L}^{\text{2nd order}}(G, B, \Gamma_\mu) = \mathcal{L}^{\text{good}} + \mathcal{L}^{\text{bad}} + O(B^4) , \tag{7}$$

with

$$\mathcal{L}^{\text{good}} = \sqrt{G}\left[R(G) - \frac{1}{12}H_{\lambda\mu\nu}H^{\lambda\mu\nu} - \frac{2}{3}B^{\mu\nu}(\partial_\mu\Gamma_\nu - \partial_\nu\Gamma_\mu)\right] , \tag{8}$$

$$\mathcal{L}^{\text{bad}} = \sqrt{G}\left[\frac{1}{4}R(G)B_{\mu\nu}B^{\mu\nu} - R_{\mu\nu\alpha\beta}(G)\, B^{\mu\alpha}B^{\nu\beta}\right] . \tag{9}$$

In Eq. (8) one has introduced the notation

$$H_{\lambda\mu\nu} = \partial_\lambda B_{\mu\nu} + \partial_\mu B_{\nu\lambda} + \partial_\nu B_{\lambda\mu} . \tag{10}$$

[Henceforth, indices are raised and lowered using the metric $G_{\mu\nu}$]. As indicated by the notation the part (8) of the action defines a mathematically and physically consistent theory. Technically speaking it would describe a gauge-fixed version (à la Nakanishi-Lautrup) of the only consistent theory describing the coupled dynamics of symmetric ($G_{\mu\nu}$) and antisymmetric ($B_{\mu\nu}$) massless tensor fields. [This theory arises when taking the low-energy limit of string theory]. The discovery a few years ago [16] that the linearization of the theory (5) around flat space (i.e. around the background $g_{\mu\nu} = f_{\mu\nu}$) was described by the "good" action (8) raised the hope that the exact theory (5) might define, after all, an interesting physical theory.

However, the presence of the "bad" coupling (9) to the curvature of the symmetric background $G_{\mu\nu}$ shatters this hope. Indeed, the part (9) of the action has two catastrophic consequences for a physical theory [14]: (i) the theory contains ghost (negative-energy) excitations that cannot be consistently separated out; and (ii) the propagation of the ghost excitations leads to the loss of any acceptable asymptotic fall-off properties for the fields.

These bad properties are all due to the coupling of $B_{\mu\nu}$ to the curvature tensor of $G_{\mu\nu}$ (last term in Eq. (9)). It is interesting to note that the presence of this coupling could have been anticipated from the mathematical study of the multiple characteristics of the exact theory (5) due to F. Maurer-Tison [17] and J. Vaillant [18]. Alas these exact mathematical results had been essentially forgotten until physically motivated perturbation calculations unraveled the inconsistencies of Einstein's last "unified theory".

References

[1] Y. Fourès-Bruhat, Acta Math. **88**, 141 (1952).

[2] D. Christodoulou and B.G. Schmidt, Commun. Math. Phys. **68**, 275 (1979).

[3] T. Damour and B. Schmidt, J. Math. Phys. **31**, 2441 (1990).

[4] Y. Choquet-Bruhat, Uspekhi Mat. Nauk. **29**, 314 (1976).

[5] A.D. Rendall, Proc. R. Soc. London Ser. A **427**, 221 (1990).

[6] H. Friedrich, Commun. Math. Phys. **91**, 445 (1983); **107**, 587 (1986).

[7] Y. Choquet-Bruhat and M. Novello, C.R. Acad. Sci. Paris **305**, série II, 155 (1987).

[8] L. Blanchet and T. Damour, Phil. Trans. R. Soc. London Ser. A **320**, 379 (1986).

[9] H. Whitney, Annals Math. **37**, 645 (1936).

[10] P.D. D'Eath, Phys. Rev. D **11**, 2183 (1975).

[11] T. Damour, in *Gravitational Radiation*, edited by N. Deruelle and T. Piran (North-Holland, Amsterdam, 1983), pp. 59-144.

[12] T. Damour, M. Soffel and C. Xu, Phys. Rev. D **43**, 3273 (1991); **45**, 1017 (1992).

[13] A. Einstein, *The Meaning of Relativity* (5th edition) (Princeton University Press, Princeton, N.J., 1955); for more references see M.A. Tonnelat, *La Théorie du Champ Unifié d'Einstein* (Gauthier-Villars, Paris, 1955).

[14] T. Damour, S. Deser and J. McCarthy, Phys. Rev. **D45**, R3289 (1992); Phys. Rev. D, 15 January 1993.

[15] A. Lichnérowicz, *Théories Relativistes de la Gravitation et de l'Electromagnétisme* (Masson, Paris, 1955).

[16] R.B. Mann and J.W. Moffat, Phys. Rev. **D31**, 2488 (1985); for more references see J.W. Moffat, in *Gravitation 1990: A Banff Summer School,* edited by R.B. Mann and P. Wesson (World Scientific, Singapore, 1991).

[17] F. Maurer-Tison, Ann. Scient. Ec. Norm. Sup. **76**, 185 (1959).

[18] J. Vaillant, C.R. Acad. Sci. Paris **253**, 231 (1961); **253**, 1909 (1961).

FUNCTIONAL INTEGRATION

A MULTIPURPOSE TOOL

Cécile DeWitt-Morette

Department of Physics and Center for Relativity.

University of Texas, Austin TX 78712.

Based on joint work with Pierre Cartier

Préambule

Je dois beaucoup à Yvonne Choquet-Bruhat. Et, s'il me fallait présenter les problèmes où elle m'a montré l'idée simple qui va au cœur du sujet, je parlerais des heures, des jours sur ... l'Analyse, les Variétés et la Physique. Mais il m'a fallu faire un choix. Pourquoi choisir l'intégration fonctionnelle? Peut-être pour marquer le cinquantième anniversaire de l'intégrale de Feynman. En réalité, en souvenir de deux dates personnelles:

En 1969, Yvonne Choquet m'invita à faire 3 conférences. Pourquoi m'invita-t-elle? Nous nous connaissions alors fort peu - encore que nous découvrîmes plus tard avoir été la même année en septième à Victor Duruy. J'ai été d'autant plus étonnée de recevoir cette invitation que je m'engourdissais alors dans une vie scientifique sclérosée par les règlements concernant les époux dans la même profession. A l'occasion de ces conférences, je revins à l'intégrale de Feynman, sujet qui m'avait intéressée vingt ans auparavant, et je repartis bon pied, bon oeil.

En 1971, Yvonne, bien que ne travaillant pas elle-même sur le sujet, réalisa qu'il me manquait un élément essentiel pour aller au-delà des idées communément acceptées, et elle me dit de lire Bourbaki, livre VI, chapitre IX, page 70 et suivantes. J'y découvrais les promesures, dans une présentation facile à généraliser pour les besoins de l'intégrale de Feynman.

M. Flato et al. (eds.), Physics on Manifolds, 67–91.

Yvonne n'a pas, sous son nom, de publication sur l'intégrale fonctionnelle mais elle est à l'origine de travaux sur le sujet. C'est avec grand plaisir que je lui offre aujourd'hui quelques progrès récents sur l'intégrale de Feynman. Ces progrès doivent beaucoup à Pierre Cartier qui a su utiliser les résultats des physiciens pour construire une axiomatique de l'intégrale de Feynman.

Abstract

The goal is to extract from work done by physicists during the last fifty years an axiomatic basis for functional integration which will provide simple and robust methods of calculation, in particular for integration by parts, change of variable of integration, sequential integrations. The mythical integrator $\mathcal{D}$ in physicists' equations such as $\int_\Phi \exp\left(\frac{i}{\hbar}\left(S(\varphi) - \hbar\,\langle J,\varphi\rangle\right)\right)\mathcal{D}\varphi = Z(J)$ is not unique ; but given two Banach spaces Φ and Φ', and two continuous bounded maps $\Theta : \Phi \times \Phi' \to \mathbb{C}$, and $Z : \Phi' \to \mathbb{C}$, one can choose an integrator $\mathcal{D}_{\Theta,Z}$ satisfying the equation $\int_\Phi \Theta(\varphi, J)\mathcal{D}_{\Theta,Z}\varphi = Z(J)$ and a related normed space $\mathcal{F}_{\Theta,Z}$ of functionals F on Φ integrable by $\mathcal{D}_{\Theta,Z}$. Prodistributions, white noise integrators, and, of course, Lebesgue measures fit with proposed scheme.

Functional integration

1. - Introduction

The subtitle on the French poster says "Un outil sûr et performant". The English subtitle says "A reliable and efficient tool". Let us say a "multipurpose tool". Given the title of the Colloquium, I have selected three applications of Feynman path integrals, one in Analysis, one in Differential Geometry, and one in Physics. In each case I shall only state the problem, indicate the key issues which have been solved, and give the answer.

No one will question the answers; they are clearly right; but are Feynman path integrals still a mathematical nonsense? The answer is "no" and I shall present a mathematical framework, nearly completed, which makes them not only efficient but also reliable. Pierre Cartier is the architect of this bridge from mathematics to physics.

2. - Prodistributions

Given a locally convex space $\mathbf{X}$, and a projective family of finite dimensional spaces $\{\mathbf{X}^\alpha\}$ with the usual coherence conditions which make it possible to reconstruct $\mathbf{X}$ from $\{\mathbf{X}^\alpha\}$, a promeasure μ on $\mathbf{X}$ is a *projective* family $\{\mu_\alpha\}$ of *bounded measures* on $\{\mathbf{X}^\alpha\}$ which satisfies coherence conditions adapted to the definition of $\{\mathbf{X}^\alpha\}$.

A topology on $\mathbf{X}$ defines the dual space $\mathbf{X}'$ of $\mathbf{X}$ and one can construct the family $\{\mathcal{F}\mu_\alpha\}$ of Fourier transforms of the promeasure $\{\mu_\alpha\}$. The coherence conditions satisfied by $\{\mathcal{F}\mu_\alpha\}$ are simpler to state than the coherence conditions satisfied by $\{\mu_\alpha\}$, namely.

$$(2.1)\qquad \begin{cases} \mathcal{F}\mu_\alpha(0) = \text{constant independent of } \alpha \\ \text{If } \Pi^{\alpha\beta} : \mathbf{X}^\beta \to \mathbf{X}^\alpha, \text{ then } \mathcal{F}\mu_\alpha = \mathcal{F}\mu_\beta \circ \widetilde{\Pi}_{\beta\alpha}, \ \widetilde{\Pi}_{\beta\alpha} : \mathbf{X}'_\alpha \to X'_\beta \end{cases}$$

where $\widetilde{\Pi}_{\beta\alpha}$ is the transpose of $\Pi^{\alpha\beta}$.

$$\begin{array}{ccccc} \mathbf{X}' & & & & \mathbf{X} \\ \widetilde{\Pi}:=\{\widetilde{\Pi}_\alpha\}\uparrow & & & & \downarrow\{\Pi^\alpha\}=:\Pi \\ \{\mathbf{X}'_\alpha\} & & & & \{\mathbf{X}^\alpha\} \\ & \mathcal{S}'(\mathbf{X}'_\alpha) & \xleftarrow{\mathcal{F}} & \mathcal{S}'(\mathbf{X}^\alpha) & \end{array}$$

Figure 1 : Prodistributions

Let

$$\widetilde{\Pi} : \bigcup_\alpha \mathbf{X}'_\alpha \to \mathbf{X}' \text{ be defined by } \widetilde{\Pi}\Big|_{\mathbf{X}'_\alpha} := \widetilde{\Pi}_\alpha. \tag{2.2}$$

A prodistribution $\mathcal{F}w$ is a family of Fourier transforms $\{\mathcal{F}w_\alpha\}$ where w_α is not necessarily a bounded measure, but where $\mathcal{F}w_\alpha$ is a continuous function on $\mathbf{X}'_\alpha$. Therefore there is not necessarily a promeasure w corresponding to a prodistribution $\mathcal{F}w$.

Does a prodistribution define an integrator ?

Is such an integrator a practical tool for Physics ?
The answer is "yes" to both questions because of the following equations :
Let P be a linear continuous map $P : \mathbf{X} \to \mathbf{Y}$. Let $\mathcal{F}w$ be a prodistribution on $\mathbf{X}$. Let $f : \mathbf{Y} \to \mathbb{C}$

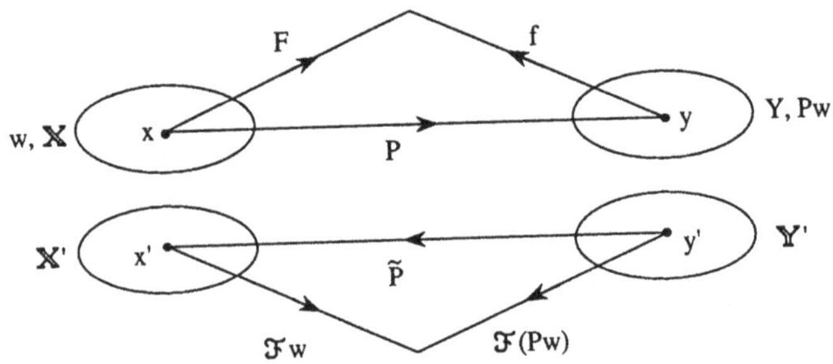

Figure 2 : Linear change of variable of integration

$$\int_{\mathbf{X}} F(x)dw(x) = \int_{\mathbf{Y}} f(y)d(Pw)(y), \tag{2.3}$$

$$\mathcal{F}(Pw) = \mathcal{F}w \circ \widetilde{P}. \tag{2.4}$$

If P maps the infinite dimensional space $\mathbf{X}$ into a finite dimensional space $\mathbf{Y}$, these equations define the functional integral over $\mathbf{X}$. If $\mathbf{Y}$ is infinite dimensional, these equations define a manipulation of the functional integral.

These two equations are sufficient to solve nontrivial problems. For instance, they have been used to compute the glory scattering cross section of classical waves by black holes. The cross section can be expressed in term of a functional integral. But this functional integral cannot be computed by any of the commonly used methods: discretization, analytical continuation, WKB expansions, for the following reasons:

a) the paths take their values in a riemannian space, and discretization of the paths is ambiguous;

b) Glory scattering is a scattering process where the final momentum is antiparallel to the initial momentum. This momentum-to-momentum transition cannot be computed from a position-to-position probability amplitude because there are no plane waves in curved spacetimes. Hence the path integral has to be set up for paths taking their values in phase space. Therefore, analytical continuation of a Wiener type integral is not an option because there are no Wiener measure for the space of paths in phase space (the positivity of the Jacobi operator in configuration space does not imply the positivity of the corresponding Jacobi operator in phase space);

c) the critical points of the action are degenerate: the paths are caustic forming in phase space. Hence the WKB approximation "breaks down".

On the other hand one can use the equations (2.3) and (2.4) to compute the glory scattering cross section. A long, but unambiguous, calculation gives, in the leading order of the semiclassical approximation

$$\frac{d\sigma}{d\Omega} = 2\pi\hbar^{-1}\left|\mathbf{p}\right| B_g^2 \frac{dB}{d\theta} J_{2s}^2\left(\hbar^{-1}\left|\mathbf{p}\right| B_g \sin\theta\right) \tag{2.5}$$

where Ω is the solid angle in the θ direction, $\mathbf{p}$ is the incoming momentum, $B(\theta)$ is the impact parameter of a particle scattered in the θ direction, B_g is the glory scattering impact parameter, $B_g = B(\pi)$, J_{2s} is the Bessel function of order $2s$ ($s = 0$ for a scalar particle, $s = 1$ for a photon, $s = 2$ for a graviton).

In the case of scattering by Schwarzschild Black Holes, the function $B(\theta)$ has been computed by Darwin. ($B(\theta) = M(3\sqrt{3} + 3.48\exp(-\theta)), M$ is the mass of the Black Hole in units where $G = c = 1$. If we use this function $B(\theta)$ in (2.5) the above cross section matches perfectly the numerical cross sections computed by Handler and Matzner who used the technique of partial wave decomposition, that is a totally different technique.

References to other applications of the prodistribution formalism are given below. I shall only mention the calculation of the propagator for the anharmonic oscillator

(2.6) $$\frac{1}{2}\dot{q}^2 + \frac{1}{2}\omega^2 q^2 + \frac{\lambda}{4}q^4 = 0,$$

because it shows that, contrary to commonly accepted ideas, the propagator is not singular in λ, but tends to the harmonic oscillator propagator when λ tends to zero.

The third application I have chosen is the expression for propagators between two points a and b of a multiply connected space M. A simple property of the path integral representation of the propagator gives :

(2.7) $$|K^{\alpha}(b, t_b; a, t_a)| = \left| \sum_{g_i \in \pi_1(M)} \chi^{\alpha}(g_i) K^i(b, t_b; a, t_a) \right|$$

where K^i is the propagator obtained by summing over all the paths in the same homotopy class, and

$$\chi^{\alpha} : \pi_1(M) \to \mathbb{C}$$

is a unitary representation of the fundamental group $\pi_1(M)$.

Prodistributions are not necessary to derive this result. I chose this example, because once more, I have to thank Yvonne Choquet-Bruhat. I had obtained the propagator (2.7) "experimentally" by studying Schulman's calculation of the path integral for a top, as a model for a path integral for spin. I knew (as a physicist knows) that the result was correct, but I knew that my proof was not convincing, to say the least. I had the opportunity to discuss the problem with Yvonne and R. Bott during a Les Houches session. When I asked Bott to help me clean up the proof, he objected strenuously to the fact that I was combining an element of $\pi_1(M)$ with a homotopy class of paths from a to $b \neq a$. But Yvonne convinced Bott not to give up the discussion. I then realised that I had to pay more than lip service to the fact that although the groups $\pi_1(M)$ based at two different points are isomorphic, they are not *canonically* isomorphic. The proof of (2.7) was then immediate : use the principle of superposition of probability amplitudes to write K as a linear combination of all the K^i's. Determine the coefficients of this linear combination by requiring that the result be independent of the base point chosen for $\pi_1(M)$ (more precisely independent of the homotopy mesh chosen to associate g_i and K^i). The linear combination is then determined up to an overall phase factor, hence the absolute value signs in (2.7). This example shows that functional integrals reflect, as could be expected, the global properties of the manifold where the paths take their values.

The problems I have mentioned were solved one at a time, the space of integrable functionals on $\mathbf{X}$ had not been identified.

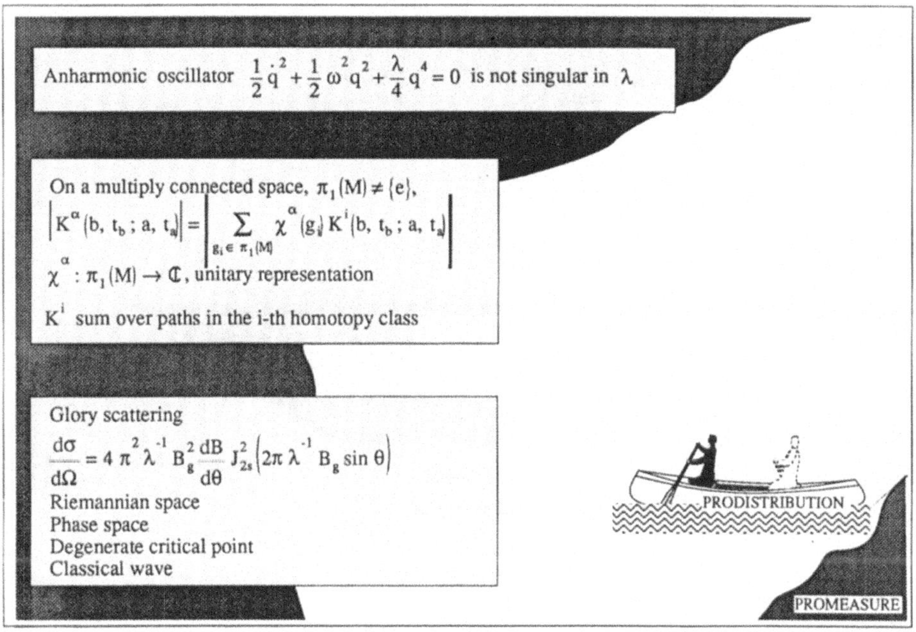

Figure 3 : Examples

3. - The Cartier Bridge

Pierre Cartier is surveying the work which has been done by physicists and by mathematicians on Feynman path integrals (see references below). Using the material at hand, he builds a bridge, now nearly completed, from mathematics to physics to make functional integration a robust tool ; in particular, a user friendly tool for

integration by parts

linear change of variable of integration

successive integrals (generalized Fubini theorem)

The bridge pillars are labelled by roman numerals, as stone pillars used to be; the roman numerals are repeated in the numbering of the equations.

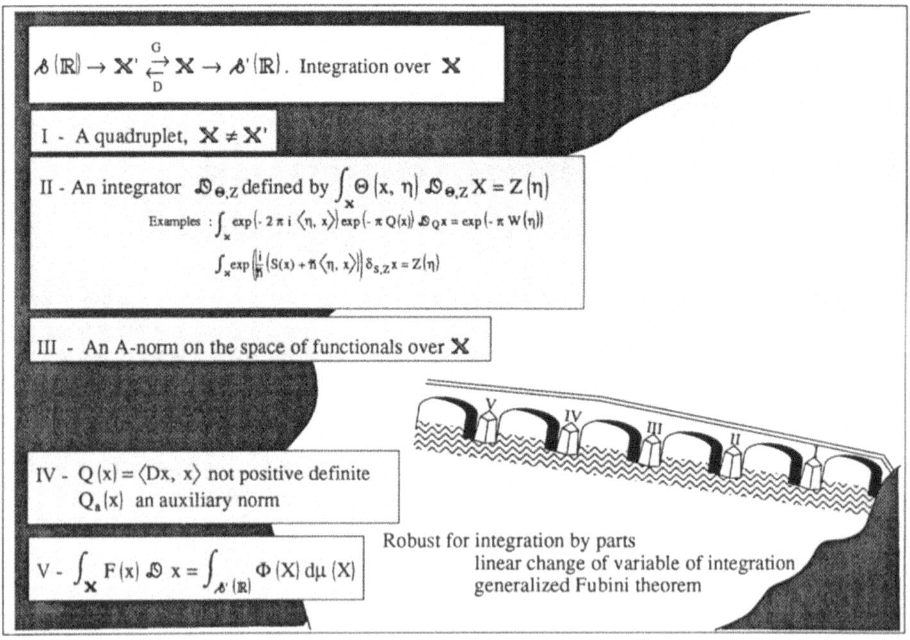

Figure 4 : The Cartier Bridge.

Pillar I : In broad outline, let $\mathbf{X}$ and $\mathbf{X}'$ be two Banach spaces in duality making together with the Schwartz spaces $\mathcal{S}(\mathbb{R})$ and $\mathcal{S}'(\mathbb{R})$ the quadruplet

$$\mathcal{S}(\mathbb{R}) \hookrightarrow \mathbf{X}' \underset{D}{\overset{G}{\rightleftarrows}} \mathbf{X} \hookrightarrow \mathcal{S}'(\mathbb{R}). \tag{3.I_1}$$

D is a differential operator acting on $\mathbf{X}$, and G is its kernel (Green function).

$$DG = \mathbb{1} \quad , \quad GD = \mathbb{1}. \tag{3.I_2}$$

The boundary conditions which define a unique kernel G are encoded in the space $\mathbf{X}$. For instance, if D is a second order differential operator on the space $\mathbf{X}$ of paths $x : [t_a, t_b] \to \mathbb{R}^d$ which vanish at t_a and t_b, $G^{\alpha\beta}(t,s)$ is the unique kernel of D which vanishes for either t or s equal to t_a or t_b.

In the case $\mathbf{X} = \mathbf{X}'$, the quadruplet reduces to a Gelfand triplet. Although the mathematics of the quadruplet is, to a large extent, similar to the mathematics of the triplet, the physics is considerably different. The physicist working with a Lagrangian needs $D : \mathbf{X} \to \mathbf{X}'$ where $\mathbf{X}'$ is not necessarily equal to $\mathbf{X}$.

Pillar II : Physicists want the action, integral of the Lagrangian, to appear explicitly. They need a path integral of the following type.

$$\int_{\mathbf{X}} \exp\left(\frac{i}{\hbar} S(x)\right) \mathcal{D}x \tag{3.II_1}$$

rather than an integral of the type

$$\int_{\mathbf{X}} \exp\left(-\frac{i}{\hbar} \int V(x(t))dt\right) dw(x). \tag{3.II_2}$$

This cannot be accomplished by a universal "Lebesgue" measure $\mathcal{D}x$ and Cartier has proposed the following framework. Let $\mathcal{D}_{\Theta,Z}$ be an integrator on $\mathbf{X}$ characterized by an equation

$$(3.II_3) \qquad \int_{\mathbf{X}} \Theta(x,\eta)\mathcal{D}_{\Theta,Z}x = Z(\eta) \quad \text{for} \quad \eta \in \mathbf{Y};$$

here $\mathbf{X}$ and $\mathbf{Y}$ are two Banach spaces, $\Theta(x,\eta)$ and $Z(\eta)$ are scalar valued functions dictated by the problem under consideration.

A simple example : with the notation of pillar I, if $\mathbf{Y} = \mathbf{X}'$, and $Q(x) = \langle Dx, x\rangle$, and $W(\eta) = \langle \eta, G\eta\rangle$,

$$(3.II_4) \qquad \int_{\mathbf{X}} \exp\left(-2\pi i \langle \eta, x\rangle\right) \exp(-\pi Q(x))\mathcal{D}_Q x = \exp(-\pi W(\eta))$$

defines an integrator $\mathcal{D}_Q$. Given the relationship between Q and W, the subscript Q is sufficient to characterize the integrator.

We note that if $\mathbf{X} = \mathbb{R}^\nu$ and $Q(x) = \sum_{i=1}^{\nu} \left(x^i\right)^2$, then $\mathcal{D}_Q x$ is the ordinary Lebesgue measure namely $dx^1 \cdots dx^\nu$.

An example from quantum physics : Let φ be a field and J a source; then the action S and the generating functional $Z(J)$ introduced by Schwinger define an integrator $\mathcal{D}_{S,Z}$ by:

$$(3.II_5) \qquad \int_{\mathbf{X}} \exp\left(\frac{i}{\hbar}\left(S(\varphi) + \hbar \langle J, \varphi\rangle\right)\right) \mathcal{D}_{S,Z}\varphi = Z(J).$$

The important point is to note that there is no universal $\mathcal{D}\varphi$ even if we abbreviate $\mathcal{D}_{\Theta,Z}\varphi$ by $\mathcal{D}\varphi$ when the context is clear.

Application : Ward-Takahashi anomalies. (See also (3. VI.$_3$)). It has often been expected that the symmetries of the generating functional $Z(J)$ are the same as the symmetries of the classical action $S(\varphi)$. Originally, the Ward-Takahashi identities expressed invariance properties of $Z(J)$ under transformations leaving $S(\varphi)$ invariant, assuming $\mathcal{D}\varphi$ (no reference to S or Z) to be invariant. When "anomalous" terms were found in the Ward-Takahashi identities, Fujikawa showed that, in the case of chiral transformations, the so called anomalous term was simply the determinant of the linear map $M : \varphi \mapsto M\varphi$, defining the chiral transformation, i.e.

$$\mathcal{D}_{\Theta,Z}(M\varphi) = \operatorname{Det} M \cdot \mathcal{D}_{\Theta,Z}(\varphi) \neq \mathcal{D}_{\Theta,Z}(\varphi). \tag{3.II_6}$$

Pillar III : Norms $\| \ \|$ on $\mathbf{X}$, and norms $\| \ \|_A$ on the space $\mathcal{F}$ of integrable functionals over $\mathbf{X}$ with respect ot $\mathcal{D}_{\Theta,Z}$.

Sobolev norms on $\mathbf{X}$ are usually suitable [1]. For instance

$$\mathbf{X} = W_2^1 \ , \ \mathbf{X}' = W_2^{-1} \tag{3.III_1}$$

where W_2^m, also labelled H^m, is the space of square integrable functions whose partial derivatives of order $\leq m$ (in the sense of distributions) are also square integrable. Its dual W_2^{-m} is the space of distributions made of derivatives of order $\leq m$ of square integrable functions.

In the case $\mathbf{X} = \mathbf{X}'$, Albeverio and Høegh-Krohn have proposed a space of integrable functionals which are Fourier transforms of bounded measures on $\mathbf{X}$. Adapting their suggestions to the integrators defined by the second pillar we consider (as a minimal choice) the space of integrable functionals to be the space $\mathcal{F}_{\Theta,Z}$ ($\mathcal{F}$ for Feynman) of functions F such that [2]

$$F(x) = \int_{\mathbf{Y}} \Theta(x,\eta) d\mu(\eta), \quad F \in \mathcal{F}_{\Theta,Z} \quad \text{(abbreviated to } \mathcal{F}) \tag{3.III_2}$$

where μ is a bounded measure on the Banach space $\mathbf{Y}$, possibly complex. This equation does not imply that there is a one-to-one correspondence between μ and F. It does not mean either that, given F, one needs to identify μ in order to compute $\int_{\mathbf{X}} F(x)\mathcal{D}x$. It only means that we can write

$$\int_{\mathbf{X}} F(x)\mathcal{D}x = \int_{\mathbf{X}} \mathcal{D}x \int_{\mathbf{Y}} \Theta(x,\eta) d\mu(\eta) \tag{3.III_3}$$

---

[1] For instance if one requires that the action of the system be finite.

[2] For simplicity we assume the functions Θ on $\mathbf{X} \times \mathbf{Y}$ and Z on $\mathbf{Y}$ to be bounded and continuous.

$$= \int_{\mathbf{Y}} d\mu(\eta) \int_{\mathbf{X}} \Theta(x,\eta)\mathcal{D}x = \int_{\mathbf{Y}} Z(\eta)d\mu(\eta)$$

and it suggests a norm on $\mathcal{F}$

$$\|F\|_A := \min_{\mu} \int_{\mathbf{Y}} |Z(\eta)|\, d\,|\mu|\,(\eta). \tag{3.III_4}$$

Although μ is not necessarily defined by (3.III$_2$), we can prove in many cases that $\int_{\mathbf{X}} F(x)\mathcal{D}x$ is well defined: assume that there exists a family $\{\lambda_n\}$ of Borel measures [3] on $\mathbf{X}$ such that

$$Z(\eta) = \lim_{n=\infty} \int_{\mathbf{X}} \Theta(x,\eta)d\lambda_n(x), \tag{3.III_5}$$

then

$$\begin{aligned} \int_{\mathbf{Y}} Z(\eta)d\mu(\eta) &= \lim_{n=\infty} \int_{\mathbf{X}} d\lambda_n(x) \int_{\mathbf{Y}} \Theta(x,\eta)d\mu(\eta) \\ &= \lim_{n=\infty} \int_{\mathbf{X}} d\lambda_n(x)F(x). \end{aligned} \tag{3.III_6}$$

Neither the measure μ in (3.III$_2$) nor the family $\{\lambda_n\}$ in (3.III$_5$) are uniquely defined. But (3.III$_5$) which is independent of the choice made for μ and (3.III$_2$) which is independent of the choice made for $\{\lambda_n\}$ lead to the equality (3.III$_6$).

Pillar IV : The axiomatic formulation of functional integrals summarized in the pillar II by $\int_{\mathbf{X}} \Theta(x,\eta)\mathcal{D}_{\Theta,Z}x = Z(\eta)$ would be nearly useless to physicists if it did not include indefinite quadratic forms. For instance, one may wish to compute (3.III$_4$) in terms of a Fresnel integrator, $\exp(-\pi Q(x))\mathcal{D}_Q x$ with $Q(x) = i\langle Dx, x\rangle$, where D is the Jacobi operator defined by the action S on $\mathbf{X}$. Therefore one needs an axiomatic framework which includes the d'Alembertian

$$D = \eta^{\mu\nu}\partial_\mu\partial_\nu. \tag{3.IV_1}$$

[3] $\mathbf{X}$ is a separable complete metric space, hence Borel subsets are defined. The λ_n are not necessarily bounded but we assume that $|\lambda_n|(B)$ is finite for $B \subset \mathbf{X}$ bounded.

That is, one needs to redefine the norm $\|F\|_A = \int |Z(\eta)|d|\mu|(\eta)$ when $Z(\eta) = \exp(-\pi W(\eta))$ with W not positive definite. Following the Gupta-Bleuler strategy, we introduce an auxiliary quadratic form Q_a positive definite [4], $Q_a(x) > 0$ for $x \neq 0$, to define an auxiliary norm $\|\ \|_a$ on $\mathbf{X}$ which defines a norm on $\mathbf{X}'$ and a norm on $\mathcal{F}$.

This can be obtained by generalizing the well-known *Sylvester decomposition* of quadratic forms. Namely, we consider real-valued quadratic forms Q on infinite-dimensional space $\mathbf{X}$ with the following property:

There exists a decomposition $\mathbf{X} = \mathbf{X}_1 \oplus \mathbf{X}_2$ into a direct sum such that

$$Q(x) = Q_1(x_1) - Q_2(x_2) \tag{3.IV$_2$}$$

for $x = x_1 + x_2$, where x_1 is in $\mathbf{X}_1$ and x_2 in $\mathbf{X}_2$. Moreover Q_1 and Q_2 are positive definite quadratic froms, and define $\mathbf{X}_1$ and $\mathbf{X}_2$ as (real) Hilbert spaces.

This being so, put

$$Q_a(x) = Q_1(x_1) + Q_2(x_2) \tag{3.IV$_3$}$$

with x, x_1, x_2 as above. Then Q_a is a positive definite quadratic from on $\mathbf{X}$, hence the norm $\|x\|_a = Q_a(x)^{1/2}$ under which $\mathbf{X}$ is complete.

The decomposition mentioned in (3.IV$_2$) is not unique but two different auxiliary norms $\|\ \|_a$ and $\|\ \|_{a'}$ are related by inequalities of the form

$$\alpha \|x\|_{a'} \leq \|x\|_a \leq \beta \|x\|_{a'} \tag{3.IV$_4$}$$

(for some constants α, β in $\mathbb{R}^+$) hence either norm can be used on $\mathbf{X}$.

---

[4] The case of a semidefinite quadratic form is not excluded but requires special consideration. The case of a degenerate quadratic form has already been investigated in the context of prodistributions.

Pillar V : On the first pillar, let the pair $\mathbf{X}, \mathbf{X}'$ in the quadruplet (3.I$_1$),

$$\mathcal{S}(\mathbb{R}) \hookrightarrow \mathbf{X}' \underset{D}{\overset{G}{\rightleftarrows}} \mathbf{X} \hookrightarrow \mathcal{S}'(\mathbb{R}), \tag{3.V_1}$$

consist of Hilbert spaces and the quadratic form $Q(x) := \langle Dx, x\rangle$ be positive definite. Then $\mathbf{X}$ is of measure zero with respect to the white noise measure on $\mathcal{S}'(\mathbb{R})$. Moreover powerful theorems have been derived for the space of Hida distributions defined on $\mathcal{S}'(\mathbb{R})$. So why work with the pair $\mathbf{X}, \mathbf{X}'$? Two reasons :

- The structure (G, D) does not exist on the pair $\mathcal{S}(\mathbb{R}), \mathcal{S}'(\mathbb{R})$, indeed $G : \mathcal{S}(\mathbb{R}) \to \mathcal{S}'(\mathbb{R})$ does not have an inverse.
- We shall show that, if $Fe^{-\pi Q}$ is in the space $\mathcal{F}$ of Feynman integrable functions, then

$$\int_{\mathbf{X}} F(x)\exp(-\pi Q(x))\mathcal{D}_Q x = \int_{\mathcal{S}'(\mathbb{R})} \widetilde{F}(X) dw_Q(X), \tag{3.V_2}$$

where the Fourier transform of dw_Q and the Fourier transform of $\exp(-\pi Q(x))\mathcal{D}_Q x$ evaluated at $\eta \in \mathcal{S}(\mathbb{R})$ are identical, and where $\widetilde{F}$ is the extension of F from $\mathbf{X}$ to $\mathcal{S}'(\mathbb{R})$ constructed as follows.

Let $\{e_i\}$ and $\{\varepsilon^i\}$ be two orthonormal bases on $\mathbf{X}$ and $\mathbf{X}'$, respectively, such that $\varepsilon^i \in \mathcal{S}(\mathbb{R})$; assume

$$De_i = \varepsilon^i \quad , \quad \langle e_i, \varepsilon^j\rangle = \delta_i^j . \tag{3.V_3}$$

Let us define

$$P_N : \mathcal{S}'(\mathbb{R}) \to \mathbf{X} \quad \text{by} \quad P_N X = \sum_{i=1}^{N} \langle X, \varepsilon^i\rangle \cdot e_i; \tag{3.V_4}$$

then the limit

$$\widetilde{F}(X) := \lim_{N=\infty} F(P_N X) \tag{3.V_5}$$

exists a.s. for w_Q and defines the extension of F from $\mathbf{X}$ to $\mathcal{S}'(\mathbb{R})$.

The privileged extension of F to $\widetilde{F}$ is analogous to the extension of a continuous function defined on the rationals to a continuous function defined on $\mathbb{R}$.

We can summarize (3.V$_2$) and (3.II$_4$) in one formula

$$(3.V_6) \qquad \int_{\mathbf{X}} F(x)\exp\left(-2\pi i\,\langle \eta, x\rangle\right)\exp(-\pi Q(x))\mathcal{D}_Q x$$

$$= \int_{\mathcal{S}'(\mathbb{R})} \widetilde{F}(X)\exp\left(-2\pi i\,\langle \eta, X\rangle\right) dw_Q(X),$$

for $\eta \in \mathcal{S}(\mathbb{R})$.

Pillar VI : Expectation values of operators. Functional integrals can be used to compute expectation values of "time ordered" products of operators. When the Cartier Bridge is completed, pillar VI will present functional integration in quantum physics as a generalisation of the relationship between the Schrödinger and the Heisenberg quantizations. It will give another justification of the well known formulae :

$$(3.VI_1) \qquad \int_{\mathbf{X}_{ba}} \exp\left(\frac{i}{\hbar}\left(S(\varphi) + \hbar\,\langle J, \varphi\rangle\right)\right)\mathcal{D}_{\Theta,Z}(\varphi)$$

$$(3.VI_2) \qquad \begin{aligned} &= \left\langle \Psi_b \middle| \mathcal{T}\exp\left(-\tfrac{i}{\hbar}\textstyle\int_{t_a}^{t_b} H\,dt\right) \middle| \Psi_a \right\rangle && \text{(Dirac bracket)} \\ &=: \langle \Psi_b, t_b | \Psi_a, t_a\rangle_J && \text{(Feynman bracket)} \end{aligned}$$

and

$$(3.VI_3) \qquad \int_{\mathbf{X}_{ba}} F(\varphi)\exp\left(\frac{i}{\hbar}\left(S(\varphi) + \hbar\,\langle J, \varphi\rangle\right)\right)\mathcal{D}_{\Theta,Z}(\varphi) = \langle \Psi_b, t_b | \mathcal{T} F(\varphi) | \Psi_a, t_a\rangle_J$$

with the following notation

- $\mathbf{X}_{ba}$ is the space of paths [or histories] with given values at times t_a and t_b corresponding respectively to the states $|\Psi_a\rangle$ and $|\Psi_b\rangle$

- H is the hamiltonian corresponding to the action $S(\varphi) + \hbar\,\langle J, \varphi\rangle$.

- $\mathcal{T}$ is a map from the space $\mathcal{F}$ of bounded functionals on $\mathbf{X}_{b,a}$ to the space of bounded operators on the Hilbert space of states of the system,

$$(3.VI_4) \qquad \mathcal{T} : \mathcal{F} \to \mathcal{B}(\mathcal{H})$$

which time-orders products of operators. For instance, in quantum mechanics,

$$(3.VI_5) \qquad \mathcal{T}\varphi_1(t)\varphi_2(s) = \theta(t-s)\widehat{\varphi}_1(t)\widehat{\varphi}_2(s) + \theta(s-t)\widehat{\varphi}_2(s)\widehat{\varphi}_1(t)$$

where $\theta(t-s) = 1$ for $t > s$, $\theta(t-s) = 0$ for $t < s$, $\theta(t-s)$ undefined for $t = s$.

The time ordering operator $\mathcal{T}$ on $\mathcal{F}$ is defined by the functional integral (3.VI$_3$) , and not by equations such as (3.VI$_5$) which can be ambiguous at equal times. Therefore, time ordering commutes with differentiation

$$(3.VI_6) \qquad \frac{d}{dt}\langle\, |\mathcal{T}\varphi(t)\varphi(s)|\,\rangle = \left\langle \left| \mathcal{T}\frac{d\varphi}{dt}\varphi(s)\right| \right\rangle$$

or in field theory

$$(3.VI_7) \qquad \frac{\partial}{\partial x^\mu}\langle\, |\mathcal{T}\varphi(x)\varphi(y)|\,\rangle = \left\langle \left| \mathcal{T}\frac{\partial}{\partial x^\mu}\varphi(x)\varphi(y)\right| \right\rangle$$

Application : Quantum Nœther's theorems.

Nœther's theorems apply to classical currents, i.e. currents which are functional of fields satisfying the Euler-Lagrange equations. There is no reason to expect that they apply to expectation values of time ordered product of currents : the left hand side of (3.VI$_3$) is, for φ a field, an integral over *all* fields with given values on an initial and a final spacelike surface. The Nœther theorems are the classical limits of "quantum Nœther's theorems" which are consequences of (3.VI$_3$) : Indeed, under a change of variable of integration $\varphi \mapsto \psi$, it follows from the defining equation of the integration $\mathcal{D}_{\Theta,Z}$ that

$$(3.VI_8) \qquad 0 = \int \Theta(\varphi, J)\mathcal{D}_{\Theta,Z}(\varphi) - \int \Theta(\psi, J)\mathcal{D}_{\Theta,Z}(\psi).$$

In case of a linear map, $\psi = M\varphi$, it follows from (3.II$_6$) that

$$0 = \int ((\text{Det}\ \mathbb{1})\Theta(\varphi, J) - (\text{Det}\ M)\Theta(M\varphi, J))\,\mathcal{D}_{\Theta,Z}(\varphi), \tag{3.VI_9}$$

and, if $\Theta(\varphi, J) = \exp\left(\frac{i}{\hbar}(S(\varphi) + \hbar\langle J, \varphi\rangle)\right)$,

$$0 = \int \Bigg(\exp\left(\frac{i}{\hbar}S(\varphi) + i\langle J, \varphi\rangle + \text{trace}\ \ln \mathbb{1}\right) - \exp\left(\frac{i}{\hbar}S(M\varphi) + i\langle J, M\varphi\rangle + \text{trace}\ \ln M\right)\Bigg)\mathcal{D}_{\Theta,Z}(\varphi). \tag{3.VI_{10}}$$

Consider the linear map

$$(M\varphi)(x) = \varphi(x) + \theta(x)\varphi(x), \tag{3.VI_{11}}$$

where θ is a continuous map on the domain of φ ; it follows from (3.VI_{10}) that

$$0 = \int \exp\left(\frac{i}{\hbar}S(\varphi) + i\langle J, \varphi\rangle\right) \frac{\delta}{\delta\theta}\left(\frac{i}{\hbar}S(M\varphi) + i\langle J, M\varphi\rangle + \text{trace}\ \ln M\right)\Bigg|_{\theta=0} \mathcal{D}_{\Theta,Z}(\varphi). \tag{3.VI_{12}}$$

In terms of time ordered product of operators (3.VI_3), this equation says

$$0 = \left\langle \mathcal{T}\frac{\delta}{\delta\theta}\left(\frac{i}{\hbar}S(M\varphi) + i\langle J, M\varphi\rangle + \text{trace}\ \ln M\right)\right\rangle_J . \tag{3.VI_{13}}$$

Repeated functional derivatives of (3.VI_{13}) with respect to J give the correct Ward-Takahashi identities. At $J = 0$ equation (3.VI_{13}) gives

$$\left\langle \mathcal{T}\frac{\delta}{\delta\theta}\left(\frac{i}{\hbar}S(M\varphi)\right)\right\rangle = -\frac{\delta}{\delta\theta}\text{trace}\ \ln M. \tag{3.VI_{14}}$$

Trace ln M is known as the anomaly function - but should not be unexpected.

The action $S(\varphi) = \int L(\varphi, \varphi_{,\mu})dx$, therefore it follows from (3.VI$_{14}$) that

$$(3.VI_{15}) \qquad \left\langle \mathcal{T} \int \left(\frac{\partial L}{\partial \varphi} - \frac{\partial}{\partial x^\mu} \left(\frac{\partial L}{\partial \varphi_{,\mu}} \right) \right) \delta\varphi + \frac{\partial}{\partial x^\mu} \left(\frac{\partial L}{\partial \varphi_{,\mu}} \delta\varphi \right) dx \right\rangle$$

$$= -i\hbar \text{ trace } \ln M.$$

Nœther's theorem says that, the current density

$$j^\mu := \left. \frac{\partial L}{\partial \varphi_{,\mu}} \delta\varphi \right|_{\varphi = \varphi_{\text{cl}}}$$

evaluated at solutions φ_{cl} of the Euler-Lagrange equation satisfies the equation

$$j^\mu{}_{,\mu} = 0$$

for actions invariant under the transformation $\varphi \mapsto \varphi + \delta\varphi$. Even if we assume (see pillar VII) that

$$\left\langle \mathcal{T} \int \left(\frac{\partial L}{\partial \varphi} - \frac{\partial}{\partial x^\mu} \left(\frac{\partial L}{\partial \varphi_{,\mu}} \right) \right) dx \right\rangle = 0,$$

equation (3.VI$_{15}$) says only that

$$\lim_{\hbar = 0} \left\langle \mathcal{T} \int j^\mu{}_{,\mu}(x) dx \right\rangle = 0.$$

Pillar VII : Renormalization.

We applied to two cases the defining equation of the integrator $\mathcal{D}_{\Theta,Z}$

$$(3.VII_1) \qquad \int_{\mathbb{X}} \Theta(\varphi, J) \mathcal{D}_{\Theta,Z}\varphi = Z(J).$$

a) The gaussian case:

$$(3.VII_2) \qquad \begin{cases} \Theta(\varphi, J) = \exp(-\pi Q(\varphi) - 2\pi i \langle J, \varphi \rangle) \\ Z(J)/Z(0) = \exp(-\pi W_Q(J)) \end{cases}$$

where the second functional derivative of Q is the inverse of the second functional derivative of W_Q :

$$(3.VII_3) \qquad \frac{\delta^2 Q(\varphi)}{\delta\varphi^2}\frac{\delta^2 W_Q(J)}{\delta J^2} = DG = \mathbb{1}.$$

b) The quantum case:

$$(3.VII_4) \qquad \begin{cases} \Theta(\varphi, J) = \exp\left(\frac{i}{\hbar}(S(\varphi) + \hbar\langle J, \varphi\rangle)\right) \\ Z(J)/Z(0) =: \exp\left(\frac{i}{\hbar}W(J)\right). \end{cases}$$

We shall now analyze the relationship between the action $S(\varphi)$, the generating functional $Z(J)/Z(0)$ and the integrator $\mathcal{D}_{\Theta,Z}$. The generating functional encodes the effect of the interaction stated by the action $S(\varphi) + \hbar\langle J, \varphi\rangle$ where φ may be an abbreviation for several interacting fields, or a self-interacting field. Experiments measure $Z(J)$ directly or indirectly. In this sense we refer to $Z(J)$ as an experimental quantity. The effect of the interactions can, in a number of cases, show up as a change in the constants used to describe the non interacting fields. One says that the (bare) constants have been renormalized (by the interactions). The ratio of the constants before and after the interactions may be infinite if we model the interactions by local actions. Various regularization techniques can be used to evaluate them; important as they are in checking the theory against experiments, they should not be given logical precedence over renormalization.

We cannot, a priori, choose the experimental generating functional, but we can assume the following property,

$$(3.VII_5) \qquad \int_{\mathbf{X}} \frac{\delta}{\delta\varphi}\left(\exp\left(\frac{i}{\hbar}(S(\varphi) + \hbar\langle J, \varphi\rangle)\right)\right)\mathcal{D}_{\Theta,Z}\varphi = 0,$$

and derive its consequences. From (3.VII$_5$) and (3. VI$_3$), it follows that

$$(3.VII_6) \qquad \left\langle \text{out}\left| T\frac{1}{\hbar}\frac{\delta S}{\delta\varphi}\right| \text{in}\right\rangle_J = -J$$

where the $|\text{in}\rangle$ and $|\text{out}\rangle$ state are the ones encoded in $\mathbf{X}$. On the other hand, having in mind an equation reminiscent of (3.VI$_3$), we note that

$$(3.VII_7) \qquad \frac{1}{\hbar}\frac{\delta W(J)}{\delta J} = \int \varphi\Theta(\varphi, J)\mathcal{D}_{\Theta,Z}\varphi = \langle \text{out}\, |\varphi|\, \text{in}\rangle_J =: \overline{\varphi}.$$

Hence the Legendre transform $\Gamma(\overline{\varphi})$ of $W(J)$,

$$(3.VII_8) \qquad \Gamma(\overline{\varphi}) = W(J) - \hbar\,\langle J, \overline{\varphi}\rangle$$

known as the *effective action*, is such that

$$(3.VII_9) \qquad \frac{1}{\hbar}\frac{\delta\Gamma}{\delta\overline{\varphi}} = -J;$$

and

$$(3.VII_{10}) \qquad \left\langle \text{out} \left| \mathcal{T}\frac{\delta S}{\delta\varphi} \right| \text{in} \right\rangle_J = \frac{\delta\Gamma}{\delta\overline{\varphi}}.$$

Moreover, the second functional derivative of the effective action is the inverse of the second functional derivative of W

$$(3.VII_{11}) \qquad \frac{\delta^2\Gamma(\overline{\varphi})}{\hbar\delta\overline{\varphi}^2}\frac{\delta^2 W(J)}{\hbar\delta J^2} = \mathbb{1}.$$

This equation can be compared with (3.VII$_3$). If the effective action $\Gamma(\overline{\varphi})$ were equal to the action $S(\varphi)$, the quantum case would be a gaussian case. Therefore we can consider the gaussian case as a non interacting quantum case, since it is a case where effective action and bare action are identical.

We note also the following consequence of (3.VII$_{4,8,9}$) :

$$(3.VII_{12}) \qquad \exp\left(\frac{i}{\hbar}\Gamma(\overline{\varphi})\right) = \frac{1}{Z(0)}\int \exp\left(\frac{i}{\hbar}\left(S(\varphi) + \left\langle\frac{\delta\Gamma(\overline{\varphi})}{\delta\overline{\varphi}}, \overline{\varphi} - \varphi\right\rangle\right)\right)\mathcal{D}_{\Theta,Z}\varphi.$$

4. - Conclusion

Given a space $\mathbf{X}$ of paths $x : [t_a, t_b] \to \mathbf{M}^d$, one can either work with

$$x \in \mathbf{X} \quad \text{or with} \quad \{x(t_1), \cdots, x(t_N)\} \in \mathbf{M}^d \times \cdots \times \mathbf{M}^d \quad (N \text{ factors}).$$

Working with $\mathbf{X}$ is admittedly more delicate than working with $\mathbf{M}^d$, but much simpler than working on $\lim_{N=\infty} \mathbf{M}^d \times \cdots \times \mathbf{M}^d$ (N factors). To mention but a few examples:

- An integrator is simpler than the family of its finite dimensional distributions.

- A linear change of variable in $\mathbf{X}$ is simpler than its discretized version on $\mathbf{M}^d \times \mathbf{M}^d \times \cdots \times \mathbf{M}^d$.

- An integration by parts is simpler on $\mathbf{X}$ than on $\mathbf{M}^d \times \cdots \times \mathbf{M}^d$ etc ...

The progress accomplished in recent years can be traced to the shift from $\mathbf{M}^d \times \mathbf{M}^d \times \cdots \times \mathbf{M}^d$ (arbitrary number of factors) to $\mathbf{X}$. This can be seen, not only in the work reported here, but in the major progress due to P. Krée, P. Malliavin, P.A. Meyer, the White Noise team, in particular, T. Hida, H.-H. Kuo, L. Streit, M. de Faria, J. Pottshoff and Khandekar.

Functional integrals are more than solutions of partial differential equations with a chosen set of boundary conditions. In particular, they provide information on the global properties of $\mathbf{M}^d$ the target space of the paths. They provide representations of expectation values of time ordered products of operators. Pierre Cartier and I plan to test the proposed axiomatic on a variety of problems to make sure that it covers, at least, the known applications of functional integrals.

REFERENCES

1. Definition, applications, review articles on prodistributions. Introduced in 1972.

[a] Cécile Morette DeWitt (1972) "Feynman's Path Integral, Definition Without Limiting Procedure", Commun. Math. Phys. **28**, 47-67.

[b] Cécile DeWitt-Morette (1974) "Feynman Path Integrals; I. Linear and Affine Techniques; II. The Feynman-Green Function", Commun. Math. Phys. **37**, 63-81.

Developed and applied in :

[c] Cécile DeWitt-Morette (1976) "The Semi-Classical Expansion", Annals of Physics **97**, 367-399.

[d] Maurice M. Mizrahi (1979) "The semiclassical expansion of the anharmonic-oscillator propagator" J. Math. Phys. **20**, 844-855.

[e] C. DeWitt-Morette, A. Maheshwari and B. Nelson (1979) "Path Integration in Non relativistic Quantum Mechanics", Physics Reports **50**, 266-372.

[f] Cécile DeWitt-Morette and Tian-Rong Zhang (1983) "Path Integrals and Conservation Laws", Phys. Rev. **D28**, 2503-2516.

[g] Cécile DeWitt-Morette and Tian-Rong Zhang (1983) "Feynman-Kac formula in phase space with application to coherent-state transitions", Phys. Rev. **D28**, 2517-2525.

[h] Cécile DeWitt-Morette, Bruce Nelson and Tian-Rong Zhang (1983) "The Caustic Problem in Quantum Mechanics with Application to Scattering Theory", Phys. Rev. **D28**, 2526-2546.

[i] Cécile DeWitt-Morette and Bruce Nelson (1984) "Glories - and other degenerate critical points of the action", Phys. Rev. **D29**, 1663-1668.

[j] P. Anninos, C. DeWitt-Morette, R. Matzner, P. Yioutas and T.-R. Zhang (1992). Orbiting cross-sections, application to black hole scattering. Armadillo preprint.

Summarized in :

[k] Cécile DeWitt-Morette (1984) "Feynman path integrals. From the Prodistribution Definition to the Calculation of Glory Scattering", in *Stochastic Methods and Computer Techniques in Quantum Dynamics*, ed. H. Mitter and L. Pittner, Acta Physica Austriaca Supp. **26**, 101-170. Reviewed in Zentralblatt für Mathematik 1985.

[l] Cécile DeWitt-Morette (1990) "Quantum Mechanics in Curved Spacetimes ; Stochastic Processes on Frame Bundles" in *Quantum Mechanics in Curved Space-Time.* Ed. by J. Audretsch and V. de Sabbata (Plenum Press, New-York).

2. The anharmonic oscillator.

See reference 1.d.

3. Paths on multiply connected spaces.

Michael G.G. Laidlaw and Cécile Morette DeWitt (1971) "Feynman Functional Integrals for Systems of Indistinguishable Particles" Phys. Rev. **D3**, 1375-1378.

4. Glory Scattering

Computed in 1.i, using results developed in 1.f, 1.g, 1.h ; summarized in 1.k.

5. Articles which helped Cartier design the bridge in paragraph 3.

[a] Richard P. Feynman (1942) "The principle of Least Action in Quantum Mechanics" (Princeton University Ph. D. Dissertation) Available from University Microfilms Inc. P.O. Box 1307, Ann Arbor, MI 48106. Tel 1-800-521 0600.

[b] Sergio A. Albeverio and Raphael J. Høegh-Krohn (1976) *Mathematical Theory of Feynman Path Integrals* (Lecture Notes in Mathematics # 523, Springer-Verlag, Berlin)

S. Albeverio, A. Boutet de Monvel-Berthier, and Z. Brzezniak (1992) "The trace formula for Schrödinger operators from infinite dimensional oscillatory integrals". Bibos preprint and references therein (submitted to Crelle's Journal).

[c] R.H. Cameron and D.A. Storvick (1983) "A simple definition of the Feynman integral, with applications" Memoirs of the American Mathematical Society #288 (Providence, Rhode Island) and references therein.

6. The Cartier bridge The construction called here the "Cartier Bridge" will be published in a Note aux Compte-Rendus de l'Académie des Sciences by P. Cartier and C. DeWitt-Morette. In the years to come the authors will write a book on functional integration and test the proposed axiomatic on many problems of physical interest.

7. A short bibliography on White Noise for Pillars I and V

[a] *General*

T. Hida, H.-H. Kuo, J. Potthoff, L. Streit. *"White Noise - An Infinite Dimensional Calculus"* monograph to appear.

H.-H. Kuo. (1991) "Lectures on White Noise Analysis", in *Proc. Preseminar Intl. Conf. on Gaussian Random Fields,* edited by T. Hida and K. Saito, Nagoya.

[b] *Characterization Theorem*

J. Potthoff, and L. Streit. (1991) "A characterization of Hida distributions", J. Funct. Analysis 101, 212.

[c] *Feynman Integrals*

T. Hida, and L. Streit. (1983) "Generalized Brownian functionals and the Feynman integral". Stoch. Proc. Appl. 16, 55-69.

M. de Faria. (1991) "White Noise Analysis and the Feynman integral". Universidade da Madeira preprint UMa-Mat 4/91, to appear in Proc. III. Intl. Conf. Stoch. Proc., Physics, and Geometry, Locarno.

M. de Faria, J. Potthoff, L. Streit. (1991). "The Feynman integrand as a Hida distribution" J. Math. Phys. 32, 2123 (1991).

D.C. Khandekar, L. Streit. (1992) "Constructing the Feynman integrand" Ann. Physik 1, 49.

L. Streit. (1992) "The Feynman Integral. Recent results" Universidade da Madeira Preprint UMa-Mat 10/92.

8. Other articles quoted in the text (by alphabetic order).

K. Bleuler (1950) Helvetica Phys. Acta **23**, 567.

S.N. Gupta (1950) Proc. Phys. Soc. **A63**, 681.

Paul Krée (1988) "La Théorie des distributions en dimension quelconque et l'Intégration stochastique" dans *Stochastic Analysis and Related Topics* Eds H. Korezlioglu and A.S. Ustunel, pp. 170-233 (Lecture Notes in Mathematics # 1316, Springer-Verlag).

P. Malliavin (1993) *Stochastic Analysis* (Grundlehren der Mathematik, Springer-Verlag, Berlin).

P.A. Meyer (1986-1987) "Eléments de probabilités quantiques", in *Séminaire des Probabilités XX* (Lecture Notes in Mathematics #1247) Eds J. Azéma, P.A. Meyer, et M. Yor (Springer-Verlag, Berlin).

L.S. Schulman (1968) "A Path Integral for Spin" Phys. Rev. **176**, 1558-1569.

ACKNOWLEDGMENTS

The recent developments reported here have been worked out during visits of Cécile DeWitt-Morette and Pierre Cartier at several institutions. Cécile DeWitt-Morette at the Institut des Hautes Etudes Scientifiques at Bures-sur-Yvette, at the Département de Mathématique et d'Informatique de l'Ecole Normale Supérieure à Paris, and at the Universidade da Madeira in Funchal ; Pierre Cartier at the Center for Relativity of the University of Texas at Austin. Both of us have appreciated the excellent working conditions and the help of colleagues and staff which have made our collaboration fruitful.

Our collaboration has been made possible by the NATO collaborative Research Grant CRG 910101.

Generalized frames of references and intrinsic Cauchy problem in General Relativity

Giorgio Ferrarese – Carlo Cattani
Dip. di Matematica "G.Castelnuovo"
Univ. di Roma "La Sapienza"
P.le A. Moro 5, 00185 Roma
Italy

Abstract

Starting from an ordinary frame of reference in General Relativity, and its anholonomic structure, we consider a review of more general quasi-product structure, in a *differentiable* (or riemannian) *manifold*, that can be deal with the same anholonomic techniques. The general case contains, in particular, the generalization of ordinary frame of reference, in the sense of polar continua, with one scalar supplementary field (non-orthogonal 1×3 structure). Finally, the anholonomic formalism is developed for the *intrinsic Cauchy problem* in General Relativity (gravitational equations), in the case of *non polar continua; properly spatial variables* are the following: *metric* γ_{ik}, *vector potential* γ_i, *deformation rate* $\widetilde{K}_{ik}$, *pure mass density* μ_0, *heat flux* $\widetilde{Q}_i$ *and temperature* T.

1. INTRODUCTION

Cattaneo's projection techniques, largely used in General Relativity with regard to the essential problem of *absolute physical laws relativization*, can be extended to more general quasi-product structures, by the systematical use of anholonomic techniques ([2] and [3], III. 2).

More generally, such structures can be extended also to *differentiable manifolds* in the sense that one can formalize, by anholonomic techniques, both the *natural decomposition*, and the *covariant derivatives adapted* to the quasi-product structure.

We have already examined both the case of a structure defined by a *distribution of hyperplanes* Σ [4], and that of a differentiable manifold with a distribution of *hyperplanes* Σ, and a *field of contravariant vectors* ([5] n. 2, and [3], IV.1). This last structure, completed by a *partial metric* g_Σ, defined only on the hyperplanes Σ, establishes the most natural geometrical scheme for *relativistic polar continua*.

From this first structure it is easy to proceed to the *general case* of a quasi product structure $p \times q$; i.e. the case of *a differential manifold* V_n, provided

M. Flato et al. (eds.), Physics on Manifolds, 93–109.

with two fields of vectorial subspaces V_p and W_q respectively of dimension p : $1 \leq p < n$, and $q = n - p$. In General Relativity, this extension includes the 2×2 structures, typical of a non singular electromagnetic field; it is essential, in this case, the adoption of *anholonomic reference frames*, since it is impossible (at least in general) to introduce *adapted coordinates* to both/either V_p and/or W_q.

The non existence of partially adapted coordinates, for $p \neq 1$, is in part balanced by an obvious *reciprocity relation* between the subspaces V_p and W_q; however this lacking in integrability does not affect the projection technic (*natural decomposition*), with all *geometrical objects and differential operators*, connected to the considered structure. More precisely, like in the riemannian case [6], if we assume an anholonomic reference frame *adapted to the quasi-product structure* $p \times q$, the anholonomy tensor decomposition points out some notable *structural invariance properties*, summarized as follows: a) there are two well-determined anti-symmetrical tensors (*vortices*), which characterize the (partial or total) integrability of the distributions V_p and W_q; b) there exist well determined *coefficients of connexion*, from which *four covariant derivatives* arise, *internal* and respectively *external* to the two subspaces V_p and W_q $(\partial V, \partial W, \partial_V W, \partial_W V)$; they are directly connected to the *Lie derivatives*, and have absolute character for the product structure, i.e. they are invariant with respect to the choice of the adapted anholonomic reference frame.

Here we will refer to the easiest case of a riemannian space-time V_4, provided with a field of *unitary time-like vectors* $\boldsymbol{\gamma}$ (ordinary case); here the product structure 1×3 is *orthogonal*, because it is locally defined by $\boldsymbol{\gamma}$ and its orthogonal hyperplane Σ.

Very interesting is also the *polar case* [7], in which Σ is not orthogonal to $\boldsymbol{\gamma}$, but to a second vector $\boldsymbol{\eta}$, which can be chosen with the condition $\boldsymbol{\gamma} \cdot \boldsymbol{\eta} = -1$: *not-orthogonal* 1×3 *structure*. Such case appears also in classic Dynamics of holonomic systems with movable constraints, where the geometrization occurs in the *riemannian space of events* [8]; the quasi-product structure $1 \times n$, defined by the *time-lines* t=var. and by the *manifolds* Σ: t=const., being not orthogonal to the previous curves.

2. REMARKS ABOUT THE PROJECTIONS TECNIQUES [9]

2.1. Quasi-natural adapted anholonomic reference frames

Let (V_4, Γ) be a space-time riemannian manifold with signature $+ + + -$, provided with a *congruence of reference* Γ, that is with the field of time-like unitary vectors $\boldsymbol{\gamma}$, tangent to Γ:

$$\boldsymbol{\gamma} \cdot \boldsymbol{\gamma} = -1. \tag{1}$$

The reference frame Γ introduces in V_4 a *time orthogonal structure* 1×3, characterized by $\boldsymbol{\gamma}$ and by the space platform Σ, orthogonal to $\boldsymbol{\gamma}$; which are respectively the *time axis* and the *physical space* locally associated to the reference frame. Since the distribution of hyperplanes Σ is not generally integrable, we are obliged to use *adapted anholonomic reference frames*, i.e. a *regular distributions of bases locally defined* , in every point of V_4, by a vector parallel to $\boldsymbol{\gamma}$ and three independent vectors of Σ.

Thus we assume the following anholonomic reference frame $\{\tilde{\mathrm{e}}_\alpha\}$:

$$\tilde{\mathrm{e}}_0 = \boldsymbol{\gamma}, \quad \tilde{\mathrm{e}}_i = \mathrm{e}_i - \frac{\gamma_i}{\gamma_0}\mathrm{e}_0 \qquad (i = 1, 2, 3). \tag{2}$$

It obviously depends on the coordinates by the natural base $\{\mathrm{e}_\alpha\}$; if the coordinates are *adapted* or *internal to the congruence* Γ, they are defined up to a *space reduced* transformation:

$$\begin{cases} y^{0'} = y^{0'}(y^0, y^1, y^2, y^3), \\ y^{i'} = y^{i'}(y^1, y^2, y^3), \qquad (i = 1, 2, 3), \end{cases} \tag{3}$$

where the functions on the right must satisfy the *equiorientation conditions* (in time and space):

$$\frac{\partial y^{0'}}{\partial y^0} > 0, \qquad \det \left\| \frac{\partial y^{i'}}{\partial y^i} \right\| > 0. \tag{4}$$

However, when the coordinates are internal to Γ, the reference frame (2) transforms as:

$$\tilde{\mathrm{e}}_{0'} = \tilde{\mathrm{e}}_0 = \text{inv.}, \quad \tilde{\mathrm{e}}_{i'} = \frac{\partial y^i}{\partial y^{i'}} \tilde{\mathrm{e}}_i; \tag{5}$$

this explains why the base $\{\tilde{\mathrm{e}}_i\}$, is named *quasi-natural base* in Σ.

In other words, the *distribution* Σ, though not integrable, i.e. lacking of adapted coordinates, *admits quasi-natural anholonomic reference frames*, in the sense that they *locally transform like the natural bases*, even if they are not derived from the coordinates.

Anyway, the conditions (2) can be inverted and, taking into account that:

$$\gamma^i = 0, \quad \gamma^0 > 0 \Rightarrow \gamma_0 \gamma^0 = -1, \tag{6}$$

they lead to the inverse formulæ:

$$\mathrm{e}_0 = -\gamma_0\, \tilde{\mathrm{e}}_0 \quad \mathrm{e}_i = \tilde{\mathrm{e}}_i - \gamma_i\, \tilde{\mathrm{e}}_0, \tag{2'}$$

For the *shifters* $\omega_{\tilde{\alpha}}{}^\beta$ we have:

$$\omega_{\tilde{0}}{}^\beta = \gamma^\beta, \qquad \omega_{\tilde{i}}{}^\beta = \delta_i{}^\beta - \frac{\gamma_i}{\gamma_0}\delta_0^\beta, \tag{7}$$

whereas the *reciprocal elements* $\omega^{\tilde{\alpha}}{}_\beta$ are

$$\omega^{\tilde{\alpha}}{}_0 = -\gamma_0 \delta^0{}_\alpha, \qquad \omega^{\tilde{\alpha}}{}_i = \delta_i{}^\alpha - \gamma_i \delta_0{}^\alpha; \tag{7'}$$

and hence there follows the (direct and inverse) relations between the *dual bases*:

$$\tilde{\mathrm{e}}^\alpha = -\boldsymbol{\gamma}\, \delta_0{}^\alpha + \delta_i{}^\alpha \mathrm{e}^i; \qquad \mathrm{e}^\beta = \gamma^\beta\, \tilde{\mathrm{e}}^0 + \left(\delta_i{}^\beta - \frac{\gamma_i}{\gamma_0}\delta_0^\beta\right) \tilde{\mathrm{e}}^i. \tag{8}$$

In anholonomic form, the components of the metric tensor are:

$$\tilde{g}_{00} = -1, \quad \tilde{g}_{0i} = 0, \quad \tilde{g}_{ik} = \gamma_{ik}; \quad \tilde{g}^{00} = -1, \quad \tilde{g}^{0i} = 0, \quad \tilde{g}^{ik} = g^{ik}; \tag{9}$$

they are determined by the *space metric tensor* γ_{ik}:

$$\gamma_{ik} \stackrel{def}{=} \tilde{\mathbf{e}}_i \cdot \tilde{\mathbf{e}}_k \qquad (i,k = 1,2,3), \tag{10}$$

or by the reciprocal $\gamma^{ik} = g^{ik}$.

The gravitational potentials $g_{\alpha\beta}$ (metric of V_4) are determined, instead, in conformity with (2′), by the $\gamma_{ik}, \gamma_0 < 0$ and γ_i as a whole; from here we have the *relative gravitational potentials*

$$\gamma_{ik}, \quad U = -c^2 \log(-\gamma_0), \quad U_i = c^2 \frac{\gamma_i}{\gamma_0} \qquad (i = 1,2,3), \tag{11}$$

which are not all effective.

2.2. Anholonomy tensor and fundamental derivatives

By using anholonomic frame of reference $\{\tilde{\mathbf{e}}_\alpha\}$ given by (2) we are led the following *Pfaff derivatives* $\tilde{\partial}_\alpha$

$$\partial = \gamma^0 \frac{\partial}{\partial y^0}, \qquad \tilde{\partial}_i = \frac{\partial}{\partial y^i} - \frac{\gamma_i}{\gamma_0} \frac{\partial}{\partial y^0}, \tag{12}$$

where for brevity

$$\partial \stackrel{def}{=} \tilde{\partial}_0; \tag{13}$$

One of the derivatives (12) has a time-like character (along Γ), whereas the remaining are space-like (in Σ). However, such derivatives do not commute:

$$\left[\tilde{\partial}_\alpha, \tilde{\partial}_\beta\right] = \tilde{A}_{\alpha\beta}{}^{\rho}\, \tilde{\partial}_\rho \neq 0, \tag{14}$$

that is the *anholonomy tensor* $\tilde{A}_{\alpha\beta}{}^{\rho}$ is *different from zero*; more precisely, we have the following *commutation formulæ*:

$$\left[\partial, \tilde{\partial}_i\right] = \tilde{C}_i \partial, \qquad \left[\tilde{\partial}_i, \tilde{\partial}_k\right] = 2\tilde{\Omega}_{ik} \partial, \tag{15}$$

which correspond to the *canonical form* of the anholonomy tensor for an orthogonal, 1×3 structure.

More generally when the coordinates are arbitrarily chosen in V_4, the components of the anholonomy tensor are [5]

$$\tilde{A}_{0i}{}^{0} = \tilde{C}_i, \quad \tilde{A}_{0i}{}^{j} = \Gamma_i{}^j; \quad \tilde{A}_{ik}{}^{0} = 2\tilde{\Omega}_{ik}, \quad \tilde{A}_{ik}{}^{j} = -2\tilde{\Omega}_{ik}\gamma^j, \tag{16}$$

$\Gamma_i{}^j$ being a *double connexion* associate to Lie derivative along Γ:

$$\Gamma_i{}^j \stackrel{def}{=} -\left(\tilde{C}_i \gamma^j + \tilde{\partial}_i\, \gamma^j\right). \tag{17}$$

Formulæ(15) put evidence on two fundamental geometric-kinematic ingredients: a) the local *curvature vector* $\mathbf{C} = \tilde{C}_i \, \tilde{\mathbf{e}}^i$ of Γ; b) the antisymmetric *tensorial field* $\tilde{\Omega}_{ik}$ (space-like), which characterizes the *olonomic distribution* Σ, i.e., in the dual point of view, the *normal congruences*.

From the kinematic point of view, the product $c\,\tilde{\Omega}_{ik}$ constitutes the *proper angular speed* of the reference fluid Γ.

Both tensors $\tilde{C}_i$ and $\tilde{\Omega}_{ik}$ *depend on* γ_α *and their first derivatives* [1]:

$$\tilde{C}_i = \gamma_0(\tilde{\partial}_i \gamma^0 + \partial \frac{\gamma_i}{\gamma_0}), \quad \tilde{\Omega}_{ik} = \frac{1}{2}\gamma_0(\tilde{\partial}_i \frac{\gamma_k}{\gamma_0} - \tilde{\partial}_k \frac{\gamma_i}{\gamma_0}); \tag{18}$$

moreover, according to (15), they are subordinate to *Jacobi identities*, which lead to the following differential relations:

$$\partial\tilde{\Omega}_{ik} = \tilde{\partial}_{[i}\tilde{C}_{k]}, \qquad \tilde{\partial}_{[i}\tilde{\Omega}_{hk]} - \tilde{C}_{[i}\tilde{\Omega}_{hk]} = 0. \tag{19}$$

The first bond is typical of Fluid dynamics (cfr. [10], 164); the second, in the framework of Kaluza-Klein 5-dimensional theory, leads to the *geometrical interpretation* of the electromagnetic field [11].

Let us consider now the *Pfaff derivatives* $\tilde{\partial}_\beta$ of the anholonomic base $\{\tilde{\mathbf{e}}_\alpha\}$; they are

$$\tilde{\partial}_\alpha \tilde{\mathbf{e}}_\beta \overset{def}{=} \tilde{\Gamma}_{\alpha\beta}{}^{\rho}\, \mathbf{e}_\rho, \tag{20}$$

$\tilde{\Gamma}_{\alpha\beta}{}^{\rho} = \tilde{\partial}_\alpha \tilde{\mathbf{e}}_\beta \cdot \tilde{\mathbf{e}}^\rho$ being the Ricci's *rotation coefficients* associated to $\{\tilde{\mathbf{e}}_\alpha\}$.

We have directly, minding also the commutation formulæ(15), the following fundamental picture:

$$\begin{cases} \partial\boldsymbol{\gamma} = \tilde{C}^i\, \mathbf{e}_i, & \tilde{\partial}_i\, \boldsymbol{\gamma} = \tilde{H}_i{}^k\, \mathbf{e}_k \\ \partial\mathbf{e}_i = \tilde{H}_i{}^k\, \tilde{\mathbf{e}}_k + \tilde{C}_i\boldsymbol{\gamma}, & \tilde{\partial}_i\, \mathbf{e}_k = \tilde{\Gamma}_{ik}{}^h\, \mathbf{e}_h + \tilde{H}_{ik}\boldsymbol{\gamma}, \end{cases} \tag{21}$$

where, together with the *curvature tensor* $\tilde{C}_i$ $(18)_1$ and the *spatial Christoffel simbols*

$$\tilde{\Gamma}_{ik}{}^{j} \overset{def}{=} \frac{1}{2}\gamma^{jh}(\tilde{\partial}_i\, \gamma_{kh} + \tilde{\partial}_k\, \gamma_{hi} - \tilde{\partial}_h\, \gamma_{ik}), \tag{22}$$

appears the spatial tensor $\tilde{H}_{ik}$, which sums up two kinematic ingredients, characteristic of the reference frame Γ; one is the *proper angular speed* $\tilde{\Omega}_{ik}$ $(18)_2$ and the other is the *proper deformation speed*:

$$\tilde{K}_{ik} = \frac{1}{2}\partial\gamma_{ik}. \tag{23}$$

In other words, we have

$$\tilde{H}_{[ik]} = \tilde{\Omega}_{ik}, \quad \tilde{H}_{(ik)} = \tilde{K}_{ik} \Longrightarrow \tilde{H}_{ik} = \tilde{\Omega}_{ik} + \tilde{K}_{ik}. \tag{24}$$

Let us notice that, in coordinates adapted to Γ, the *longitudinal derivative* $\partial = \gamma^0 \dfrac{\partial}{\partial y^0}$ is invariant with respect to every transformations internal to the reference frame, and coincides with its tensorial extension (Lie derivative along Γ); if the coordinates chosen in V_4 instead, are not adapted to Γ, so that $\partial \equiv \gamma^\alpha \frac{\partial}{\partial x^\alpha}$, then the tensorial extension of that derivative can be obtained from the double connexion $\Gamma_i{}^k$ (17). In such a case, (23) turns into the following:

$$\tilde K_{ik} = \frac{1}{2}\left(\partial\gamma_{ik} - \Gamma_i{}^j\gamma_{jk} - \gamma_k{}^j\gamma_{ij}\right). \tag{23'}$$

3. NATURAL DECOMPOSITION OF THE CURVATURE TENSOR

By now we have got all the elements for calculating the anholonomic components of the curvature tensor, which corresponds to its natural decomposition; the following *general formula* [12] is involved:

$$\tilde{\mathbf{R}}_{\alpha\beta\rho} \equiv \tilde R_{\alpha\beta\rho}{}^{\sigma}\,\tilde{\mathbf{e}}_\sigma = \left[\tilde\partial_\alpha, \tilde\partial_\beta\right]\tilde{\mathbf{e}}_\rho - \tilde A_{\alpha\beta}{}^{\sigma}\,\tilde\partial_\sigma\,\tilde{\mathbf{e}}_\rho, \tag{25}$$

$\tilde A_{\alpha\beta}{}^{\sigma}$ being the *anholonomy tensor* of the considered reference frame. In our case, according to (15), the only non-vanishing components are the following:

$$\tilde A_{0i}{}^{0} = \tilde C_i, \qquad \tilde A_{ik}{}^{0} = 2\tilde\Omega_{ik}. \tag{26}$$

Thus from $(21)_4$, for all space indexes, we have:

$$\tilde\partial_i\,(\tilde\partial_k\tilde{\mathbf{e}}_h) = \tilde\partial_i\,\tilde\Gamma_{kh}{}^{j}\,\tilde{\mathbf{e}}_j + \tilde\Gamma_{kh}{}^{j}\,(\tilde\Gamma_{ij}{}^{l}\,\tilde{\mathbf{e}}_l + \tilde H_{ij}\boldsymbol{\gamma}) + \tilde\partial_i\,\tilde H_{kh}\boldsymbol{\gamma} + \tilde H_{kh}\tilde H_i{}^{j}\,\tilde{\mathbf{e}}_j,$$

and

$$\left[\tilde\partial_i, \tilde\partial_k\right]\tilde{\mathbf{e}}_h = (\tilde P_{ikh}{}^{j} + \tilde H_{kh}\tilde H_i{}^{j} - \tilde H_{ih}\tilde H_k{}^{j})\,\tilde{\mathbf{e}}_j + (\tilde\nabla_i\,\tilde H_{kh} - \tilde\nabla_k\,\tilde H_{ih})\boldsymbol{\gamma}. \tag{27}$$

In the above formula, $\tilde P_{ikh}{}^{j}$ indicates the *spatial curvature tensor*

$$\tilde P_{ikh}{}^{j} \overset{def}{=} \tilde\partial_i\,\tilde\Gamma_{kh}{}^{j} - \tilde\partial_k\,\tilde\Gamma_{ih}{}^{j} + \tilde\Gamma_{kh}{}^{l}\,\tilde\Gamma_{il}{}^{j} - \tilde\Gamma_{ih}{}^{l}\,\tilde\Gamma_{kl}{}^{j}, \tag{28}$$

which, though it has the typical expression of a curvature tensor, does not own all its typical algebraic properties; $\tilde\nabla_i$ is the *covariant extension* (in Σ) of *Pfaff's derivative* $\tilde\partial_i$, obtained with the spatial Christoffel simbols (Cattaneo's *transversal derivative* [1]).

Analogously, we have

$$\begin{cases} \partial\,\tilde\partial_i\,\boldsymbol{\gamma} = \partial\tilde H_i{}^{j}\,\tilde{\mathbf{e}}_j + \tilde H_i{}^{j}\,(\tilde H_j{}^{k}\,\tilde{\mathbf{e}}_k + \tilde C_j\boldsymbol{\gamma}) \\ \tilde\partial_i\,\partial\boldsymbol{\gamma} = \tilde\partial_i\,\tilde C^{j}\,\tilde{\mathbf{e}}_j + \tilde C^{j}\,(\tilde\Gamma_{ij}{}^{l}\,\tilde{\mathbf{e}}_l + \tilde H_{ij}\boldsymbol{\gamma}), \end{cases}$$

whence

$$\left[\partial, \tilde\partial_i\right]\boldsymbol{\gamma} - (\partial\tilde H_i{}^{j} + \tilde H_i{}^{k}\,\tilde H_k{}^{j} - \tilde\nabla_i\,\tilde C^{j})\,\tilde{\mathbf{e}}_j,$$

and finally, thanks to the following identity:

$$\partial \tilde{H}_i{}^j = \gamma^{jk} \partial \tilde{H}_{ik} - 2\tilde{K}^{jk} \tilde{H}_{ik},$$

we obtain the commutation formula:

$$\left[\partial, \tilde{\partial}_i\right] \gamma = (\partial \tilde{H}_{ik} - \tilde{H}_i{}^j \tilde{H}_{kj} - \tilde{\nabla}_i \tilde{C}_k) \tilde{\mathrm{e}}^k . \tag{29}$$

Formulæ(27) and (28) lead us, according to (25), to the following *anholonomic components of the curvature tensor*

$$\begin{cases} \tilde{R}_{ikh}{}^j = \tilde{P}_{ikh}{}^j + \tilde{H}_{kh} \tilde{H}_i{}^j - \tilde{H}_{ih} \tilde{H}_k{}^j - 2\tilde{\Omega}_{ik} \tilde{H}_h{}^j \\ \tilde{R}_{ikh}{}^0 = B_{ikh}, \qquad \tilde{R}_{0i0k} = C_{ik}, \end{cases} \tag{30}$$

where, for the sake of brevity,

$$\begin{cases} B_{ikh} \stackrel{def}{=} \tilde{\nabla}_i \tilde{H}_{kh} - \tilde{\nabla}_k \tilde{H}_{ih} - 2\tilde{\Omega}_{ik} \tilde{C}_h \\ C_{ik} \stackrel{def}{=} \partial \tilde{H}_{ik} - (\tilde{\nabla}_i + \tilde{C}_i) \tilde{C}_k - \tilde{H}_i{}^j \tilde{H}_{kj}. \end{cases} \tag{31}$$

Only the independent components are involved, because the others are obtained from these by by taking into account the algebraic properties of the curvature tensor.

In other words, *the decomposition of the curvature tensor is characterized by three spatial tensor* $\tilde{R}_{ikhj}, \tilde{B}_{ikh}, \tilde{C}_{ik}$, since only three are the independent natural projections [13]-[14].

The four indices tensor $\tilde{R}_{ikhj}$ obviously satisfies all algebraic properties of a curvature tensor; $\tilde{B}_{ikh}$ is anti-symmetric with respect to its first two indices, and verifies the cyclic property, according to $(19)_2$. Finally $\tilde{C}_{ik}$ is a symmetric tensor, according to $(19)_1$.

4. INTRINSIC FORMULATION OF CAUCHY PROBLEM IN GENERAL RELATIVITY

4.1. Relative form of the gravitational equations

According to (30), we can decompose the *Ricci tensor* $R_{\alpha\beta} \stackrel{def}{=} R_{\alpha\rho\beta}{}^\rho$ as follows:

$$\tilde{R}_{ih} = \tilde{R}_{ijh}{}^j - \tilde{C}_{ih}, \quad \tilde{R}_{i0} = \tilde{B}_{ij}{}^j, \quad R_{00} = \tilde{C}_i{}^i; \tag{32}$$

from here we obtain the curvature scalar $R = g^{\alpha\beta} R_{\alpha\beta}$

$$R = \tilde{R}_{ij}{}^{ij} - 2\tilde{C}_i{}^i, \tag{33}$$

and the *gravitational tensor* $S_{\alpha\beta}$

$$\tilde{S}_{ik} = \tilde{R}_{ik} - \frac{1}{2} R \gamma_{ik}, \quad \tilde{S}_{i0} = \tilde{B}_{ij}{}^j \quad \tilde{S}_{00} = \frac{1}{2} \tilde{R}_{ij}{}^{ij} . \tag{34}$$

First of all, let us examine *the case of the vacuum*; the conditions $\tilde{R}_{ik} = 0, \tilde{R}_{\alpha 0} = 0$ being equivalent to $\tilde{R}_{ik} = 0, \tilde{S}_{\alpha 0} = 0$, the gravitational equations are equivalent to

$$\tilde{R}_{ik} = 0, \quad \tilde{S}_{\alpha 0} = 0, \tag{35}$$

with the additional conditions, from (19):

$$\partial \tilde{\Omega}_{ik} = \tilde{\nabla}_{[i} \tilde{C}_{k]}, \quad \tilde{\nabla}_{[i} \tilde{\Omega}_{kh]} - \tilde{C}_{[i} \tilde{\Omega}_{kh]} = 0. \tag{36}$$

Also in anholonomic terms, like in terms of coordinates [15], the two groups of equations (35) have a different role; we mean that: a) *the conditions* $\tilde{R}_{ik} = 0$ *constitute the only field equations*; b) *the remaining conditions* $\tilde{S}_{\alpha 0} = 0$ *constrain the initial data*, because they do not contain second timelike derivatives.

More precisely, $(35)_1$ and (23), lead us to the following *differential system of first order (in time) for the variables* γ_{ik} *and* $\tilde{H}_{ik} = \tilde{\Omega}_{ik} + \tilde{K}_{ik}$:

$$\begin{cases} \partial \gamma_{ik} = 2\tilde{H}_{(ik)} \\ \partial \tilde{H}_{ik} = \tilde{P}_{ik} + \tilde{H}_i^{\ j} \tilde{H}_{jk} + \tilde{H}_k^{\ j} \tilde{H}_{ji} - \tilde{H}_j^{\ j} \tilde{H}_{ik} + (\tilde{\nabla}_i + \tilde{C}_i)\tilde{C}_k, \end{cases} \tag{37}$$

being $\tilde{P}_{ik}$ the *spatial Ricci tensor* :

$$\tilde{P}_{ik} \stackrel{def}{=} \tilde{P}_{ijk}^{\quad j}. \tag{38}$$

Together with (37), we have to consider the further conditions (36); but $(36)_1$ already appears in the differential system (37). In fact, equation (37), because of the decomposition (24), and of the symmetry of the spatial tensor $\tilde{Q}_{ik}$:

$$\tilde{Q}_{ik} \stackrel{def}{=} \tilde{P}_{ik} - \tilde{H}_j^{\ j} \tilde{H}_{ik}, \tag{39}$$

can be written in the equivalent form

$$\begin{cases} \partial \gamma_{ik} = 2\tilde{K}_{ik} \qquad \partial \tilde{\Omega}_{ik} = \tilde{\nabla}_{[i} \tilde{C}_{k]} \\ \partial \tilde{K}_{ik} = \tilde{Q}_{ik} + \tilde{H}_i^{\ j} \tilde{H}_{jk} + \tilde{H}_k^{\ j} \tilde{H}_{ji} + \tilde{\nabla}_{(i} \tilde{C}_{k)} + \tilde{C}_i \tilde{C}_k. \end{cases} \tag{40}$$

Thus, *the effective constraints for the field variables* $\tilde{\gamma}_{ik}$ *and* $\tilde{H}_{ik}$ *are given by* $(35)_2$ *and* $(36)_2$ *as a whole*:

$$\begin{cases} \tilde{B}_{ij}^{\ \ j} \equiv \tilde{\nabla}_i \tilde{H}_j^{\ j} - \tilde{\nabla}_j \tilde{H}_i^{\ j} - 2\tilde{\Omega}_{ij} \tilde{C}^j = 0 \\ \tilde{R}_{ij}^{\ \ ij} \equiv \tilde{Q} + \tilde{H}_i^{\ j} \tilde{H}_j^{\ i} - 2\tilde{\Omega}_{ij} \tilde{H}^{ij} = 0 \\ \frac{1}{2} \tilde{B}_{[ikh]} \equiv \tilde{\nabla}_{[i} \tilde{\Omega}_{kh]} - \tilde{C}_{[i} \tilde{\Omega}_{kh]} = 0, \end{cases} \tag{41}$$

where $\tilde{Q}$ is connected to *the spatial curvature scalar*:

$$\tilde{Q} \stackrel{def}{=} \gamma^{ik} \tilde{P}_{ik} - (\tilde{H}_j^{\ j})^2 \equiv \tilde{P}_{ij}^{\ \ ij}. \tag{42}$$

These are *constraints in involution, in the sense of Cartan, as a consequence of Bianchi's identities*, including the conservative character of the gravitational tensor $S_{\alpha\beta}$ [5] and [16]; they constitute *limitations only for the initial data*, which are not yet precised. About these, as always occurs, the classical situation should inspire us, where it deals with the equations of evolution of a generic continuous system, translated in terms of acceleration, angular speed and deformation speed [17]; here the point is stating the *initial configuration, initial act of motion* and finally *initial distributions of angular and deformation speeds*, these last being constrained to satisfy the *linearized congruence conditions*.

From the relativistic point of view, this corresponds to stating the *initial hypersurface* Σ_0 with the relative *metric*, and on it the *distributions of linear 4-speed and angular and deformation speeds*, which, like in the classic situation, are not free, but subordinate to certain constraints, according to (41); *they also involve the accelerations*, which is a typical relativistic effect [18].

But what role do the accelerations, we mean $\widetilde{C}_i$, play in (40) and (41)? They are not further field variables (to be added to γ_{ik} and $\widetilde{H}_{ik}$), because $\widetilde{C}_i$, as a consequence of the conservation equations, are *subordinate to Galileo's principle*; i.e. they are well determined functions of the *thermodynamic variables* of the continuum associated with the reference frame Γ. All this loses effective meaning in the case of the vacuum. However, it is worthwhile to underline that, for any choice of the reference frame Γ, the *formulation* of the evolution problem for the space time V_4, in the differential terms (40)-(41), has *an invariant character for transformations* (3), and so it has got an intrinsic meaning in the considered reference frame; this is a substantial difference from the formulation in terms of coordinates, generally lacking of invariance properties.

4.2. Case of the vacuum

The case of the vacuum, though extremely important for what concerns the analitic study of gravitational equations, is strictly different from the case of a space-time with presence of matter. In fact, according to Mach's principle, in the absence of matter, there are no privileged reference frames (galileian or not).

In other words, *in the case of the vacuum, the reference frame Γ cannot have more than a purely geometrical-kinematic meaning;* it is completely disposable, but deductively there are no privileged choices due to the physical reasons. We will be authorized to make any simplifying choice, for what concernes the gravitational equations, possibly suggested by the initial conditions.

Let us first consider the *general case*, in which deductively there are no restrictions at all for the space-time V_4; without loosing generality, we can suppose that the coordinates adapted to Γ satisfy the condition:

$$\gamma_0 = -1 \quad \sim \quad g_{00} = -1. \tag{43}$$

This condition is not only always compatible, but is invariant for coordinate transformations of the kind:

$$y^{0'} = y^0 + \psi(y^1, y^2, y^3), \quad y^{i'} = y^{i'}(y^1, y^2, y^3), \tag{44}$$

where ψ and $y^{i'}$ are arbitrary functions of the spatial coordinates.

Let Γ_0 be the congruence defined by the *geodesics* from the points of the initial surface Σ_0 and *orthogonal* to it; this *congruence* is not only geodesic but also *normal*, because condition (43) and the geodesic character, reduce $(18)_1$ to the form[1]

$$\partial\gamma_i = 0 \Longrightarrow \gamma_i = \gamma_i\,|_{\Sigma_0} = 0.$$

On the other hand, in *a generic geodesic congruence*, we have necessarily, according to $(36)_1$: $\partial\tilde{\Omega}_{ik} = 0$; that is the vortex $\tilde{\Omega}_{ik} = 0$ remains along the lines of the congruence. In the case of Γ_0, we have the condition $\tilde{\Omega}_{ik}\,|_{\Sigma_0} = 0$, because in every point of Σ_0 the facets orthogonal to the congruence joint together on Σ_0 itself.

Let *us now consider, as reference frame* Γ, *the congruence* Γ_0; then we have, identically in V_4:

$$\gamma_0 = -1, \quad \gamma_i = 0, \quad \tilde{C}_i = 0, \quad \tilde{\Omega}_{ik} = 0. \tag{45}$$

Condition $\gamma_i = 0$ proves that *the space-time V_4 has a time orthogonal structure*; moreover we must have $\tilde{\partial}_i = \partial_i$, so that $\tilde{P}_{ik}$ *coincides with the ordinary Ricci tensor relative to the metric* γ_{ik}; analogously, the coefficients $\tilde{\Gamma}_{ik}^{\ \ j}$ reduce to ordinary Christoffel symbols of second species relative to the same metric. Thus, system (40) takes the reduced form

$$\partial\gamma_{ik} = 2K_{ik}, \quad \partial K_{ik} = P_{ik} + 2K_i^{\ j}K_{jk} - K_j^{\ j}K_{ik}, \quad (\partial \equiv \partial/\partial y^0) \tag{46}$$

where we omitted the "tilde" for simplicity; it leads to a *well determined Cauchy problem of the first order, in the variables* γ_{ik} and K_{ik} (of second order in the γ_{ik} alone).

The initial data $\gamma_{ik}{}^0$ and $K_{ik}{}^0$ (first and second quadratic form of Σ_0) to be jointed to the differential system (46) *are not free*; they are instead subordinated to (41)

$$\nabla i K_j^{\ j} - \nabla j K_i^{\ j} = 0, \quad P + K_{ij}K^{ij} - (K_j^{\ j})^2 = 0. \tag{47}$$

We can see well right from here that, if Σ_0 is a strictly euclidean space, in which case we must agree that $K_{ij} = 0, P_{ij} = 0$ on Σ_0, system (46)-(47) admits the solution $\gamma_{ik} = const., K_{ik} = 0$, and so, minding the uniqueness theorem [19], the manifold V_4 is a Minkowsky space.

The general case, considered before, does not include peculiar situations of the space-time, that should require the existence deductively of special congruences or coordinates, or Killing fields; as, for example, under the (geometrical) assumption of stationarity. This assumption is characterized by the existence, in V_4, of a one-parameter group of isometries; that is by an orthogonal 1×3 structure, where the congruence Γ_s admits internal coordinates y^α for which the ten gravitational potentials (absolute or relative) are independent from $y^0 = ct$:

$$\partial\gamma_{ik} = 0, \qquad \partial\gamma_\alpha = 0. \tag{48}$$

[1] We admit that Σ_0 is representable by $y^0 = 0$.

These coordinates (*adapted to both Γ_s and the stationarity*) are not univocally defined, but up to a transformation of the kind (44); it follows that in general the stationarity condition is not compatible with[2] (43). In fact (43) presupposes the existence of coordinates $y^{\alpha'}$ with $y^{0'}(y)$ solution of the first order differential equation:

$$\partial_0 y^{0'} = -\gamma_0(y) \neq 1; \tag{49}$$

so that $y^{0'}$ is not included in (44).

Formula (48) implies that the trasversal derivatives $\tilde{\partial}_i$, and the ordinary ∂_i coincide, when they operate on gravitational potentials. It follows that: a) spatial Christoffel symbols $\tilde{\Gamma}_{ik}^{\ \ j}$ turn into ordinary Christoffel symbols for the metric $\gamma_{ik}(y^1, y^2, y^3)$; b) the tensor $\tilde{P}_{ihk}^{\ \ \ j}$ coincides with the *curvature tensor* of the riemannian manifold V_3, with metric γ_{ik} and coordinates y^i : *quotient manifold of V_4 with respect to the presupposed isometry group*. At the same time (18) turns into:

$$\tilde{C}_i = \partial_i \varphi, \qquad \tilde{\Omega}_{ik} = -\frac{1}{2} e^{\varphi} (\partial_i \varphi_k - \partial_k \varphi_i), \tag{50}$$

where φ and φ_i are respectively the scalar and vector potential:

$$\varphi \stackrel{def}{=} \log(-\gamma_0), \qquad \varphi_i \stackrel{def}{=} \frac{\gamma_i}{\gamma_0}. \tag{51}$$

Finally, with the usual notations

$$\xi = -\gamma_0 > 0, \qquad \varphi_i = \frac{\gamma_i}{\gamma_0}, \quad F_{ik} = \partial_i \varphi_k - \partial_k \varphi_i, \tag{52}$$

the *gravitational field relative to the reference frame* Γ_s is of the following kind:

$$K_{ik} = 0, \quad \Omega_{ik} = \frac{1}{2} \xi F_{ik}, \quad C_i = \frac{1}{\xi} \partial_i \xi, \tag{53}$$

where, for simplicity, we have omitted the inessential "tilde" in the first member.

At the same time it results

$$\left(\tilde{\nabla}_i + \tilde{C}_i\right) \tilde{C}_k = \frac{1}{\xi} \nabla_i (\partial_k \xi), \tag{54}$$

and so equations (40)-(41), all valid on V_3, take the form

$$\left\{ \begin{array}{l} P_{ik} + \frac{1}{2} \xi^2 F_i^{\ j} F_{jk} + \frac{1}{\xi} \nabla_i (\partial_k \xi) = 0 \\ \xi \nabla_j F_i^{\ j} + 3 F_i^{\ j} \partial_j \xi = 0 \\ P = \frac{3}{4} \xi^2 F_{ij} F^{ij}. \end{array} \right. \tag{55}$$

P_{ik} being Ricci tensor of the quotient manifold V_3. So we obtain, in V_3, *ten differential conditions of the second order as many as the unknown quantities*

[2]This could be possible only in some particular cases.

γ_{ik}, ξ and F_{ik}, all function of the y^i alone; conditions that, eliminating P from $(55)_{1,3}$, can be written under the form:

$$\begin{cases} \xi^4 P_{ik} + \frac{1}{2}(f_i f_k - f_l f^l \gamma_{ik}) + \xi^3 \nabla_i(\partial_k \xi) = 0 \\ \nabla_{[j} f_{i]} = 0, \qquad \frac{1}{2} f_l f^l + \xi^3 \nabla^i(\partial_i \xi) = 0, \end{cases} \tag{56}$$

where f_i is the *adjoint vector* of tensor $\xi^3 F_{ik}$:

$$f_i \stackrel{def}{=} \frac{1}{2}\xi^3 \eta_{ihk} F^{hk} \Longrightarrow \xi^3 F^{hk} = \eta^{hki} f_i. \tag{57}$$

It follows that f_i *is a gradient*:

$$f_i = grad\varphi, \tag{58}$$

and φ *is an auxiliary scalar field not well determined.*

Once the static case which corresponds to $\varphi =$ const. , is excluded *the solving equation* is

$$\nabla^i(\partial_i \xi) = -\frac{(grad\varphi)^2}{2\xi^3} < 0 \tag{59}$$

and, in order to integrate, together with φ we must specify the "boundary conditions".

Although the two cases: general and stationary, are somehow complementary, (one implies $\tilde{C}_i = 0$, and $\tilde{\Omega}_{ik} = 0$, and the other $\tilde{K}_{ik} = 0$) they have completely different characteristics; the first one has an *evolutive character*, while the second does not, presupposing, for V_4, an *isometric leafletness*, not orthogonal to the congruence of stationarity Γ_s .

We are dealing about the vacuum, that is a limit case in which Physics, completely lacking, cannot give any suggestion concerning the choice of the reference frame Γ. Things go in an enterily different way in presence of matter.

4.3. Disgregate matter - Gauss-Poisson equation

For estabilishing our key points, we will consider here the simplest case of disgregate matter, that is the case of a material system lacking of 4-tensions (absence of *internal energy, mechanical stress and thermical flow*); it is a case that, in order to equalizate, does not need constitutive equations.

In such a case, the congruence Γ_m of the material lines of the continuous medium is *necessarily geodetic*. So, if we assume that the reference frame Γ coincides with Γ_m, and suppose that initial 4-speeds $\mathbf{V}_0$ are orthogonal to[3] Σ_0, we fall in previous (45); of course the Cauchy problem modifies for the presence of the *term of pure matter* μ_0.

---

[3]If this condition is not satisfied, γ_i and $\tilde{\Omega}_{ik}$, independent from time but not null, *are determined by initial conditions*:

$$\gamma_i = \frac{1}{c} V_i^0(y^1, y^2, y^3), \quad \tilde{\Omega}_{ik} = \frac{1}{2}(\partial_i \gamma_k - \partial_k \gamma_i)$$

More precisely, we have the following differential system

$$\begin{cases} \partial\gamma_{ik} = 2K_{ik}, \quad \partial\mu_0 = -K\mu_0 \\ \partial K_{ik} = P_{ik} + 2K_i{}^j K_{jk} - KK_{ik} + \frac{1}{2}\chi\mu_0 c^2\gamma_{ik}, \end{cases} \tag{60}$$

with initial constraints of the type:

$$\begin{cases} \nabla_j(K\delta_i{}^j - K_i{}^j) = 0 \\ P + K_{ij}K^{ij} - K^2 + 2\chi\mu_0 c^2 = 0, \end{cases} \tag{61}$$

K being the *cubic dilatation rate* of the physical space Σ joined to normal reference frame Γ_m:

$$K \overset{def}{=} K_j{}^j = Div\mathbf{V}. \tag{62}$$

For what concerns $(60)_3$, saturating with γ^{ik}, we have directly the *expansion condition of the physical space* Σ:

$$\partial K = -K_{ij}K^{ij} - \frac{1}{2}\chi\mu_0 c^2 \leq 0 \qquad (\partial = \partial_0). \tag{63}$$

Let us notice that the assumption that Σ_0 is an euclidean space ($P_{ik} = 0, K_{ik} = 0$ in Σ_0) because of $(61)_2$, is not compatible with the presence of matter; moreover, the condition of stationarity of V_4 with respect to Γ_m (rest reference frame of the material system) *is not compatible either with the assumption of disgregated matter*, because from (63) we would have again $\mu_0 = 0$.

The previous argument agrees with the common assertion that "the presence of matter bends the space".

Now let us suppose V_4 *stationary*, without asking the reference frame Γ_s to be also of rest for the disgregated system: $\Gamma_s \neq \Gamma_m$.

Taking Γ_s as reference frame, the assumption of stationarity is represented by the conditions (48). Moreover, by choosing as parameter along the time lines of the continuous medium, Cattaneo's *standard time*:

$$dT = -\frac{1}{c}\gamma_\alpha dy^\alpha, \tag{64}$$

the 4-*speed* V^α is decomposable in the following manner:

$$V^\alpha = \eta(v^\alpha + c\gamma^\alpha), \quad v^\alpha \in \Sigma, \quad \eta^{-1} \overset{def}{=} \sqrt{1 - v^2/c^2}; \tag{65}$$

at the same time, the anholonomic components of the *energy tensor* $M^{\alpha\beta} = \mu_0 V^\alpha V^\beta$ become:

$$\tilde{M}{}^{ik} = \mu \tilde{v}{}^i \tilde{v}{}^k, \quad \tilde{M}{}^{i0} = \mu c \tilde{v}{}^i, \quad \tilde{M}{}^{00} = \mu c^2, \tag{66}$$

μ being the *material density relative to* Γ_s:

$$\mu = \eta^2\mu_0. \tag{67}$$

Therefore, the previous (60)-(61) picture here results completely modified. More precisely, the evolution equations $\nabla_\beta M^{\alpha\beta} = 0$, omitting the "tilde" for brevity, turn into

$$\begin{cases} \frac{1}{c}\partial(\mu v_i) = -\mu(\delta_i{}^k + \frac{1}{c^2}v_i v^k)C_k + \frac{2}{c}\mu\Omega_{ik}v^k - \frac{1}{c^2}\nabla_k(\mu v_i v^k) \\ \partial\mu = -\frac{1}{c}(\nabla_i + 2C_i)\mu v^i; \end{cases} \tag{68}$$

on the contrary, the Cauchy problem for the gravitational equations *looses his evolutive character* and, with notations (52), is represented by the following *differential conditions on the initial surface* Σ_0:

$$\begin{cases} P_{ik} + \frac{1}{2}\xi^2 F_{ij}F^j{}_k + \frac{1}{\chi}\nabla_i(\partial_k\xi) + \mu\chi\left[v_i v_k - \frac{1}{2}(v^2 - c^2)\gamma_{ik}\right] = 0 \\ \frac{1}{\xi^2}\nabla_j(\xi^3 F_i{}^j) + 2\mu\chi c v_i = 0 \\ P - \frac{3}{4}\xi^2 F_{ij}F^{ij} + 2\chi\mu c^2 = 0. \end{cases} \tag{69}$$

It follows that the determination of the unknown quantities v_i and μ (the only to have an evolutive character) needs the resolution of the differential system (68), in which *the tensor* $C_i = \dfrac{1}{\xi}\partial_i\xi$ *and* $\Omega_{ik} = \frac{1}{2}\xi F_{ik}$, *both independent of time, appear as initial data*, even if subordinate to equations (69); in these, of course, together with ξ, F_{ik} and γ_{ik}, the speeds v_i also appear as initial data.

Now it is worthwhile to deal with Gauss-Poisson equation; it is obtained, in the current case, by eliminating *the spatial curvature scalar* P from $(69)_{1,3}$. It gives rise to Gauss-Poisson *generalized equation*

$$\frac{1}{\xi}\nabla^i(\partial_i\xi) + \frac{1}{4}\xi^2 F_{ij}F^{ij} - \frac{1}{2}\chi\mu c^2(1 + \frac{v^2}{c^2}) = 0. \tag{70}$$

This denomination has good foundations [21]; in fact, by using (53) and (54), (70) it is equivalent to

$$(\nabla^i + C^i)C_i + \Omega_{ij}\Omega^{ij} - \frac{1}{2}\mu c^2\chi(1 + \frac{v^2}{c^2}) = 0. \tag{71}$$

On the other hand, instead of C_i, we can let the dragging (that is of rest) gravitational field appear:

$$G_i = -c^2 C_i \equiv \partial_i U, \tag{72}$$

U being Cattaneo's *standard relative gravitational potential*:

$$U = -c^2 \log\xi. \tag{73}$$

That being stated, minding that the first term in (71) coincides with the space-time divergence of the curvature vector:

$$(\nabla^i + C^i)C_i = \overset{\sim}{\nabla}{}^\alpha \tilde{C}_\alpha = \nabla^\alpha C_\alpha \equiv -\frac{1}{c^2}\Box_g U \tag{74}$$

multiplying (71) by $-c^2$, we have directly

$$Div\mathbf{G} \equiv \Box_g U = 2\omega^2 - \frac{1}{2}\mu c^4 \chi (1 + \frac{v^2}{c^2}); \tag{75}$$

here $\omega^2 = \gamma_{ik}\omega^i\omega^k$ is the *proper angular speed of the reference frame* $\Gamma \equiv \Gamma_s$ (not to be confused with that of the continuum relative to Γ):

$$\omega^i = \frac{c}{2}\eta^{ijk}\Omega_{jk} \Longrightarrow \omega^2 = \frac{1}{2}\Omega_{ij}\Omega^{ij}. \tag{76}$$

Equation (75) is, at sight, the *Gauss-Poisson generalized equation, for the disgregated matter, in motion with respect to the stationarity reference frame of* V_4.

In the case of slow motion: $\frac{v^2}{c^2} \ll 1$, (75) coincides with the classic Gauss-Poisson equation relative to a stationary matter distribution; in the same approximation, but in the static case: $\Omega_{ik} = 0$, we recognize the *ordinary Gauss-Poisson equation*

$$\Delta_2 U = -4\pi f \mu \qquad (\chi = \frac{8\pi f}{c^4}), \tag{77}$$

with the further assumption of a *weak gravitational field:* $U/c^2 \ll 1$, that is neglecting the product $\frac{1}{c^2}G_i G^i$.

Of course, proceeding to the case of a more general continuum, the Cauchy problem, the expansion equation (63) and the Gauss-Poisson equation (75), are more general. For example, for a *generic ordinary continuum* ($\partial\gamma_i \neq 0$), in the stationary case, (63) can be generalized with the analogous equation [16]:

$$\partial K = \Omega_{ij}\Omega^{ij} - K_{ij}K^{ij} + (\nabla^i + C^i)C_i - \frac{1}{2}\chi(X + \rho_0 c^2), \tag{78}$$

where X is the *track of the proper mechanical stress* and ρ_0 the *proper total density of energy*

$$X \stackrel{def}{=} \gamma^{ik}X_{ik}, \quad \rho_0 \stackrel{def}{=} \mu_0 + \frac{\varepsilon_0}{c^2}. \tag{79}$$

At the same time, for what concerns (75), the Gauss-Poisson equation holds in the form

$$Div\mathbf{G} \equiv (\nabla^i + C^i)\partial_i U = 2\omega^2 - \frac{\chi c^2}{2}(X + \rho_0 c^2); \tag{80}$$

here does not appear the *termic flow*, while it appears if the continuum *is not in rest* in the stationarity reference frame [16]. We can suppose $\gamma_0 = -1$, and the Cauchy problem is of the form

$$\begin{cases} \partial\gamma_{ik} &= 2\tilde{K}_{ik}, \quad \partial\gamma_i = \tilde{C}_i \\ \partial\tilde{K}_{ik} &= \tilde{P}_{ik} + \tilde{H}_i{}^j \tilde{H}_{jk} - \tilde{H}_k{}^j \tilde{H}_{ji} - \tilde{K}\,\tilde{H}_{ik} + \tilde{\nabla}_{(i}\,\tilde{C}_{k)} + \\ &+ \tilde{C}_i\tilde{C}_k + \chi[\tilde{X}_{ik} - \frac{1}{2}(\tilde{X} - \rho_0 c^2)\gamma_{ik}] \\ \partial\mu_0 &= -\tilde{K}\,\mu_0 - \frac{1}{c^2}[\partial\varepsilon_0 + \tilde{K}\,\varepsilon + \tilde{K}_{ik}\tilde{X}^{ik} + \frac{1}{c}(\tilde{\nabla}_i + 2\tilde{C}_i)\,\tilde{Q}^i] \\ \partial Q_i &= -\frac{hc}{\varepsilon_0}(\tilde{Q}_i + k\,\tilde{\partial}_i\,T) \\ \partial T &= -T\frac{\partial}{\partial T}(\tilde{K}_{ik}\tilde{X}^{ik}) - \frac{1}{c}(\tilde{\nabla}_i + \tilde{C}_i)\,\tilde{Q}^i \end{cases}$$

where, at second member, (inside $\widetilde{H}_{ik}$) we must think $\widetilde{\Omega}_{ik}$ to be of the kind

$$\widetilde{\Omega}_{ik} = \frac{1}{2}(\partial_i\gamma_k - \partial_k\gamma_i) + \gamma_i\widetilde{C}_k - \gamma_k\widetilde{C}_i;$$

at the same time, the field $\widetilde{C}_i$ is given by the *evolution equations*, in terms of thermodinamical variables, after giving the *constitutive equations*:

$$(\rho_0\delta_i{}^k + \frac{1}{c^2}\widetilde{X}_i{}^k)\widetilde{C}_k = -\frac{1}{c^2}[\widetilde{\nabla}_k\, \widetilde{X}_i{}^k - \frac{h}{\varepsilon}(\widetilde{Q}_i + k\,\widetilde{\partial}_i\, T) - \frac{1}{c}(K\,\widetilde{Q}_i + 2\widetilde{\Omega}_{ji}\,\widetilde{Q}^j)].$$

REFERENCES

[1] Cattaneo, C. (1959) "Proiezioni naturali e derivazione trasversa in una varietà a metrica iperbolica normale", Ann. Matem. P. Appl. (IV), XLVIII, 361-386.

[2] Schouten, A. (1954) Ricci calculus, 2a Ed., Springer-Verlag, Berlino.

[3] Ferrarese, G. (1979) Introduzione alla dinamica Riemanniana dei sistemi continui, Vol. I, Ed. Pitagora, Bologna.

[4] Ferrarese, G. (1964) "Proprietà di 2° ordine di un generico riferimento fisico in Relatività Generale", Seminario al I Corso C.I.M.E., 1-18.

[5] Ferrarese, G. (1965) "Proprietà di 2° ordine di un generico riferimento fisico in Relatività Generale", Rendic. Matem. Roma, 24, 57-100.

[6] Ferrarese, G. and Cattani, C. (1980) "Tecnica delle proiezioni e derivazioni covarianti adattate ad una struttura riemanniana quasi-prodotto", Rendic. Matem. Roma,VI, 13, 135-51.

[7] Ferrarese, G. and Bini, D. (1992) "Riferimenti generalizzati di tipo polare in Relatività generale", Ricerche di Matem., Napoli,vol. in memoria di C.Tolotti.

[8] Ferrarese, G. (1963) "Sulle equazioni di moto di un sistema soggetto ad un vincolo anolonomo mobile", Rendic. Matem. Roma, 22, 1-20.

[9] Ferrarese, G. (1988) "Formulation intrinsèque du problème de Cauchy en relativité générale", C.R. Acad. Sci. Paris, t. 307, serie I, 107-110; (1988) "Formulazione intrinseca del Problema di Cauchy in Relatività Generale", Rendic. Acc. Lincei, 4.

[10] Ferrarese, G. (1965) Lezioni di Meccanica relativistica I, Ed. Pitagora, Bologna, .

[11] Ferrarese, G. (1989) "Kaluza-Klein theories and projection techniques", Atti del Simposio internazionale "Classical Mechanics and Relativity: Relationship and Consistency", Ed. Bibliopolis, 9-19; idem (1990) Rend. Matem. 10, 103-117.

[12] Choquet-Bruhat, Y. (1968) Geometrie differentielle et systemes exterieurs, Dunod, Paris.

[**13**] Cattaneo-Gasparini, I. (1961) "Projections naturelles des tenseurs de courbure d'une varieté V_{n+1} a metrique hyperbolique normale", C.R. Acad. Sc., 252, 3722-3724; (1963) "Proiezioni dei tensori di curvatura di una varietà riemanniana a metrica iperbolica normale", Rend. di Matem. (1-2), XXII, 127-146.

[**14**] Ferrarese, G. (1963) "Contributi alla tecnica delle proiezioni in una varietà a metrica iperbolica normale", Rendic.Matem., 22, 147-168.

[**15**] Lichnerowicz, A. (1954) Thèories relativistes de la Gravitation et de l' E'lectromagnétisme, Masson, Paris.

[**16**] Ferrarese, G. (1987) "Intrinsic formulation in relativistic continuum mechanics", Proceedings dell'Italian-Polish Meeting on "Selected Problems of Modern Continuum Theory", Bologna 3-6 Giugno ,49-58, Pitagora editrice, Bologna. Idem: 1987, Atti del Seminario matem. e Fisico di Modena, 36, 289-302., v.a. (1975) "Sulla formulazione intrinseca della dinamica dei continui iperelastici", Rend. Matem., Roma, 8,1, 285-49.

[**17**] Ferrarese, G. (1991) "Compatibility Conditions in Relativistic Continuum Dynamics", Atti del Convegno internazionale "Advances in Modern Continuum Dynamics", in memoria di A.Signorini, Elba, 6-12 giugno.

[**18**] Choquet-Bruhat, Y. (1952) "Theoreme d'existence pour certains systemes d'equations aux derivees partielles non lineaires", Acta Math., 88, 141-225.

[**19**] Cattaneo, C. (1958) "General Relativity: relative standard mass, momentum, energy and gravitational field in a general system of reference", Il Nuovo Cimento, 10, 318-337.

[**20**] Cattaneo, C. (1961) "Equation de Gauss-Poisson dans les Univers stationnaires", C.R. Acad. Sc. 252, 3748-50.

Reducing Einstein's Equations to an Unconstrained Hamiltonian System on the Cotangent Bundle of Teichmüller Space

Arthur E. Fischer and Vincent Moncrief

Department of Mathematics
University of California
Santa Cruz, California 95064
Email: aef@cats.ucsc.edu

Departments of Mathematics and Physics
Yale University
New Haven, Connecticut 06511
Email: moncrief@yalph2.bitnet

Dedicated to Yvonne-Choquet Bruhat

To Yvonne: Scholar, Co-author, Friend

M. Flato et al. (eds.), Physics on Manifolds, 111–151.

Abstract

A program for reducing Einstein's vacuum equations to an *unconstrained* dynamical system where the variables are the *true degrees of freedom* of the gravitational field is presented. The reduced phase space on which this dynamics occurs is the cotangent bundle $T^*\mathcal{T}$ of

$$\mathcal{T} \equiv \frac{\mathcal{M}/\mathcal{P}}{\mathcal{D}_0}$$

the $\mathcal{D}_0$-restricted conformal superspace of a compact connected orientable 3-manifold M, which we also refer to as the Teichmüller space of conformal structures on M. From results regarding linearization stability of spacetimes, we argue that the program of reduction should be successful for spacetimes that admit constant mean curvature continuously non-symmetric (or $\deg(M) = 0$) non-F_6 compact spacelike Cauchy hypersurfaces, inasmuch as the space of solutions to Einstein's equations is a manifold in a neighborhood of such spacetimes. Modulo the Poincaré conjecture and a conjecture regarding finite free actions on S^3, the topological types of all such spacelike Cauchy hypersurfaces M are identified.

Restricting to the case of continuously non-symmetric Yamabe type -1 manifolds M, $\mathcal{T}$ can be represented as

$$\mathcal{T} \approx \mathcal{M}_{-1}/\mathcal{D}_0$$

and the full Einstein dynamics can be reduced to an implicitly defined time-dependent Hamiltonian system on the cotangent bundle $T^*\mathcal{T}$. The parameter of evolution for this system is the parameter of a monotonically increasing constant mean curvature slicing by spacelike Cauchy hypersurfaces in a neighborhood of the given one, and the Hamiltonian is the volume functional of these hypersurfaces.

Some possible approaches to the more general case of continuously non-symmetric Yamabe type $+1$ manifolds are proposed. It is speculated that if conformal methods are not successful in treating this case, then, since linearization stability arguments imply that a reduction *does* exist (modulo discrete isometry groups of the spacetime), there must be another method of reduction that does work.

1 Introduction, notation, and background

Of the innumerable contributions of Yvonne Choquet-Bruhat to the mathematical development of general relativity, one of special significance for the authors has been her treatment of the initial value problem for Einstein's equations. Her work in this area has extended to fundamental theorems of existence, uniqueness, and stability of Einstein's equations in harmonic coordinates, to the existence of maximal Cauchy developments for the initial value problem, to the use of conformal methods to solve the constraint equations in the Hamiltonian formulation, and to the establishment of existence theorems regarding maximal Cauchy hypersurfaces and linearization stability for asymptotically flat spacetimes. Yvonne's work has involved a powerful combined use of tools from topology, geometry, and analysis, finely tuned and brought to bear with surgical precision on understanding the complicated systems of non-linear equations that arise in mathematical physics. The methods and tools that she has crafted over the years have been adopted by many others to extend our domain of knowledge of the universe. It is a great pleasure to contribute this article to this volume in honor of her work.

The problem that we address in this paper is how to write Einstein's vacuum equations as an *unconstrained* dynamical system where the variables are the *true degrees of freedom* of the gravitational field. This program, which we refer to as *reduction*, consists of four components. The first component consists of identifying the reduced phase space on which the dynamics will occur. The second component consists of finding the proper Hamiltonian function for the unconstrained variables. The third component consists of reconstructing the spacetime from the Hamiltonian flow. The fourth component consists of studying the nature and general properties of the Hamiltonian flow. Our intention here is only to outline this program of reduction; further details will be presented elsewhere.

The first three components of this program have been carried out in $2+1$ dimensions (Moncrief [1989], [1990]). In the analysis there, the spacelike Cauchy hypersurfaces are compact connected orientable 2-manifolds Σ_p of genus p. Let $\mathcal{T}$ denote the Teichmüller space of Σ_p. It is shown that in the case $p \geq 2$ the full Einstein dynamics can be reduced to an implicitly defined time-dependent Hamiltonian system on the cotangent bundle $T^*\mathcal{T}$, a $(12p-12)$-dimensional space. The parameter of evolution for this system, i.e., the time, is the parameter of a monotonically increasing constant mean curvature slicing by spacelike Cauchy hypersurfaces in a neighborhood of the given one, and the Hamiltonian in this slicing is the area functional of these hypersurfaces.

In this paper, we outline an extension of this program to $3+1$ dimensions. Our first task is to identify the 3-manifolds that are the analogues of Σ_p, $p \geq 2$. For this task, we take our lead from results regarding linearization stability of spacetimes. We consider spatially compact globally hyperbolic Ricci-flat spacetimes (V, g_V) with a constant mean curvature spacelike Cauchy hypersurface M, and find sufficient topological conditions on M that insures that the spacetime is linearization stable. The topological conditions that we find are that M is a compact connected orientable 3-manifold that is *continuously non-symmetric* (i.e., $\deg(M) = 0$, or equivalently, every Riemannian metric on M has a finite isometry group), and that M is not diffeomorphic F_6, a particular manifold of compact orientable flat type that is the underlying manifold of one of the flat affine diffeomorphism classes in 3-dimensions. When these conditions are met, (V, g_V) does not have any Killing vector fields and must be linearization stable. Moreover, the space of spatially compact globally hyperbolic solutions $\mathcal{E}^{sc}_{gh}(V)$ to Einstein's equations is a manifold in a neighborhood of g_V. Thus, in particular, modulo discrete isometries, the space of spatially compact globally hyperbolic Ricci-flat geometries $\mathcal{E}^{sc}_{gh}(V)/\mathcal{D}(V)$ has a manifold structure in the neighborhood of $[g_V] \in \mathcal{E}^{sc}_{gh}(V)/\mathcal{D}(V)$.

Because of this result, we conjecture that the program of reduction should be successful for spatially compact globally hyperbolic Ricci-flat spacetimes that admit constant mean curvature continuously non-symmetric non-F_6 spacelike Cauchy hypersurfaces.

Subject to the Poincaré conjecture and a conjecture regarding free actions of finite groups on S^3, we are able to identify the continuously non-symmetric non-F_6 manifolds explicitly. These manifolds come in two varieties, Yamabe type -1 and Yamabe type $+1$, where M is of Yamabe type -1 if it *has no* Riemannian metric with scalar curvature $R(g) = 0$, and is of Yamabe type $+1$ if it *has* a Riemannian metric with scalar curvature $R(g) = 1$.

Restricting to Yamabe type -1 manifolds, we are able to show that the $\mathcal{D}_0$-restricted conformal superspace, which we also refer to as the Teichmüller space of conformal structures, can be represented as

$$\mathcal{T} \equiv \frac{\mathcal{M}/\mathcal{P}}{\mathcal{D}_0} \approx \mathcal{M}_{-1}/\mathcal{D}_0$$

and that $\mathcal{T}$ has at most orbifold-type singularities. Modulo these orbifold singularities and the use of local cross-sections rather than a global cross-section, the analysis for $3+1$ reduction can be completed much as in the $2+1$ dimensional case. Thus one gets as the reduced phase space the cotangent bundle $T^*\mathcal{T}$ of the Teichmüller space of conformal structures, and one gets reduction of the full classical non-reduced Hamiltonian formulation to a reduced dynamical system on $T^*\mathcal{T}$, where the time parameter is the parameter of a family of monotonically increasing constant mean curvature spacelike Cauchy hypersurfaces in a neighborhood of the given one, and the Hamiltonian is the volume functional on these hypersurfaces (see Arnowitt, Deser, and Misner [1962] for a review of the classical Hamiltonian formulation).

For continuously non-symmetric Yamabe type $+1$ manifolds, the situation is complicated by two problems. The first is that unlike $\mathcal{M}_{-1}$, the space $\mathcal{M}_{+1}$ may have non-manifold singular points. This situation arises because the derivative of the scalar curvature function is not surjective at certain positive scalar curvature metrics. This complication causes a difficulty in finding a subspace of $\mathcal{M}$ to represent $\mathcal{M}/\mathcal{P}$.

The second problem is that there is a certain amount of non-uniqueness in the construction of positive constant scalar curvature metrics in a given conformal class. Thus there may exist $p_1 g$, $p_2 g$, $p_1 \neq p_2$, $p_1, p_2 > 0$, with $R(p_1 g) = R(p_2 g) = 1$. This non-uniqueness compounds the problems involved in representing $\mathcal{M}/\mathcal{P}$. It would be of interest if these two difficulties were related, with perhaps the non-uniqueness problem "causing" the non-manifold problem of $\mathcal{M}_{+1}$.

Some possible approaches to these difficulties are discussed. Again, our feeling is that the continuously non-symmetric Yamabe type $+1$ case should be reducible, inasmuch as the results from linearization stability analysis show that this case is not singular and that in principle, a reduction does exist. Of course, this does not prove that the conformal method of reduction that works in the Yamabe type -1 case need work in the Yamabe type $+1$ case. However, if it doesn't, then the question becomes what method does?

2 Some topological and geometrical preliminaries

In this section we present some topological and geometrical background information that we shall need. Throughout this paper, an *n-manifold* M will mean a smooth (C^∞) Hausdorff second countable n-dimensional manifold without boundary. All tensor fields, including all Riemannian and Lorentz metrics, all maps between manifolds, and all group actions will be smooth.

Let M be a connected n-manifold. We let

$$
\begin{aligned}
\mathcal{M} &= \text{Riem}(M) = \text{ the space of smooth } (C^\infty) \text{ Riemannian metrics on } M \\
\mathcal{D} &= \text{Diff}(M) = \text{ the group of smooth diffeomorphisms of } M \\
\mathcal{X} &= \text{the space of vector fields on } M \\
\Omega^k &= \text{the space of k-forms on } M,\ 0 \le k \le n \\
S_2 &= \text{the space of symmetric 2-covariant tensor fields on } M \\
S^2 &= \text{the space of symmetric 2-contravariant tensor fields on } M \\
\mathcal{F} &= C^\infty(M, \boldsymbol{R}) = \Omega^0 = \text{the space of smooth real-valued functions on } M \\
\mathcal{P} &= \text{the space of smooth real-valued positive functions on } M
\end{aligned}
$$

If M is compact, each of these spaces can be given the structure of a smooth infinite dimensional ILH-manifold (Inverse Limit Hilbert) (see Omori [1970], Ebin [1970], or Fischer-Marsden [1975a] for more details regarding these spaces).

For a Riemannian metric $g \in \mathcal{M}$, we let

$$
\begin{aligned}
(M, g) &= \text{the corresponding Riemannian manifold} \\
\text{Riem}(g) &= \text{the Riemann-Christoffel curvature tensor of } g
\end{aligned}
$$

$$\begin{aligned}\mathrm{Ric}(g) &= \text{the Ricci curvature tensor of } g\\ \mathrm{R}(g) &= \text{the scalar curvature tensor of } g\\ \mathrm{K}(g) &= \text{the sectional curvature of } g\end{aligned}$$

Our sign conventions on these curvature tensors are as in Kobyashi-Nomizu [1963].

If (M,g) is an oriented Riemannian manifold, we let μ_g denote the unique volume element on M determined by g and the orientation of M. If, additionally, M is compact, we let

$$\mathrm{vol}(M,g) = \int_M \mu_g$$

denote the volume of (M,g).

For (M,g) a Riemannian manifold, we let

$$I_g(M) = \mathrm{Isom}_g(M) = \{f \in \mathcal{D} \mid f^*g = g\}$$

denote the group of isometries of the Riemannian metric g, we let $I_g^0(M)$ denote the connected component of the identity, and we let

$$\mathcal{I}_g(M) = \{X \in \mathcal{X} \mid L_X g = 0\}$$

denote the Killing vector fields of (M,g). Here $L_X g$ denotes the Lie derivative of g with respect to X.

A useful concept for a connected n-manifold M is its degree of symmetry (Hsiang [1967a], [1967b], and [1971]).

Definition 2.1 *For M a connected n-manifold, the* **degree of symmetry** *of M is defined by the non-negative integer*

$$\deg(M) = \max\{\dim I_g(M) \mid g \in \mathcal{M}\} \leq \frac{1}{2}n(n+1)$$

Note that if M is compact, then

$$\deg(M) = \max\{\dim G \mid G \text{ is a compact Lie group acting smoothly and effectively on } M\}$$

Note also that $\deg(M)$ is independent of any Riemannian metric on M and thus is a differential-topological invariant, rather than a geometrical invariant. Thus the degree of symmetry is a property of the manifold structure of M.

The following definition is useful:

Definition 2.2 *A connected n-manifold M is*

1. **symmetric** *if there exists a $g \in \mathcal{M}$ such that $I_g(M) \neq \{\mathrm{id}_M\}$;*
2. **continuously symmetric** *if* $\deg(M) > 0$ *(equivalently, if there exists a $g \in \mathcal{M}$ such that* $\dim I_g(M) > 0$*);*

3. **non-symmetric** *if for all* $g \in \mathcal{M}$, $I_g(M) = \{\mathrm{id}_M\}$;

4. **continuously non-symmetric** *if* $\deg(M) = 0$ *(equivalently, if for all* $g \in \mathcal{M}$ $\dim I_g(M) = 0$).

Thus M being symmetric and non-symmetric are mutually exclusive, and M being continuously symmetric and continuously non-symmetric are mutually exclusive. However, M is continuously non-symmetric if and only if for all $g \in \mathcal{M}$, $I_g(M)$ is a discrete group (since $\dim I_g(M) = 0$), so a continuously non-symmetric M is either non-symmetric or is symmetric with discrete isometry group for all g. If M is continuously non-symmetric and compact, then for all $g \in \mathcal{M}$, $I_g(M)$ is a finite group (see Fischer [1970] for an application of these concepts to the construction of superspace).

For compact manifolds, there is a useful necessary and sufficient condition for a manifold to be continuously symmetric (or continuously non-symmetric).

Proposition 2.3 *Let M be a compact connected n-manifold. Then M is continuously symmetric if and only if M admits an effective* $\boldsymbol{SO}(2)$*-action.*

Proof: If M admits an effective $\boldsymbol{SO}(2)$-action, then averaging an arbitrary Riemannian metric g with respect to this action yields an averaged metric $\overline{g}$ whose isometry group $I_{\overline{g}}(M) \supseteq \boldsymbol{SO}(2)$, and hence $\deg(M) \geq \dim I_{\overline{g}}(M) \geq 1$.

Conversely, if $\deg(M) \geq 1$, then there is a Riemannian metric g such that $\dim I_g(M) \geq 1$. Thus $\dim I_g^0(M) = \dim I_g(M) \geq 1$, and so a maximal torus in the compact connected Lie group $I_g^0(M)$ contains an effective $\boldsymbol{SO}(2)$-action on M. ∎

Recall that a *space form* is a complete connected Riemannian manifold of constant sectional curvature. The space form is *Euclidean* if the sectional curvature is zero, *spherical* if the sectional curvature is positive, and *hyperbolic* if the sectional curvature is negative.

The following definition is useful:

Definition 2.4 *A manifold M is of* **flat type** *(respectively,* **spherical type, hyperbolic type***) if M is diffeomorphic to a Euclidean space form (respectively, a spherical form, a hyperbolic space form).*

Note that in the above definition, we do not require that M have a Riemannian metric on it, nor if it does, need it be isometric to a one of the space forms.

In three dimensions, the isometric classification of Euclidean space forms is known (see Wolf [1972]). We shall need the affine classification for the compact orientable case:

Theorem 2.5 *Let (M, g_F) be a compact connected orientable flat Riemannian 3-manifold. Then (M, g_F) is affinely diffeomorphic to one of six torsion-free affinely* *flat model spaces, which we denote by* $\{\mathcal{F}_1,\ldots,\mathcal{F}_6\}$, *where* $\mathcal{F}_i = (F_i,\nabla_i)$, $1 \le i \le 6$, *and where* F_i *is the underlying manifold (of flat type) of* $\mathcal{F}_i$, *and* ∇_i *is the flat affine connection on* F_i.

Here $\mathcal{F}_1$ is affinely diffeomorphic to any flat 3-torus $\boldsymbol{T}^3 = \boldsymbol{R}^3/(\boldsymbol{Z}\times\boldsymbol{Z}\times\boldsymbol{Z})$, and the other $\mathcal{F}_i$, $2 \le i \le 6$, are affinely diffeomorphic to the Euclidean space forms $\boldsymbol{R}^3/\boldsymbol{\Gamma}_i$, where the $\boldsymbol{\Gamma}_i$ are described in Wolf [1972] (see also Ellis [1971] for a description of the F_i). For our purposes, an important point to note is that F_1, F_2, F_3, F_4, and F_5 all fiber over the circle with a 2-torus as a fiber, but that F_6 does not. Thus for $1 \le i \le 5$, $\deg(F_i) \ge 1$, and $\deg(F_6) = 0$. These degrees can be computed more specifically as follows:

Proposition 2.6 *The degrees of the manifolds* F_i, $1 \le i \le 6$, *are given as follows:*

1. $\deg(F_1) = 3$;
2. $\deg(F_2) = \deg(F_3) = \deg(F_4) = \deg(F_5) = 1$;
3. $\deg(F_6) = 0$.

Proof: From Bochner's theorem [1946], a Killing vector field on a compact connected flat Riemannian n-manifold (F, g_F) must be parallel and hence harmonic. Conversely, a harmonic vector field is parallel, and hence Killing (see also Fischer and Wolf [1974], [1975]). Thus

$$\{\text{Killing vector fields}\} = \{\text{parallel vector fields}\} = \{\text{harmonic vector fields}\}$$

so that by Hodge's Theorem, $\deg(F) \ge \dim I_{g_F}(F) = \dim(\ker \Delta_{g_F}) = b_1(F)$, where Δ_{g_F} is the Laplacian on vector fields (or 1-forms), and $b_1(F)$ is the first Betti number of M. Now the first integral homology groups of the affine diffeomorphism classes $F_i, 1 \le i \le 6$ are given in Wolf [1972], from which we see that $b_1(F_1) = 3$, $b_1(F_2) = b_1(F_3) = b_1(F_4) = b_1(F_5) = 1$, and $b_1(F_6) = 0$. Now from the results in Fischer [1970], these manifolds admit no group actions with groups whose dimension is higher than $b_1(F_i)$, $1 \le i \le 6$. Thus the flat metrics on these manifolds are the most symmetric, and so we have equality of the degree of the manifold with the first Betti number. ∎

For each of the affine diffeomorphism classes above, the finer isometric classification is known (see Wolf [1972]).

Now we recall the following regarding Eilenberg-MacLane spaces (see for example Spanier [1966]).

Definition 2.7 *Let* i *be an integer* ≥ 1 *and let* π *be a group, Abelian if* $i \ge 2$. *A topological path-connected pointed space* (X, x_0) *is an* **Eilenberg-MacLane space of type** (π, i), *or a* $K(\pi,i)$**-space**, *if the* i^{th} *homotopy group* $\pi_i(X, x_0) = \pi$, *and* $\pi_j(X, x_0) = 0$ *for* $j \ne i$. *A connected manifold* M *is a* $K(\pi,i)$**-manifold** *if* M *is a* $K(\pi,i)$*-space.*

Thus $K(\pi, i)$-spaces are spaces all of whose homotopy is concentrated in a single group. The usefulness of such spaces is that their homotopy is relatively simple, and thus they can be used as a set of building blocks for understanding more complicated spaces.

As a simple example, note that $\pi_i(S^1) = 0$ for $i \geq 2$, and so S^1 is a $K(Z, 1)$-space. Note however that since $\pi_n(S^n) = Z$ for $n \geq 1$ and $\pi_{n+1}(S^n) \neq 0$ for $n \geq 2$ (see for example Spanier [1966]), the spheres S^n for $n \geq 2$ are not Eilenberg-MacLane spaces.

Also, for connected spaces M and N, the homotopy of the product space $M \times N$ is given by

$$\pi_i(M \times N) = \begin{cases} \pi_i(M) \oplus \pi_i(N) & \text{for } i = 1 \\ \pi_i(M) \times \pi_i(N) & \text{for } i \geq 2 \end{cases}$$

where the right hand side denotes the direct sum of groups for $i = 1$ and the direct product of groups for $i \geq 2$ (see e.g., Hu [1959]). It follows that the finite Cartesian product of $K(\pi, i)$-spaces is also a $K(\pi, i)$-space. Thus $T^n = S^1 \times \ldots \times S^1$ is a $K(\pi, 1)$-manifold with $\pi = Z + \ldots + Z$.

A nice class of examples of $K(\pi, 1)$-manifolds is given by the following:

Proposition 2.8 *Let M be a manifold of either flat or hyperbolic type. Then M is a $K(\pi, 1)$-manifold.*

Proof: If M is of flat type, then from the classical theorem of Killing and Hopf (see Wolf [1972]), M is diffeomorphic to R^n/Γ where Γ is a subgroup of the Euclidean group acting freely and properly discontinuously on R^n. Since R^n is contractible and since the orbit projection map $R^n \to R^n/\Gamma$ is a covering map, $\pi_1(R^n/\Gamma) = \Gamma$. Since the homotopy sequence of a fibration is exact, $\pi_i(R^n/\Gamma) = \pi_i(R^n) = 0$ for $i \geq 2$. Thus $M \approx R^n/\Gamma$ is a $K(\Gamma, 1)$-manifold.

If M is of hyperbolic type, then M is diffeomorphic to H^n/Γ, where H^n is unit hyperbolic space, and where Γ is a subgroup of the isometry group $I(H^n)$ acting freely and properly discontinuously on H^n. Since H^n is diffeomorphic to R^n, the same argument applies. ∎

The connected sum $M_1 \# M_2$ of two connected orientable n-manifolds M_1 and M_2 is defined by removing the interior of a solid n-ball from each and then matching the resulting boundaries using an orientation-reversing diffeomorphism (e.g., see Bröcker and Jänich [1982]). The resulting manifold does not depend on which balls are removed. This operation is well-defined up to diffeomorphism, is associative and commutative (up to diffeomorphism), and the n-sphere S^n serves as an identity element.

Definition 2.9 *A connected orientable n-manifold M is* **non-trivial** *if it is not diffeomorphic to S^n. A non-trivial manifold M is* **prime** *if there is no decomposition $M = M_1 \# M_2$ with both M_1 and M_2 non-trivial, and is* **non-prime** *if there is a decomposition $M = M_1 \# M_2$ with both M_1 and M_2 non-trivial (i.e., if it is not prime).*

Note that from this definition, $\boldsymbol{S}^n$ is neither prime nor non-prime, being somewhat of an exception throughout this discussion inasmuch as it serves as an identity element for connected sums. Thus there are three diffeomorphism classes of connected orientable n-manifolds, namely, trivial, prime, and non-prime.

If two manifolds M_1 and M_2 are diffeomorphic, we denote this by $M_1 \approx M_2$. We shall also be considering orbit spaces which may only be topological spaces. For these spaces "$\approx$" shall denote homeomorphic.

Now we recall the prime decomposition theorem for compact connected orientable 3-manifolds.

Theorem 2.10 (Kneser (1929), Milnor (1961)) *Let M be a non-trivial compact connected orientable 3-manifold. Then M is diffeomorphic to a finite connected sum of prime manifolds*

$$M \approx P_1 \# \cdots \# P_k \ ,$$

where the summands are uniquely determined up to order and diffeomorphism.

The existence part of the above Theorem was proved by Kneser [1929] and the uniqueness part was proved by Milnor [1961].

Let M be a compact connected orientable 3-manifold. Then if M is prime, M is diffeomorphic to either a handle $\boldsymbol{S}^1 \times \boldsymbol{S}^2$, or to a $K(\boldsymbol{\pi}, 1)$-manifold, or else M is non-trivial and is finitely covered by a homotopy 3-sphere (Milnor [1961]). Thus, together with Theorem 2.10, we have

Theorem 2.11 *Let M be a non-trivial compact connected orientable 3-manifold. Then M is diffeomorphic to a finite connected sum of prime manifolds*

$$M \approx \Sigma_1 \# \ldots \# \Sigma_k \# (\boldsymbol{S}^1 \times \boldsymbol{S}^2)_1 \# \ldots \# (\boldsymbol{S}^1 \times \boldsymbol{S}^2)_l \# K_1 \# \ldots \# K_m$$

where k, l, and m are non-negative integers, $k + l + m \geq 1$, and where each Σ_i for $1 \leq i \leq k$ is non-trivial and is finitely covered by a homotopy 3-sphere, where each $K_j = K(\boldsymbol{\pi}_j, 1)$ for $1 \leq j \leq m$ is a compact orientable $K(\boldsymbol{\pi}_j, 1)$-manifold, and where the summands are uniquely determined up to order and diffeomorphism.

In the above theorem and throughout this paper, we adopt the convention that if any of the integers k, l, or m are zero, then terms of that type do not appear.

Now suppose that the Poincaré conjecture is true. Then every homotopy 3-sphere is a true 3-sphere, so each Σ_i is finitely covered by $\boldsymbol{S}^3$. Let $\boldsymbol{\Gamma}_i = \pi_1(\Sigma_i)$, so that $\boldsymbol{\Gamma}_i$ is a finite group acting freely on $\boldsymbol{S}^3$, and Σ_i is diffeomorphic to $\boldsymbol{S}^3/\boldsymbol{\Gamma}_i$.

However, there is a question of whether or not a free action of a finite group acting on $\boldsymbol{S}^3$ is equivalent to a standard orthogonal action (see the review by Edmonds [1985] and the references therein). We shall refer to this question as the following:

Conjecture B *Every finite group acting freely on S^3 is equivalent to a standard orthogonal action of a subgroup of $O(4)$ acting freely on S^3.*

Thus if the Poincaré and B conjectures are true, then each Σ_i in the prime factorization given in Theorem 2.11 is diffeomorphic to S^3/Γ_i, where Γ_i is a non-trivial finite subgroup of $SO(4)$ (since M is non-trivial and orientable) acting freely and orthogonally on S^3. Thus in this case, each $\Sigma_i \approx S^3/\Gamma_i$ is non-trivial and of orientable spherical type.

3 Scalar curvature on compact manifolds

We now review the results regarding the scalar curvature of Riemannian metrics on compact connected n-manifolds, $n \geq 3$, that we shall need. We begin by making the following two remarks.

Remark 1 Aubin ([1970], [1982]) has shown that every compact connected n-manifold, $n \geq 3$, admits a Riemannian metric g_{-1} with $R(g_{-1}) = -1$. Since for c a positive real constant, the homothetically deformed Riemannian metric cg has scalar curvature $R(cg) = c^{-1}R(g)$, among compact n-manifolds, $n \geq 3$, there is no topological obstruction to having any constant negative scalar curvature.

Remark 2 Using Aubin's result and a continuity argument (see Kazdan and Warner [1975a], [1985]), it then follows that if M admits a metric g_1 with $R(g_1) = 1$, then M admits a metric g_0 such that $R(g_0) = 0$.

Because of these two remarks, the following definition partitions the class of compact n-manifolds, $n \geq 3$, into three classes. We introduce the terminology **Yamabe-type** as follows.

Definition 3.1 *Let M be a compact connected n-manifold, $n \geq 3$.*

1. *M is of* **Yamabe-type -1** *if M admits no metric with $R(g) = 0$.*
2. *M is of* **Yamabe-type 0** *if M admits a metric with $R(g) = 0$, but no metric with $R(g) = 1$.*
3. *M is of* **Yamabe-type +1** *if M admits a metric with $R(g) = 1$.*

Manifolds of Yamabe-type -1 and Yamabe-type 0, and manifolds of Yamabe-type 0 and Yamabe-type $+1$ are disjoint by definition. Manifolds of Yamabe-type -1 and Yamabe-type $+1$ are disjoint by Remark 2 above. Thus these three classes are mutually exclusive. Since every M either admits a metric with $R(g) = 0$ (in which case M is of Yamabe-type either 0 or $+1$) or doesn't (in which case M is of Yamabe-type -1), the definition implies that these classes are exhaustive. Thus these three classes are mutually exclusive and exhaustive, and thus partition the set of compact connected n-manifolds, $n \geq 3$, into these three types.

The following definition is useful:

Definition 3.2 *A function $\rho \in \mathcal{F} = C^\infty(M, \boldsymbol{R})$ is a* **scalar curvature function** *if there exists a metric $g \in \mathcal{M}$ such that $R(g) = \rho$.*

The relationship of the Yamabe-type of a manifold to the possible scalar curvature functions that it can support is given by the following:

Theorem 3.3 (Kazdan-Warner (1975a,b), (1985)) *Let M be a compact connected n-manifold, $n \geq 3$.*

1. *M is of Yamabe-type -1 if and only if every scalar curvature function is negative somewhere.*

2. *M is of Yamabe-type 0 if and only if every scalar curvature function is either negative somewhere or is identically zero.*

3. *M is of Yamabe-type $+1$ if and only if every function is a scalar curvature function.*

We now summarize the results of Gromov and Lawson ([1980a], [1980b], [1983]) that we shall need regarding manifolds which can and cannot support positive scalar curvature Riemannian metrics (see also Schoen and Yau [1979a], [1979b]).

Firstly, Gromov and Lawson have the following positive result:

Theorem 3.4 (Gromov-Lawson (1980a,b)) *Let M be a compact connected orientable n-manifold, $n \geq 3$, of the form*

$$M \approx \boldsymbol{S}^n/\boldsymbol{\Gamma}_1 \# \ldots \# \boldsymbol{S}^n/\boldsymbol{\Gamma}_k \# (\boldsymbol{S}^1 \times \boldsymbol{S}^{n-1})_1 \# \ldots \# (\boldsymbol{S}^1 \times \boldsymbol{S}^{n-1})_l \,,$$

where k, l are non-negative integers, $k + l \geq 1$, and where for $1 \leq i \leq k$, $\boldsymbol{\Gamma}_i \subset \boldsymbol{SO}(n+1)$ is a finite group acting freely on $\boldsymbol{S}^n$ as a group of isometries. (Thus each $\boldsymbol{S}^n/\boldsymbol{\Gamma}_i$ is an orientable spherical space form.) Then there exists a Riemannian metric g on M with scalar curvature $R(g) > 0$.

Secondly, in dimension $n = 3$, Gromov and Lawson have very specific negative results:

Theorem 3.5 (Gromov-Lawson ((1980a,b), (1983))) *Let M be a compact connected orientable 3-manifold. If M has a $K(\pi, 1)$-factor in its prime decomposition, then M has no Riemannian metric with scalar curvature $R(g) > 0$. Furthermore, any metric with $R(g) \geq 0$ must be flat.*

In terms of the Yamabe types, Theorems 3.3, 3.4, and 3.5 can be rephrased as follows:

Theorem 3.6 *Let M be a compact connected orientable 3-manifold. If either*

1. *M is a $K(\pi,1)$-manifold not of flat type, or*

2. *M is non-prime and has a $K(\pi,1)$-factor in its prime decomposition (which may be of flat type),*

then M is of Yamabe-type -1. Conversely, assuming that both the Poincaré and B conjectures are true, if M is of Yamabe type -1, then M is of type (1) or (2) above.

If M is of flat type, then M is of Yamabe type 0. Conversely, assuming that both the Poincaré and B conjectures are true, if M is of Yamabe type 0, then M is $K(\pi,1)$-manifold of flat type.

If M is diffeomorphic to a finite connected sum of orientable spherical space forms and handles, then M is of Yamabe type $+1$. Conversely, assuming that both the Poincaré and B conjectures are true, if M is of Yamabe type $+1$, then M is diffeomorphic to a finite connected sum of orientable spherical space forms and handles.

Proof: In either case (1) or (2) above, M has a $K(\pi,1)$-factor in its prime decomposition, so by Theorem 3.5, M has no metric g with $R(g) > 0$, and so M is not of Yamabe-type $+1$. If M were of Yamabe-type 0, then M would have a metric g_0 such that $R(g_0) = 0$, and so by Theorem 3.5 again, g_0 would be flat, and so M would be a $K(\pi,1)$-manifold of flat type, contradicting (1). Thus M is of Yamabe-type -1.

Conversely, assume that both the Poincaré and B conjectures are true, and that M is of Yamabe type -1. If M has no $K(\pi,1)$-factor in its prime decomposition, then from the Prime Decomposition Theorem 2.11 (and the remarks following the theorem), M must be a connected sum of spherical space forms and handles, and so from Theorem 3.4, M admits a metric g with $R(g) > 0$, contradicting the assumption that M is of Yamabe type -1. Thus M must have a $K(\pi,1)$-factor in its prime decomposition. Now M cannot be a $K(\pi,1)$-manifold of flat type, since then it would admit a metric g with $R(g) = 0$ and thus would be of Yamabe type 0. Thus M must either be a $K(\pi,1)$-manifold not of flat type, or be non-prime with a $K(\pi,1)$-factor (which may be of flat type).

If M is of flat type, then M admits a flat metric g, and so M is of Yamabe type either 0 or $+1$. But from Proposition 2.8, M is a $K(\pi,1)$-manifold, and so from Theorem 3.5, M has no metric g with $R(g) > 0$. Thus M is of Yamabe type 0.

Conversely, assuming the Poincaré and B conjectures are true, if M is of Yamabe type 0, then M must have a $K(\pi,1)$-factor in its prime decomposition (otherwise M would be of Yamabe type $+1$, as above). Since M has a metric g with $R(g) = 0$, then from Theorem 3.5 again, g is flat, and so M is a $K(\pi,1)$-manifold of flat type.

If M is diffeomorphic to a finite connected sum of orientable spherical space forms and handles, then from Theorem 3.4, M admits a Riemannian metric g such that

$R(g) > 0$. Then from Theorem 3.3, M is of Yamabe type $+1$. Conversely, if M is of Yamabe type $+1$, from Theorem 3.3, M admits a Riemannian metric g with $R(g) > 0$. Thus from Theorem 3.5, M has no $K(\pi, 1)$-factor in its prime decomposition. Thus, assuming both the Poincaré and B conjectures are true, from the prime decomposition theorem, M is diffeomorphic to a finite connected sum of orientable spherical space forms and handles. ∎

Thus assuming the Poincaré and B conjectures, under the hypothesis of Theorem 3.6, the Gromov-Lawson results give necessary and sufficient conditions for M to be of Yamabe type -1 (conditions (1) or (2) above), of Yamabe type 0 (M is a $K(\pi, 1)$-manifold of flat type), or of Yamabe type $+1$ (M is a finite connected sum of orientable spherical space forms and handles). We shall return to this result later.

4 Continuously non-symmetric 3-manifolds

The results of Raymond [1968] and Orlik and Raymond [1968] give a complete topological and equivariant classification of effective $\boldsymbol{SO}(2)$-actions on compact 3-manifolds. Thus by Proposition 2.3 their results give a list of compact 3-manifolds with $\deg(M) > 0$ (the continuously symmetric case). We are only interested in the topological classification and the orientable case, but the equivariant classification and the non-orientable case are treated there as well.

Here is their list for the orientable case (without the actions):

Theorem 4.1 (Orlik-Raymond (1968)) *Let M be a compact connected orientable 3-manifold such that M admits an effective $\boldsymbol{SO}(2)$-action. Then M is diffeomorphic to one of the following manifolds:*

1. $\boldsymbol{S}^3$, $\boldsymbol{S}^1 \times \boldsymbol{S}^2$, *or a lens space* $L(p,q) = \boldsymbol{S}^3/\boldsymbol{\Gamma}(p,q)$, $p > q > 0$, p, q *relatively prime.*

2. *A connected sum of the above.*

3. *A quotient of $\boldsymbol{SO}(3)$ or $\boldsymbol{Spin}(1)$ by a finite, non-abelian, discrete subgroup (see Remark 4 below).*

4. *A $K(\pi, 1)$-manifold whose fundamental group π has infinite cyclic center (provided it is not the 3-dimensional torus) (see Remark 5 below).*

Remarks:

1. See Fischer ([1970]) for an application of these results to superspace, where also the degree of symmetry of each of the above manifolds is given.

2. In the construction of the lens spaces $L(p,q) = \boldsymbol{S}^3/\boldsymbol{\Gamma}(p,q)$ in (1) above, $\boldsymbol{S}^3$ is viewed as the unit quaternions, i.e., the unit sphere in $\boldsymbol{C} \times \boldsymbol{C}$, with multiplicative group structure induced from $\boldsymbol{C} \times \boldsymbol{C}$ (making $\boldsymbol{S}^3$ isomorphic to $\boldsymbol{SU}(2) = \boldsymbol{Spin}(1)$). The group $\Gamma(p,q) \subset \boldsymbol{SO}(2) \times \boldsymbol{SO}(2) \subset \boldsymbol{SO}(4)$ denotes the cyclic group of order p generated by $(e^{2\pi i/p}, e^{2\pi iq/p})$, acting on the unit sphere in $\boldsymbol{C} \times \boldsymbol{C}$.

3. The manifolds in (1), (3), and (4) are all prime, whereas the manifolds in (2) are non-prime.

4. Their item (3) above must be interpreted as "M is a quotient of $\boldsymbol{S}^3$ by a finite non-abelian group $\boldsymbol{\Gamma}$ acting freely and orthogonally on $\boldsymbol{S}^3$", since such manifolds admit $\boldsymbol{SO}(2)$-actions. Thus $M \approx \boldsymbol{S}^3/\boldsymbol{\Gamma}$ is a spherical space form that is not a lens space (since $\boldsymbol{\Gamma}$ is non-Abelian). The lens spaces are included in item (1) so as to be able to be part of the connected sums in item (2).

5. Their item (4) above must be interpreted as "M is diffeomorphic to either the 3-dimensional torus, or to a $K(\boldsymbol{\pi},1)$-manifold whose fundamental group π has infinite cyclic center", so that the 3-torus is included in this item. The 3-torus is an exception because it is a $K(\boldsymbol{\pi},1)$-manifold whose fundamental group $\pi_1(\boldsymbol{T}^3) = \boldsymbol{Z} \oplus \boldsymbol{Z} \oplus \boldsymbol{Z}$ does not have infinite cyclic center.

6. Now further partition the $K(\boldsymbol{\pi},1)$-manifolds whose fundamental groups have infinite cyclic centers into those of flat types and those of non-flat types. Examining the fundamental groups of the flat types $\{F_2, F_3, F_4, F_5, F_6\}$ shows that only the fundamental groups of the flat types $\{F_2, F_3, F_4, F_5\}$ have infinite cyclic center, so that F_6 does not admit an $\boldsymbol{SO}(2)$-action (see also Proposition 2.6 and the paragraph preceding that proposition). Now $F_1 = \boldsymbol{T}^3$ is also a $K(\boldsymbol{\pi},1)$-manifold which occurs in their item (4) (as revised by Remark 5), but whose fundamental group does not have infinite cyclic center. Thus their item (4) can be expressed as "Either M is a $K(\boldsymbol{\pi},1)$-manifold of flat type diffeomorphic to either F_1, F_2, F_3, F_4, or F_5 (F_6 is excluded), or M is a $K(\boldsymbol{\pi},1)$-manifold not of flat type whose fundamental group has infinite cyclic center". This further refinement of their item (4) is useful since in considering results involving scalar curvature, it is natural to further divide the $K(\boldsymbol{\pi},1)$-manifolds into those of flat and non-flat types (e.g., see Theorem 4.2 below).

The Orlik-Raymond results list the manifolds M with $\deg(M) > 0$. We shall be mostly interested in the complement to this list, namely, those M with $\deg(M) = 0$ (i.e., the continuously non-symmetric case).

Theorem 4.2 *Let M be a compact connected orientable 3-manifold. Assume the following:*

1. *If M is prime, then M is diffeomorphic to either*

(a) *a compact orientable $K(\pi, 1)$-manifold not of flat type whose fundamental group π does not have an infinite cyclic center, or*

(b) *F_6 (a $K(\pi, 1)$-manifold of flat type).*

2. *If M is non-prime, then in the prime decomposition of M, either*

   (a) *there is a compact orientable $K(\pi, 1)$-factor, or*

   (b) *there is a factor Σ_i (finitely covered by a homotopy 3-sphere) such that Σ_i is not a lens space.*

Then $\deg(M) = 0$.

Conversely, assume that the Poincaré and B conjectures are true and that $\deg(M) = 0$. *Then M is diffeomorphic to one of the manifolds in (1a,1b,2a,2b), and in the case (2b), the factor Σ_i is of spherical type but is not diffeomorphic to a lens space.*

Proof: If M is diffeomorphic to any of the manifolds in (1a,1b,2a,2b) then M does not appear on the Orlik-Raymond list, and so $\deg(M) = 0$.

Conversely, assume that the Poincaré and B conjectures are true and that $\deg(M) = 0$. If M is prime, then M is diffeomorphic to either a spherical space form S^3/Γ (not a sphere), a handle, or a $K(\pi, 1)$-manifold. Since a S^3/Γ or a handle have $\deg(M) > 0$, M must be a $K(\pi, 1)$-manifold. If M is a $K(\pi, 1)$-manifold of flat type, then from Theorem 4.1 (see Remark 6 following that Theorem), M must be diffeomorphic to F_6 (since the other F_i's have degree > 0). If M is a $K(\pi, 1)$-manifold not of flat type, then from Theorem 4.1 again, π cannot have an infinite cyclic center.

Now assume that M is non-prime and that $\deg(M) = 0$. From the Orlik-Raymond list, the only non-prime manifolds with $\deg(M) > 0$ are connected sums of handles and lens spaces. Thus either M has a $K(\pi, 1)$-factor in its prime decomposition, or else it has a spherical space form factor S^3/Γ_i that is not a lens space. ∎

Thus assuming the Poincaré and B conjectures are true, a compact connected orientable 3-manifold M is prime and has $\deg(M) = 0$ if and only if either (1a) or (1b) is true, and is non-prime and has $\deg(M) = 0$ if and only if and only if M is diffeomorphic to

$$S^3/\Gamma_1 \# \ldots \# S^3/\Gamma_k \# (S^1 \times S^2)_1 \# \ldots \# (S^1 \times S^2)_l \# K_1 \# \ldots \# K_m$$

where k, l, and m are non-negative integers, $k + l + m \geq 2$ (since M is non-prime), where for $1 \leq i \leq k$, Γ_i is a non-trivial finite group acting freely and orthogonally on S^3, and for $1 \leq j \leq m$, $K_j = K(\pi_j, 1)$ is a compact orientable $K(\pi_j, 1)$-manifold, and where either $m \geq 1$ or $k \geq 1$ and at least one of the S^3/Γ_i is not a lens space.

We also remark that from Theorem 3.6, the class of manifolds occurring in (1a) or (2a) are of Yamabe type -1, in the class (1b), $M \approx F_6$ is of Yamabe type 0, and in the class (2b) the manifolds are connected sums of spherical space forms and handles, where at least one of the spherical space forms is not a lens space. From Theorem

3.4, this latter class is of Yamabe type +1. Thus deg(M) = 0 manifolds contain all three Yamabe types.

5 The space of conformal structures

In this section we summarize the results from Fischer-Marsden ([1975a], [1977]) and Fischer-Tromba ([1984a,b], [1987]) that we shall need.

Let M be a compact connected n-manifold. Then the infinite-dimensional Abelian group of positive functions $\mathcal{P}$ on M acts on $\mathcal{M}$ by pointwise multiplication,

$$\mathcal{P} \times \mathcal{M} \longrightarrow \mathcal{M}; \quad (p, g) \longmapsto pg$$

For $g \in \mathcal{M}$, we let

$$\langle g \rangle = \{pg \mid p \in \mathcal{P}\}$$

denote the orbit through g, so that $\langle g \rangle$ is the pointwise conformal class of g. The resulting orbit space

$$\mathcal{M}/\mathcal{P} = \{\langle g \rangle \mid g \in \mathcal{M}\}$$

is the space of *pointwise conformal structures on M.* In a natural way $\mathcal{M}/\mathcal{P}$ has the structure of a smooth ILH-manifold, and the projection

$$\pi : \mathcal{M} \longrightarrow \mathcal{M}/\mathcal{P}; \quad g \longmapsto \langle g \rangle$$

is a $\mathcal{P}$-principle fiber bundle over $\mathcal{M}/\mathcal{P}$ (see Fischer-Marsden [1977] and Fischer-Tromba [1984a]).

The group $\mathcal{D}$ acts on $\mathcal{M}/\mathcal{P}$ on the right by pull-back,

$$\mathcal{M}/\mathcal{P} \times \mathcal{D} \longrightarrow \mathcal{M}/\mathcal{P}; \quad (\langle g \rangle, f) \longmapsto \langle f^* g \rangle$$

For $\langle g \rangle \in \mathcal{M}/\mathcal{P}$, we let

$$[\langle g \rangle] = \{f^* \langle g \rangle \mid f \in \mathcal{D}\}$$

denote the orbit through g, so that $[\langle g \rangle]$ is the set of metrics conformally equivalent to g. The resulting orbit space

$$\frac{\mathcal{M}/\mathcal{P}}{\mathcal{D}} = \{[\langle g \rangle] \mid \langle g \rangle \in \mathcal{M}/\mathcal{P}\}$$

is the space of *conformal structures on M*, or, in relativity terms, *conformal superspace.* For comparison, recall that superspace itself, the space of geometries on M, is $\mathcal{M}/\mathcal{D}$.

The orbit space of conformal superspace can be approached in a slightly different manner by considering the semi-direct product $\mathcal{D} \dot{\times} \mathcal{P}$ of $\mathcal{D}$ with $\mathcal{P}$, namely the product space $\mathcal{D} \times \mathcal{P}$ with semi-direct product group structure

$$(f_1, p_1) \cdot (f_2, p_2) = (f_1 \circ f_2, p_2(f_2^* p_1)) = (f_1 \circ f_2, p_2(p_1 \circ f_2))$$

(see Fischer-Marsden [1977]). Note that this construction is analogous to the construction of the Euclidean group $\boldsymbol{E(n)} = \boldsymbol{O(n)} \dot{\times} \boldsymbol{R}^n$ as a semi-direct product of the orthogonal group $\boldsymbol{O(n)}$ and the translation group $\boldsymbol{R}^n$ (here $\mathcal{D}$ plays the role of the orthogonal group, and $\mathcal{P}$ plays the role of the translation group). The resulting orbit space

$$\mathcal{M}/(\mathcal{D} \dot{\times} \mathcal{P}) = \frac{\mathcal{M}/\mathcal{P}}{\mathcal{D}}$$

then gives directly the space of conformal structures. However, in studying this space, it is often more useful to consider the two-step procedure above.

Let $\mathcal{D}_0$ denote the connected component of the identity of $\mathcal{D}$. Our main interest will be in the space

$$\mathcal{T} \equiv \frac{\mathcal{M}/\mathcal{P}}{\mathcal{D}_0}$$

In 2-dimensions this space is the Teichmüller space of the manifold (see Fischer-Tromba [1984a]). In higher dimensions we continue with this terminology and refer to $\mathcal{T}$ as the *Teichmüller space (of conformal structures) of M*. However, contrary to the 2-dimensional case (with genus ≥ 2), in general neither $\mathcal{D}_0$ acts freely on $\mathcal{M}/\mathcal{P}$ nor is $\frac{\mathcal{M}/\mathcal{P}}{\mathcal{D}_0}$ a manifold (see Sections 8 and 9). However, in those cases in which $\mathcal{D}_0$ does act freely and $\frac{\mathcal{M}/\mathcal{P}}{\mathcal{D}_0}$ is a manifold, the projection

$$\pi : \mathcal{M}/\mathcal{P} \longrightarrow \mathcal{T} = \frac{\mathcal{M}/\mathcal{P}}{\mathcal{D}_0}; \quad \langle g \rangle \longmapsto [\langle g \rangle]$$

is a $\mathcal{D}_0$-principle fiber bundle over $\mathcal{T}$.

The space $\mathcal{T} = \frac{\mathcal{M}/\mathcal{P}}{\mathcal{D}_0}$ will be important for us as the reduced configuration space for the Hamiltonian formulation of general relativity. In this regard, our ultimate goals are threefold:

1. Represent $\frac{\mathcal{M}}{\mathcal{P}}$ as a subspace of $\mathcal{M}$

2. Find the circumstances under which $\mathcal{D}_0$ acts freely on $\mathcal{M}/\mathcal{P}$ and $\mathcal{T} = \frac{\mathcal{M}/\mathcal{P}}{\mathcal{D}_0}$ is a manifold, and at those points where $\mathcal{D}_0$ does not act freely, find the nature of the non-manifold points in the orbit space?

3. Under the circumstances that $\mathcal{D}_0$ acts freely and $\mathcal{T}$ is a manifold, find further conditions on M for the $\mathcal{D}_0$-principle fiber bundle $\mathcal{M}/\mathcal{P} \to \mathcal{T}$ to have a global cross-section (rendering it isomorphic to a product bundle)?

In regard to the first question, for $\rho \in \mathcal{F}$, let

$$\mathcal{M}_\rho = \{g \in \mathcal{M} \mid R(g) = \rho\}$$

denote the subspace of Riemannian metrics with prescribed scalar curvature ρ.

Then we have the following result:

Theorem 5.1 (Fischer-Marsden (1975a))*Let M be a compact connected n-manifold, $n \geq 2$. Let $\rho \in \mathcal{F}$ be such that ρ is not a constant ≥ 0. Then $\mathcal{M}_\rho$ is a smooth closed ILH-submanifold of $\mathcal{M}$.*

Note that for certain M and certain ρ, $\mathcal{M}_\rho$ may be empty (for example if $\rho > 0$ and M is of Yamabe type -1).

Combining Theorem 5.1 with Aubin's result (see Remark 1 at the beginning of Section 3) and the methods used in Fischer-Tromba [1984a], we have the following:

Theorem 5.2 *Let M be a compact connected orientable n-manifold, with either $n = 2$, and* genus$(M) \geq 2$, *or $n \geq 3$. Then $\mathcal{M}_{-1}$ is a smooth closed non-empty ILH-submanifold of $\mathcal{M}$.*

If, under the additional assumption in the case $n \geq 3$, that M is of Yamabe type -1, then the projection map

$$\pi_{-1} : \mathcal{M}_{-1} \longrightarrow \mathcal{M}/\mathcal{P}; \quad g \longmapsto [g]$$

is a is a $\mathcal{D}$-equivariant diffeomorphism. Thus

$$\mathcal{M}_{-1} \approx \mathcal{M}/\mathcal{P}$$

and so

$$\mathcal{M}_{-1}/\mathcal{D} \approx \frac{\mathcal{M}/\mathcal{P}}{\mathcal{D}}$$

and

$$\mathcal{M}_{-1}/\mathcal{D}_0 \approx \frac{\mathcal{M}/\mathcal{P}}{\mathcal{D}_0}$$

In the above theorem, $\approx$ means isomorphic as spaces, which in the first case means diffeomorphic, and in the next two cases means homeomorphic, depending on whether the spaces are manifolds or topological spaces.

Thus for $n \geq 3$, the condition that M be of Yamabe type -1 is sufficient to be able to represent $\mathcal{M}/\mathcal{P}$ as the submanifold $\mathcal{M}_{-1}$.

Now we consider our second question regarding the circumstances under which $\mathcal{D}_0$ acts freely on $\mathcal{M}/\mathcal{P}$ and $\mathcal{T} = \frac{\mathcal{M}/\mathcal{P}}{\mathcal{D}_0}$ is a manifold. For $g \in \mathcal{M}$, let

$$C_g(M) = \{f \in \mathcal{D} \mid f^*g = pg \text{ for some } p \in \mathcal{P}\}$$

denote the *conformal group* of (M, g). Note that $C_g(M) = C_{g'}(M)$ for all g' in the conformal class $\langle g \rangle$. Now note that in the right action of pull-back of $\mathcal{D}$ on $\mathcal{M}/\mathcal{P}$, the isotropy group for this action at $\langle g \rangle \in \mathcal{M}/\mathcal{P}$ is given by

$$\text{Isotropy}_{\langle g \rangle} = \{f \in \mathcal{D} \mid f^*\langle g \rangle = \langle g \rangle\} = C_{g'}(M)$$

for $g' \in \langle g \rangle$.

Now we have to consider infinite dimensional orbifolds (see, for example, Thurston for a finite dimensional version).

An **orbifold** is a second countable Hausdorff space which is locally homeomorphic to the quotient space of a local (perhaps infinite dimensional) manifold by the action of a finite group.

As an example, if a group (perhaps infinite dimensional) acts on a manifold (also perhaps infinite dimensional) such that at each point in the manifold the action has a slice and the isotropy group is *finite*, then the orbit space of the action is an orbifold. Here the slice is the local manifold, and the finite isotropy group at each point then implies that the orbit space is an orbifold.

Now we have the following:

Theorem 5.3 *Let M be a compact connected n-manifold, $n \geq 3$, and with* $\deg(M) = 0$. *Then superspace*

$$\mathcal{M}/\mathcal{D}$$

conformal superspace

$$\frac{\mathcal{M}/\mathcal{P}}{\mathcal{D}}$$

and the Teichmüller space of conformal structures

$$\mathcal{T} \equiv \frac{\mathcal{M}/\mathcal{P}}{\mathcal{D}_0}$$

are orbifolds.

Proof: Since M is compact and $\deg(M) = 0$, for all $g \in \mathcal{M}$, $I_g(M)$ is a finite group. Since the isotropy groups of the action of $\mathcal{D}$ on $\mathcal{M}$ are the isometry groups of the metrics in $\mathcal{M}$ (see Fischer [1970] for a more extensive use of this point of view in the study of superspace), the action of $\mathcal{D}$ on $\mathcal{M}$ has finite isotropy groups for each $g \in \mathcal{M}$. Since this action also has a slice at each g (Ebin [1970]), $\mathcal{M}/\mathcal{D}$ is an orbifold.

For the case of conformal superspace and Teichmüller space, a theorem of Lelong-Ferrand ([1969], [1971]) states that if (M, g) is a compact connected Riemannian n-manifold, $n \geq 3$, then the conformal group $C_g(M)$ is compact if and only if (M, g) is not conformally equivalent to the standard sphere. Another theorem of Lelong-Ferrand ([1969], [1971]) and Obata [1971] states that under the same conditions on (M, g), the identity component $C_g^0(M)$ of $C_g(M)$ is a subgroup of the isometry group $I_{pg}(M)$ of some conformally related metric pg (for some $p \in \mathcal{P}$) if and only if (M, g) is not conformally equivalent to the standard sphere. Thus if $\deg(M) = 0$, M cannot be diffeomorphic to an n-sphere, and (M, g) cannot be conformally equivalent to a standard n-sphere. Thus applying both theorems to this case, for every $g \in \mathcal{M}$, by the first theorem $C_g(M)$ is compact, and by the second theorem, $C_g^0(M) \subseteq I_{pg}(M)$ for some $p \in \mathcal{P}$, and hence $C_g^0(M) \subseteq I_{pg}^0(M)$. Since, as above, $I_{pg}(M)$ is a finite group, $C_g^0(M) = I_{pg}^0(M) = \{\mathrm{id}_M\}$. Thus $C_g(M)$ is discrete and compact, and hence finite.

Thus any metric on a compact $\deg(M) = 0$ manifold must have a finite conformal group as well as a finite isometry group. Since the isotropy groups of the action of $\mathcal{D}$ on $\mathcal{M}/\mathcal{P}$ are the conformal groups of the conformal classes in $\mathcal{M}/\mathcal{P}$, and since this action has a slice (see Fischer-Marsden [1977]), both conformal superspace and Teichmüller space are orbifolds. ∎

Continuing with the case of $\deg(M) = 0$, an important question regarding $\mathcal{T} \equiv \frac{\mathcal{M}/\mathcal{P}}{\mathcal{D}_0}$ is under what circumstances "orbifold-type" singularities do not occur. In this regard the following is important:

Theorem 5.4 (Frankel (1966)) *Let (M, g) be a compact connected Riemannian n-manifold, $n \geq 2$, with sectional curvature $K(g) \leq 0$, and with Ricci curvature $Ric(g)$ negative definite. Then $I_g(M)$ is a finite group, and*

$$I_g(M) \cap \mathcal{D}_0 = \{id_M\}$$

Thus by Frankel's theorem, for 2-manifolds Σ_p, $p \geq 2$, $\mathcal{D}_0$ acts freely on $\mathcal{M}_{-1}$, and so Teichmüller space $\mathcal{T} \equiv \frac{\mathcal{M}/\mathcal{P}}{\mathcal{D}_0} \approx \mathcal{M}_{-1}/\mathcal{D}_0$ is a manifold. Unfortunately, for compact connected orientable n-manifolds M, $n \geq 3$, $\deg(M) = 0$, and M of Yamabe type -1, then $\mathcal{T} \approx \mathcal{M}_{-1}/\mathcal{D}_0$, but the scalar curvature condition $R(g) = -1$ may not be sufficient to imply that non-trivial discrete isometries of g are not isotopic to the identity. Thus in this case, $\mathcal{M}_{-1}/\mathcal{D}_0$ could have orbifold type singularities.

Nevertheless, our candidate manifolds for reduction will be compact connected orientable 3-manifolds M of $\deg(M) = 0$ and Yamabe type -1. Summarizing, the condition that M be of Yamabe type -1 lets us represent $\mathcal{M}/\mathcal{P}$ by $\mathcal{M}_{-1}$, and thus $\mathcal{T} \approx \mathcal{M}_{-1}/\mathcal{D}_0$, and the condition that $\deg(M) = 0$ implies that $\mathcal{T} = \frac{\mathcal{M}/\mathcal{P}}{\mathcal{D}_0}$ has at most orbifold-type singularities.

Now we consider another natural representation of $\mathcal{M}/\mathcal{P}$ in the Yamabe type -1 case. We are looking for a subspace of $\mathcal{M}$ that can replace $\mathcal{M}_{-1}$ and that may generalize to manifolds of Yamabe type $+1$.

For M a compact connected orientable n-manifold, $n \geq 3$, of any Yamabe type, let

$$\mathcal{M}^1_{\boldsymbol{R}} = \{g \in \mathcal{M} \mid R(g) = \text{constant and } \mathrm{vol}(M, g) = 1\}$$

denote the space of constant scalar curvature unit volume Riemannian metrics on M. Then by Yamabe's theorem (see Lee and Parker [1987] for a history of the Yamabe problem), the projection

$$\pi^1_{\boldsymbol{R}} : \mathcal{M}^1_{\boldsymbol{R}} \longrightarrow \mathcal{M}/\mathcal{P}; \qquad g \longmapsto \langle g \rangle$$

is surjective (note that the volume normalization can always be achieved by a homothetic rescaling of g). Let

$$\mathcal{M}^1_{\boldsymbol{R}^-} = \{g \in \mathcal{M} \mid R(g) = \text{constant} < 0 \text{ and } \mathrm{vol}(\mathrm{M}, \mathrm{g}) = 1\}$$

Then analogous to Theorems 5.1 and 5.2 we have:

Theorem 5.5 *Let M be a compact connected orientable n-manifold, $n \geq 3$. Then $\mathcal{M}^1_{\boldsymbol{R}^-}$ is a smooth closed non-empty submanifold of $\mathcal{M}$, and the map*

$$\Psi : \mathcal{M}^1_{\boldsymbol{R}^-} \longrightarrow \mathcal{M}_{-1}; \quad g \longmapsto |R(g)|g$$

is a diffeomorphism of $\mathcal{M}^1_{\boldsymbol{R}^-}$ with $\mathcal{M}_{-1}$, with inverse $\Psi^{-1}(g) = (\mathrm{vol(M,g)})^{-2/\mathrm{n}}\mathrm{g}$.

If M is of Yamabe type -1, then $\mathcal{M}^1_{\boldsymbol{R}} = \mathcal{M}^1_{\boldsymbol{R}^-}$ and the projection map

$$\pi^1_{\boldsymbol{R}^-} : \mathcal{M}^1_{\boldsymbol{R}^-} \longrightarrow \mathcal{M}/\mathcal{P}; \quad g \longmapsto [g]$$

is a diffeomorphism.

Note that relative to $\mathcal{M}_{-1}$, the additional constraint in $\mathcal{M}^1_{\boldsymbol{R}^-}$ of unit volume is offset by expanding $\mathcal{M}_{-1}$ to $\mathcal{M}_{\boldsymbol{R}^-} = \boldsymbol{R}^+ \cdot \mathcal{M}_{-1} = \{cg \mid g \in \mathcal{M}_{-1}, c = \text{constant} > 0\} = \{g \in \mathcal{M} \mid R(g) = \text{constant} < 0\}$.

The reason that we introduce the space $\mathcal{M}^1_{\boldsymbol{R}}$ is that we wish to find a representation of $\mathcal{M}/\mathcal{P}$ as a subset of $\mathcal{M}$ for manifolds of Yamabe type $+1$, and $\mathcal{M}^1_{\boldsymbol{R}}$ is a natural extension of $\mathcal{M}^1_{\boldsymbol{R}^-}$. Unfortunately, $\mathcal{M}^1_{\boldsymbol{R}}$ need not be a manifold (see Section 9, where we discuss the possibility that a subset of $\mathcal{M}^1_{\boldsymbol{R}}$ is a manifold diffeomorphic to $\mathcal{M}/\mathcal{P}$). In this regard, note that $\mathcal{M}^1_{\boldsymbol{R}}$ is invariant under the action of $\mathcal{D}_0$ (and of $\mathcal{D}$), so that the quotient space

$$\mathcal{M}^1_{\boldsymbol{R}}/\mathcal{D}_0$$

is well-defined. Thus a subspace of this quotient space is a possible candidate for a representation of the Teichmüller space of a manifold of Yamabe type $+1$ (see Section 9).

6 Some applications to linearization stability

In this section we apply some of the results from Sections 3 and 4 to considerations involving the linearization stability of Einstein's empty space field equations. These results will also give us a lead on what class of spacetimes we can expect reduction to be successful.

By a *Lorentz manifold* we shall mean a pair (V, g_V), where V is a $(n+1)$-manifold, and g_V is a Lorentz metric on V with signature $(-1, \underbrace{+1, \ldots, +1}_{n})$. By a *spacetime* we shall mean a Lorentz manifold (V, g_V), where V is a *connected* $(n+1)$-manifold. Our main consideration will be the $3+1$ case, but we shall also discuss briefly the $2+1$ case (see Section 7).

For a spacetime (V, g_V), we let $\mathrm{Ric}(g_V)$ denote the Ricci curvature tensor, $\mathrm{R}(g_V)$ the scalar curvature, and $G\ \mathrm{Ein}(g_V) = \mathrm{Ric}(g_V) - \frac{1}{2}R(g_V)g_V$ the Einstein tensor of the Lorentz metric g_V.

We shall only be considering the Einstein field equations without sources and without cosmological constant, so that the empty space field equations are

$$\mathrm{Ein}(g_V) = 0 \Longleftrightarrow \mathrm{Ric}(g_V) = 0$$

Thus a solution (V, g_V) to these equations is a Ricci-flat spacetime, and in the $2+1$ case, the solutions are flat spacetimes. We shall be interested in when these empty space equations are linearization stable, i.e., when infinitesimal deformations of these equations can be integrated to finite deformations (see Fischer-Marsden ([1975a,b]) for more details).

We say that a globally hyperbolic spacetime (V, g_V) is *spatially compact* if there exists a compact spacelike hypersurface (M, i) (without boundary). It follows from Budic, Isenberg, Lindblom, and Yasskin [1978] that $S = i(M)$ is a Cauchy hypersurface. Also, from the global hyperbolicity and connectedness of V, V has topology $\boldsymbol{R} \times M$, and M (and S) are connected.

For V a connected 4-manifold, we let

$$\mathcal{L}(V) = \{g_V \mid g_V \text{ is a Lorentz metric on } V\}$$

denote the space of Lorentz metrics on V so that $\mathcal{L}(V)$ is the "space" of spacetimes with underlying manifold V. Let

$$\mathcal{E}^{sc}_{gh}(V) = \{g_V \in \mathcal{L} \mid \mathrm{Ric}(g_V) = 0 \text{ and } (V, g_V) \text{ is globally hyperbolic and spatially compact}\}$$

denote the subset of spatially compact globally hyperbolic Ricci-flat spacetimes on V, and let

$$\mathcal{E}^{sc}_{gh}(V)/\mathcal{D}(V)$$

denote the orbit space of spatially compact globally hyperbolic Ricci-flat geometries on V.

One goal of the linearization stability analysis is to determine which Ricci-flat spacetimes $g_V \in \mathcal{E}^{sc}_{gh}(V)$ and geometries $[g_V] \in \mathcal{E}^{sc}_{gh}(V)/\mathcal{D}(V)$ have smooth manifold neighborhoods in $\mathcal{E}^{sc}_{gh}(V)$ and $\mathcal{E}^{sc}_{gh}(V)/\mathcal{D}(V)$, respectively.

Let (V, g_V) be a Ricci-flat spacetime, and let (M, i) be an oriented spacelike hypersurface in (V, g_V). Then a classical calculation using the Gauss-Codazzi equations (see, for example, Arnowitt, Deser, and Misner [1962], Hawking-Ellis [1973], or Fischer-Marsden [1972], [1979a]), shows that the Cauchy data (g, π) on M must satisfy the Hamiltonian and divergence constraint equations

$$\mathcal{H}(g, \pi) \equiv (-\pi' \cdot \pi' + \frac{1}{2}(\mathrm{tr}_g \pi')^2 + R(g))\mu_g = 0$$

$$\delta(g, \pi) \equiv \delta_g \pi = 0$$

where μ_g is the unique volume element induced on M by the metric g and the orientation of M, π' is the tensor part of the momentum density $\pi = \pi' \otimes \mu_g = \pi' \mu_g$

(suppressing the tensor product symbol), and $\pi' \cdot \pi' = \langle \pi', \pi' \rangle_g = \|\pi'\|_g^2$ is the square of the norm of π' (in the pointwise metric induced by g on the 2-contravariant symmetric tensor bundle over M), $R(g)$ is the scalar curvature of the Riemannian metric g, $\delta_g \pi$ denotes the divergence of the symmetric tensor density π. The momentum density π is related to the second fundamental form k by

$$\pi = (k - (\mathrm{tr}_g k)g)^\sharp \mu_g$$

so that $\mathrm{tr}_g \pi' = -2\mathrm{tr}_g k$.

In local coordinates $x = (x^i)$ on M, $1 \leq i \leq 3$, and $y = (y^\mu)$ on V, $0 \leq \mu \leq 3$, letting "$_|$" denote covariant differentiation on (M, g), these equations are

$$(-(\pi')^{ij}(\pi')_{ij} + \frac{1}{2}(g^{ij}\pi'_{ij})^2 + R(g))(\det g_{ij})^{1/2} = 0$$
$$-\pi^{ij}{}_{|k} = 0$$

We let

$$\mathcal{C}_{\mathcal{H}} = \{(g, \pi) \in T^*\mathcal{M} \mid \mathcal{H}(g, \pi) = 0\}$$
$$\mathcal{C}_\delta = \{(g, \pi) \in T^*\mathcal{M} \mid \delta_g \pi = 0\}$$

denote the solution subspace for each of the constraint equations, and let

$$\mathcal{C} = \mathcal{C}_{\mathcal{H}} \cap \mathcal{C}_\delta$$

denote the joint constraint space.

We will need to consider the following conditions on pairs $(g, \pi) \in T^*\mathcal{M}$:

$\mathrm{C}_{\mathcal{H}}$: If $\pi = 0$, then g is not flat;
C_δ : If for a vector field $X \in \mathcal{X}, L_X g = 0$ and $L_X \pi = 0$, then $X = 0$;
C_{tr}: $\mathrm{tr}_g \pi' =$ constant on M.

If $\mathrm{tr}_g \pi' =$ constant, we say that the Cauchy data (g, π) has *constant mean curvature.*

Let

$$\Phi = (\mathcal{H}, \delta) : T^*\mathcal{M} \longrightarrow \mathcal{F}_d \times \mathcal{X}_d; \quad (g, \pi) \longmapsto (\mathcal{H}(g, \pi), \delta(g, \pi))$$

denote the joint constraint map, where $\mathcal{F}_d = \Omega^3$ is the space of scalar densities, or 3-forms, on M, and $\mathcal{X}_d$ is the space of vector field densities on M. Let $(D\mathcal{H}(g, \pi))^*$, $(D\delta(g, \pi))^*$,and $(D\Phi(g, \pi))^*$ denote the L_2-adjoint of the derivatives of the maps $\mathcal{H}$, δ, and Φ, respectively, at (g, π) (see Fischer-Marsden ([1975b], [1979a,b]) for details).

Now we have the following results:

Theorem 6.1 (Fischer-Marsden (1975b), (1979a,b)) *Let M be a compact connected orientable n-manifold, $n \geq 3$.*

Let $(g, \pi) \in \mathcal{C}_{\mathcal{H}}$. The following are equivalent:

(1a) (g, π) satisfies condition $\mathrm{C}_{\mathcal{H}}$.

(1b) $D\mathcal{H}(g,\pi)$ *is surjective.*

(1c) $\ker(D\mathcal{H}(g,\pi))^* = 0$.

(1d) The Hamiltonian equation $\mathcal{H}(g,\pi) = 0$ *is linearization stable at* (g,π).

If any of these conditions are satisfied, then $C_{\mathcal{H}}$ *is a smooth ILH-submanifold of* $T^*\mathcal{M}$ *in a neighborhood of* (g,π).

Let $(g,\pi) \in C_\delta$. *Then the following are equivalent:*

(2a) (g,π) *satisfies condition* C_δ.

(2b) $D\delta(g,\pi)$ *is surjective.*

(2c) $\ker(D\delta(g,\pi))^* = 0$.

(2d) The divergence equation $\delta(g,\pi) = 0$ *is linearization stable at* (g,π).

If any of these conditions are satisfied, then C_δ *is a smooth submanifold of* $T^*\mathcal{M}$ *in a neighborhood of* (g,π).

Let $(g,\pi) \in C$. *Then the following are equivalent:*

(3a) $D\Phi(g,\pi)$ *is surjective.*

(3b) $\ker(D\Phi(g,\pi))^* = 0$.

(3c) The joint constraint equations $\Phi(g,\pi) = (\mathcal{H}(g,\pi), \delta(g,\pi)) = 0$ *are linearization stable at* (g,π).

If either (3a), (3b), or (3c) is satisfied, then C *is a smooth submanifold of* $T^*\mathcal{M}$ *in a neighborhood of* (g,π).

If (g,π) *satisfies conditions* $C_{\mathcal{H}}, C_\delta$, *and* C_{tr}, *then (3a), (3b), and (3c) are satisfied, and* C *is a smooth submanifold of* $T^*\mathcal{M}$ *in a neighborhood of* (g,π).

The conditions $C_{\mathcal{H}}, C_\delta$, and C_{tr} give sufficient conditions for a spacetime to be linearization stable.

Theorem 6.2 (Fischer-Marsden (1975b), (1979a,b)) *Let* (V, g_V) *be a maximally developed spatially compact Ricci-flat spacetime. Let* (M, i) *be a compact orientable spacelike Cauchy hypersurface in* (V, g_V), *and let* $(g,\pi) \in C$ *be the induced Cauchy data on* M. *Then the following are equivalent:*

1. $D\Phi(g,\pi)$ *is surjective.*

2. $\ker(D\Phi(g,\pi))^* = 0$.

3. *The joint constraint equations*

$$\Phi(g,\pi) = (\mathcal{H}(g,\pi), \delta(g,\pi)) = 0$$

are linearization stable at (g,π).

4. *The spacetime* (V, g_V) *is linearization stable (i.e., Einstein's empty space equations are linearization stable at* g_V*).*

5. *the space* $\mathcal{E}^{sc}_{gh}(V)$ *of spatially compact globally hyperbolic Ricci-flat spacetimes on* V *is a manifold in a neighborhood of* g_V *with tangent space equal to*

$$T_{g_V}\mathcal{E}^{sc}_{gh} = \ker(D\mathrm{Ric}(g_V)) \approx \ker(D\Phi(g,\pi))$$

If (g,π) *satisfies conditions* $\mathrm{C}_{\mathcal{H}}, \mathrm{C}_{\delta}$, *and* C_{tr}, *then* $\ker(D\Phi(g,\pi))^* = 0$, *and so conclusions (1), (3), (4), and (5) also hold.*

Note that $\ker(D\mathrm{Ric}(g_V))$ is equal to the solution space of the linearized empty space Einstein equations.

The conditions $\mathrm{C}_{\mathcal{H}}, \mathrm{C}_{\delta}$, and C_{tr} are sufficient conditions to assure linearization stability of the ambient spacetime. The trace condition C_{tr} does not seem to be intrinsic but seems to be needed to show that the joint constraint equations are linearization stable. Thus it would be nice to have more intrinsic necessary and sufficient conditions for a spacetime to be linearization stable. Such conditions have been found by Moncrief ([1975a,b], [1976]).

For a spacetime (V, g_V), let

$$\mathcal{I}_{g_V}(V) = \{W \in \mathcal{X}(V) \mid L_W g_V = 0\}$$

denote the Lie algebra of Killing vector fields.

Theorem 6.3 (Moncrief ((1975a,b), (1976))) *Let* (V, g_V) *be a spatially compact globally hyperbolic Ricci-flat spacetime. Let* (M, i) *be a compact orientable spacelike Cauchy hypersurface in* (V, g_V), *and let* $(g,\pi) \in C$ *be the induced Cauchy data on* M. *Then*

$$\ker(D\Phi(g,\pi))^* \approx \mathcal{I}_{g_V}(V)$$

Thus (V, g_V) *is linearization stable if and only if* (V, g_V) *has no Killing vector fields.*

This gives a very natural condition for linearization stability in terms of the 4-geometry rather than in terms of the geometry of some arbitrarily chosen hypersurface. In particular, encoded in the above is the natural result that the dimension of the kernel of the adjoint map is independent of the choice of hypersurface on which it is evaluated, being equal to the dimension of the space of Killing vector fields of (V, g_V).

With these results in mind, we look for topologies of M which are incompatible with global continuous isometries for Ricci-flat spacetimes. Thus we are trying to link the topology of M with the manifold structure of $\mathcal{E}^{sc}_{gh}(V)$.

Theorem 6.4 *Let (V, g_V) be a spatially compact globally hyperbolic Ricci-flat spacetime. Let (M, i) be a compact orientable spacelike Cauchy hypersurface in (V, g_V), and let $(g, \pi) \in C$ be the induced Cauchy data on M. Assume that either of the conditions $C_{\mathcal{H}}$ or C_δ does not hold, i.e., that either:*

1. *g is flat and $\pi = 0$, in which case the spacetime is static; or*

2. *there exists a non-zero vector field X on M such that $L_X g = 0$ and $L_X \pi = 0$.*

Then (V, g_V) admits a Killing vector field, and thus is not linearization stable.

Conversely, assume that (V, g_V) admits a Killing vector field, and that (V, g_V) admits a compact orientable spacelike Cauchy hypersurface (M, i_o) with constant mean curvature. Then either (1) or (2) above is true.

Proof: From Theorem 6.1, if (1) holds, then $\ker(D\mathcal{H}(g,\pi)^*) \neq 0$, and if (2) holds, then $\ker(D\delta(g,\pi)^*) \neq 0$. But

$$\ker(D\mathcal{H}(g,\pi)^*) \times \ker(D\delta(g,\pi)^*) \subseteq \ker(D\Phi(g,\pi)^*)$$

so that in either case, $\ker(D\Phi(g,\pi)^*) \neq 0$. Thus from Theorem 6.3, (V, g_V) has a Killing vector field.

Conversely, if (V, g_V) has a Killing vector field, then on *any* compact orientable spacelike Cauchy hypersurface (M, i) with induced Cauchy data (g, π), $\ker(D\Phi(g,\pi)^*) \neq 0$. Thus if (V, g_V) has a constant mean curvature hypersurface (M, i_o), then for the Cauchy data (g_o, π_o) on this hypersurface, $\ker(D\Phi(g_o, \pi_o)^*) \neq 0$. Since on this hypersurface $\mathrm{tr}_{g_o}\pi_o' =$ constant, from Theorem 6.1, at least one of the conditions $C_{\mathcal{H}}$ or C_δ must fail. Thus either (1) or (2) above must hold. ∎

Remarks:

1. If (1) occurs, (M, g) is a compact connected orientable flat Riemannian 3-manifold, and hence a compact orientable Euclidean space form. In this case M is of flat type, and is diffeomorphic to one of the underlying manifolds $\{F_1, F_2, F_3, F_4, F_5, F_6\}$ of the six affine diffeomorphism classes of compact orientable Euclidean space forms.

2. If (2) occurs, then $\deg(M) \geq \dim \mathcal{I}_g(M) = \dim I_g(M) \geq 1$, and so from Proposition 2.3, M must admit an $\boldsymbol{SO}(2)$-action.

3. Cases (1) and (2) may both occur simultaneously. Under this circumstance, (M, g) is a compact orientable Euclidean space form with $\deg M \geq 1$. Then from Proposition 2.6, M must be diffeomorphic to one of the manifolds $F_i, 1 \leq i \leq 5$.

Now we can give sufficient topological conditions on a hypersurface M for a spacetime to be linearization stable. First we consider the constraint equations.

Theorem 6.5 *Let M be a compact connected orientable 3-manifold.*
If M is not of flat type, then

(1a) for all Cauchy data $(g,\pi) \in \mathcal{C}_{\mathcal{H}}$, condition $\mathrm{C}_{\mathcal{H}}$ is satisfied, and so conditions (1b), (1c), and (1d) of Theorem 6.1 are satisfied.

(1b) $\mathcal{C}_{\mathcal{H}}$ is a smooth submanifold of $T^*\mathcal{M}$.

If $\deg M = 0$, *then*

(2a) for all Cauchy data $(g,\pi) \in \mathcal{C}_{\delta}$, condition C_{δ} is satisfied, and thus conditions (2b), (2c), and (2d) of Theorem 6.1 are satisfied.

(2b) $\mathcal{C}_{\delta}$ is a smooth submanifold of $T^*\mathcal{M}$.

Suppose $\deg(M) = 0$ *and M is not diffeomorphic to F_6. Then for all Cauchy data $(g,\pi) \in \mathcal{C}$, the conditions $\mathrm{C}_{\mathcal{H}}$ and C_{δ} are satisfied. Thus if, additionally, $tr_g\pi' =$ constant, then*

(3a) the conditions (3a), (3b), and (3c) of Theorem 6.1 are satisfied.

(3b) $\mathcal{C}$ is a smooth submanifold of $T^*\mathcal{M}$ in a neighborhood of (g,π).

Proof: If M is not of flat type, then condition $\mathrm{C}_{\mathcal{H}}$ must be satisfied globally on $T^*\mathcal{M}$. Thus conditions (1b), (1c), and (1d) of Theorem 6.1 are satisfied, and $\mathcal{C}_{\mathcal{H}}$ is globally a smooth submanifold of $T^*\mathcal{M}$.

If $\deg(M) = 0$, then condition C_{δ} is satisfied globally on $T^*\mathcal{M}$. Thus conditions (12b), (2c), and (2d) of Theorem 6.1 are satisfied, and $\mathcal{C}_{\delta}$ is globally a smooth submanifold of $T^*\mathcal{M}$. Moreover, from Proposition 2.6, M cannot be diffeomorphic to F_i, $1 \leq i \leq 5$. Thus if $\deg(M) = 0$ and M is not diffeomorphic to F_6, then M is not of flat type, and so both conditions $\mathrm{C}_{\mathcal{H}}$ and C_{δ} are satisfied globally on $T^*\mathcal{M}$. Thus if $\mathrm{tr}_g\pi' =$ constant, then conditions $\mathrm{C}_{\mathcal{H}}$, C_{δ}, and C_{tr} are satisfied at (g,π), and so (3a), (3b), and (3c) of Theorem 6.1 are satisfied, and so $\mathcal{C}$ is a smooth submanifold of $T^*\mathcal{M}$ in a neighborhood of (g,π). ∎

What is somewhat remarkable is that topological conditions on M force the constraint subsets $\mathcal{C}_{\mathcal{H}}$, $\mathcal{C}_{\delta}$, and (locally) $\mathcal{C}$ of $T^*\mathcal{M}$ to be smooth submanifolds. The $\deg(M) = 0$ manifolds are identified in Theorem 4.2.

Now we can give sufficient topological conditions on a spacelike Cauchy hypersurface to insure that a spacetime is linearization stable.

Theorem 6.6 *Let (V, g_V) be a spatially compact globally hyperbolic Ricci-flat spacetime that admits a compact orientable spacelike Cauchy hypersurface (M, i_o) of constant mean curvature. Suppose that* $\deg(M) = 0$ *and that M is not diffeomorphic to F_6. Then*

1. $\mathcal{I}_{g_V}(V) = 0$

2. (V, g_V) *is linearization stable*

3. $\mathcal{E}^{sc}_{gh}(V)$ *is a smooth ILH-manifold in a neighborhood of* g_V.

Proof: From Theorem 6.5, the Cauchy data (g_o, π_o) on (M, i_o) satisfy conditions $C_{\mathcal{H}}$, C_{δ}, and C_{tr}, so from Theorem 6.2, the result follows. ∎

Thus we have found that the topological conditions deg(M) $= 0$ and M not diffeomorphic to F_6 are sufficient to exclude continuous isometries of Ricci-flat spacetimes that admit constant mean curvature Cauchy hypersurfaces with the topology of M, and which, in turn, insures that these spacetimes are linearization stable. We now go on to examine if these topological conditions are sufficient to do reduction of the dynamical formulation for these spacetimes.

7 Reduction in the case of $2+1$ spacetimes

In this section we review the process of reduction as it applies to $2+1$ spacetimes as given in Moncrief ([1989], [1990]). This case will serve as a model for our considerations regarding the $3+1$ case, discussed in the next section.

In $2+1$ dimensions, the Ricci-flat spacetimes are flat, but there is still considerable non-triviality of such spacetimes (see Moncrief [1989] for a discussion). The $2+1$ spacetimes that we shall be considering are thus globally hyperbolic spatially compact flat spacetimes with a constant mean curvature orientable spacelike Cauchy hypersurface. Thus the Cauchy hypersurfaces in these spacetimes are compact connected orientable 2-manifolds, and so their topological types consist of three classes, namely, the topology of the sphere $\boldsymbol{S}^2$, the torus $\boldsymbol{T}^2$, and all higher genus orientable surfaces $\boldsymbol{\Sigma}_p$, where $p \geq 2$ is the genus of the surface.

The first two classes require special treatment, as the sphere case is vacuous, as there are no globally hyperbolic flat spacetimes containing a constant mean curvature spacelike Cauchy hypersurface with topology $\boldsymbol{S}^2$. In the torus case, there are such globally hyperbolic flat spacetimes containing a constant mean curvature spacelike Cauchy hypersurface with topology $\boldsymbol{T}^2$, but in this case the isometry groups of different solutions change in dimension. Thus the generic solution has isometry group whose connected component of the identity is $\boldsymbol{U}(1) \times \boldsymbol{U}(1)$, but there also exist exceptional stationary solutions whose isometry group has connected component of the identity $\boldsymbol{U}(1) \times \boldsymbol{U}(1) \times \boldsymbol{R}$. Such a jump in dimension in the isometry group causes problems in the construction of the reduced phase space for the torus. These problems are analogous to the problems that arise in the construction of superspace (see Fischer [1970]).

However, these problems can be overcome, and the reduced phase space for the torus case can be constructed explicitly. The resulting reduced phase space includes conical singularities at those points which correspond to Cauchy data for solutions with exceptional symmetry. This is the analogue in $2+1$ dimensions of the *linearization instability* result for Einstein's equations, which asserts that the constraint subset admits conical singularities at points whose neighborhoods contain solutions of lower dimensional symmetry (see Fischer, Marsden, and Moncrief [1980] and Arms, Marsden, and Moncrief [1982]). In the above example, every solution having exceptional symmetry lies at the boundary of a manifold of solutions having generic symmetry. The rigidity of Einstein's field equations in $2+1$ dimensions, which forces all solutions to be flat, prohibits a further breaking of this symmetry down from the isometry group $\boldsymbol{U}(1) \times \boldsymbol{U}(1)$, which is shared by all the flat spacetimes that admit a spacelike Cauchy hypersurface with $\boldsymbol{T}^2$ topology.

Now we consider the third class consisting of all higher genus surfaces $\boldsymbol{\Sigma}_p$, $p \geq 2$. By contrast with the first two classes, the isometry group of every flat spacetime on $\boldsymbol{\Sigma}_p \times \boldsymbol{R}$ can be shown to have no non-zero Killing vector fields, and thus its isometry group is discrete. Consequently, all of the higher genus surfaces can be treated in a uniform fashion. The absence of such continuous symmetries guarantees that the solution set of the constraint equations is everywhere a manifold and leads one to anticipate that the reduced phase space is a manifold as well.

Thus we consider the Teichmüller space of conformal structures

$$\frac{\mathcal{M}/\mathcal{P}}{\mathcal{D}_0} \approx \mathcal{M}_{-1}/\mathcal{D}_0$$

for $\boldsymbol{\Sigma}_p$, $p \geq 2$. Since in two dimensions $R(g) = -1$ implies that g has constant negative sectional curvature, from Frankel's Theorem 5.4, $\mathcal{D}_0$ acts freely on $\mathcal{M}_{-1}$. Thus $\mathcal{M}_{-1}/\mathcal{D}_0$ is a smooth manifold, diffeomorphic to $\mathcal{T}$, and the projection

$$\pi_{-1} : \mathcal{M}_{-1} \longrightarrow \mathcal{M}_{-1}/\mathcal{D}_0$$

is a $\mathcal{D}_0$-principle fiber bundle over $\mathcal{M}_{-1}/\mathcal{D}_0$. Moreover, continuing with this Riemannian approach to Teichmüller theory, one can show that $\mathcal{M}_{-1}/\mathcal{D}_0$ is a $6p-6$-dimensional Euclidean cell (see Eells and Earle [1969], Fischer-Tromba ([1984a,b,c], [1987]), and Tromba [1992] for more information regarding this Riemannian approach to Teichmüller theory). Thus the principle fiber bundle

$$\mathcal{M}_{-1} \longrightarrow \mathcal{M}_{-1}/\mathcal{D}_0$$

admits a global cross section

$$\chi : \mathcal{M}_{-1}/\mathcal{D}_0 \longrightarrow \mathcal{M}_{-1}$$

which can, in principle be constructed through the use of the existence theory for harmonic maps between compact Riemannian surfaces (see Eells and Earle [1969]). Let

$$\tilde{\mathcal{T}} \equiv \chi(\mathcal{M}_{-1}/\mathcal{D}_0) \subseteq \mathcal{M}_{-1}$$

denote the image of χ in $\mathcal{M}_{-1}$. Then $\tilde{\mathcal{T}}$ is a smooth submanifold of $\mathcal{M}_{-1}$ diffeomorphic to $\mathcal{T} \approx \mathcal{M}_{-1}/\mathcal{D}_0$.

Now $\mathcal{T}$ is the configuration space that is of interest in reduction in the 2 + 1-dimensional case, and its cotangent bundle $T^*\mathcal{T}$ plays the role of the reduced phase space. Now $T^*\mathcal{T}$ is bundle-isomorphic to the cotangent bundle $T^*\tilde{\mathcal{T}}$ of $\tilde{\mathcal{T}}$. This latter cotangent bundle consists of all pairs (g, π) where $g \in \tilde{\mathcal{T}}$ and where $\pi \in (S^2_d)^{TT}_g$, the space of 2-contravariant symmetric tensor densities that are transverse (i.e., divergence-free) and traceless with respect to g. Thus

$$T^*\mathcal{T} \approx T^*\tilde{\mathcal{T}} = \bigcup_{g \in \tilde{\mathcal{T}}} (S^2_d)^{TT}_g$$

Using these facts together with the Lichnerowicz, Choquet-Bruhat, York conformal method (see Choquet-Bruhat and York [1980]) for solving the constraints on a constant mean curvature spacelike Cauchy hypersurface, one shows that the equivalence classes of solutions to the constraint equations, where equivalence is taken with respect to gauge transformations, is represented by a point (g, π) in $T^*\tilde{\mathcal{T}}$, and that this bijection is unique up to possible discrete isometries not isotopic to the identity diffeomorphism of Σ_p. Here the mean curvature constant condition fixes the time gauge (i.e., the choice of initial slice), whereas taking the $\mathcal{D}_0$-quotient mods out the spatial gauge transformations.

We remark that had the quotient of $\mathcal{M}_{-1}$ been taken with respect to $\mathcal{D}$ rather than $\mathcal{D}_0$, we would have been led to Riemann moduli space in place of Teichmüller space as the reduced configuration space. However, Riemann moduli space is not a manifold inasmuch as it has "orbifold-type" singularities at those $[g]$ admitting discrete isometries not isotopic to the identity. Thus from a mathematical point of view, working with Teichmüller space gives better results than working with Riemann moduli space. This advantage however may disappear in the 3 + 1 case (see Section 8). We also remark that there are also physical reasons to prefer working with Teichmüller space rather than Riemann moduli space (see the discussion in Friedman and Witt [1986], Witt [1986]), although these reasons are related to quantum rather than classical general relativity.

If a vacuum spacetime admits a compact spacelike Cauchy hypersurface of constant mean curvature, then in a tubular neighborhood of this initial slice, it admits a foliation by spacelike Cauchy surfaces of constant mean curvature (Choquet-Bruhat, Fischer, and Marsden [1979], Choquet-Bruhat and York [1980]). When the spacetime is not static (a case which cannot occur for the higher genus surfaces), the mean curvature varies monotonically throughout the foliation and thus can be chosen as a convenient time function. The (time-dependent) Hamiltonian which generates the corresponding flow on the reduced phase space $T^*\tilde{\mathcal{T}}$ can easily be shown to be the area functional of the mean curvature constant hypersurfaces re-expressed in terms of the cotangent bundle coordinates through the solution of the constraints (see Moncrief ([1989], [1990]) for details).

8 Reduction in the case of $3+1$ spacetimes

In this section we sketch a program on how the reduction in $2+1$ dimensions can be generalized to $3+1$ dimensions. The first consideration is to find the analogue to the surfaces Σ_p, $p \geq 2$. These surfaces have three desirable properties. Firstly, they are "topologically" incompatible with global continuous isometries for the solutions of the vacuum Einstein equations on $\Sigma_p \times \mathbf{R}$. This fact insured that the solution set of globally hyperbolic flat spacetimes were linearization stable. Secondly, their Teichmüller space $\mathcal{T} \approx \mathcal{M}_{-1}/\mathcal{D}_0$ is a manifold. Thirdly, the bundle $\pi_{-1} : \mathcal{M}_{-1} \to \mathcal{T}$ has a global cross-section χ, so that $\mathcal{T}$ could be represented as a submanifold $\tilde{\mathcal{T}} = \chi(\mathcal{T})$ of $\mathcal{M}_{-1}$.

Thus we are looking to find topological conditions on 3-manifolds M that reproduce as far as possible these optimal properties of Σ_p, $p \geq 2$.

As we have seen in Theorem 6.6, as far as linearization stability considerations go, the class of compact orientable connected 3-manifolds with $\deg(M) = 0$ and $M \not\approx F_6$ generates constant mean curvature spacetimes which are linearization stable, and which are points of regularity in $\mathcal{E}^{sc}_{gh}(V)$. Thus we suspect that these topological conditions on M should be sufficient to carry out reduction as in the $2+1$ dimensional case. However, two immediate problems arise inasmuch as the class of 3-manifolds with $\deg(M) = 0$ and $M \not\approx F_6$ contains both Yamabe type -1 and $+1$ manifolds (see the remarks following Theorem 4.2). For example, from Theorems 4.2 and 3.4, a non-prime connected sum of spherical space forms and handles containing at least one spherical space form that is not a lens space has $\deg(M) = 0$ and is of Yamabe type $+1$. On the other hand, under the assumption of the Poincaré and B conjectures, the only $\deg(M) = 0$ Yamabe type 0 manifold is F_6, so by excluding F_6, we are excluding Yamabe type 0 manifolds (see below). The first problem that arises is that for manifolds of Yamabe type $+1$, if a metric is pointwise conformal to a metric of scalar curvature $+1$, the choice of the conformal representative is not in general unique. Consequently, there is currently no way to represent $\mathcal{M}/\mathcal{P}$ for Yamabe type $+1$ manifolds as a submanifold of $\mathcal{M}$.

The second difficulty is that the space $\mathcal{M}_{+1}$ may have certain non-manifold points at those g such that $\ker(DR(g))^* \neq 0$ (see Fischer-Marsden [1975a] for details). These problems are presumably related, with the non-uniqueness in the conformal representation "causing" the occurrence of non-manifold points of $\mathcal{M}_{+1}$.

While there may be a way of avoiding these problems (see Section 9), here we temporarily exclude manifolds of Yamabe-type $+1$. The manifolds which remain are then the $\deg(M) = 0$ non-F_6 manifolds not of Yamabe type $+1$. Assuming the Poincaré and B conjectures, F_6 is the only $\deg(M) = 0$ manifold of Yamabe type 0, so this class is precisely the class of $\deg(M) = 0$ manifolds of Yamabe type -1.

For this class of manifolds, we have seen (Theorem 5.2) that the Teichmüller space of conformal structures $\mathcal{T} \equiv \frac{\mathcal{M}/\mathcal{P}}{\mathcal{D}_0}$ is homeomorphic to $\mathcal{M}_{-1}/\mathcal{D}_0$ and is an orbifold.

Thus for these manifolds we recover some of the optimal properties that occur in the 2-dimensional case.

Thus currently our candidate manifolds are compact connected orientable 3-manifolds M with $\deg(M) = 0$ and Yamabe type -1. More explicitly, combining Theorems 3.6 and 4.2, we have the following description of the candidate manifolds:

Theorem 8.1 *Let M be a compact connected orientable 3-manifold with* $\deg(M) = 0$ *and of Yamabe type -1. Assume that the Poincaré and B conjectures are true. Then either*

1. *M is a compact orientable $K(\pi, 1)$-manifold not of flat type and whose fundamental group π does not have an infinite cyclic center (such a manifold is a prime manifold); or*

2. *M is a non-prime manifold containing a compact orientable $K(\pi, 1)$-factor.*

Proof: The manifolds in (1) and (2) are the overlap of the manifolds of Yamabe type -1 and $\deg(M) = 0$ (Theorems 3.6 and 4.2). ∎

Thus for a compact connected orientable manifold M of Yamabe type -1 and with $\deg(M) = 0$, any metric g on M is uniquely conformal to a metric pg having scalar curvature $R(pg) = -1$, and any globally hyperbolic spatially compact Ricci-flat spacetime containing M as a spacelike Cauchy hypersurface of constant mean curvature must be linearization stable (Theorem 6.6).

For such manifolds, we are left with two questions:

1. Under what circumstances is $\mathcal{T}$ a manifold?

2. When $\mathcal{T}$ is a manifold, under what circumstances does the $\mathcal{D}_0$-principle fiber bundle $\pi_{-1} : \mathcal{M}_{-1} \to \mathcal{M}_{-1}/\mathcal{D}_0$ have a global cross-section.

Regarding the first question, the "orbifold-type" singularities that may occur in $\mathcal{M}_{-1}/\mathcal{D}_0$ are due to the occurrence of non-trivial discrete isometries of $g \in \mathcal{M}_{-1}$ lying in $\mathcal{D}_0$, i.e., they are due to the possibility that

$$I_g(M) \cap \mathcal{D}_0 \neq \emptyset$$

By Frankel's Theorem 5.4, this does not occur if g has sectional curvature $K(g) \leq 0$ and Ricci curvature $\mathrm{Ric}(g)$ negative definite. But for more general $g \in \mathcal{M}_{-1}$ that do not satisfy these inequalities, such a result is not known.

Thus, in contrast with the 2-dimensional case Σ_p, $p \geq 2$, the reduced configuration space $\mathcal{T}$ for $\deg(M) = 0$ and Yamabe class -1 need not be a manifold. Generically, however, on an open dense subset of the submanifold $\mathcal{M}_{-1}$, discrete isometries will be absent and the quotient space will be well-behaved. In particular it will be non-singular throughout the subspace satisfying Frankel's inequalities.

We remark that if discrete isometries isotopic to the identity do occur for such manifolds, then the mathematical advantage of working with $\mathcal{T} \approx \mathcal{M}_{-1}/\mathcal{D}_0$ rather than the more complicated space $\mathcal{M}_{-1}/\mathcal{D}$ is diminished, and one might then prefer to accept $\mathcal{M}_{-1}/\mathcal{D}$ as the reduced configuration space, thereby working with an analogue of Riemann moduli space rather than with Teichmüller space.

Regarding the second question above, in the 2-dimensional case of Σ_p, $p \geq 2$, the triviality of the $\mathcal{D}_0$-principle fiber bundle $\mathcal{M}_{-1} \to \mathcal{M}_{-1}/\mathcal{D}_0$ was established by the use of harmonic maps (see Eells-Earle [1969], and a private communication by Sampson) which in that case were shown to be global diffeomorphisms. In the 3-dimensional case, even if $\mathcal{T}$ were a manifold, an analogous approach would need additional curvature conditions to even get the existence of a unique harmonic map, and so the success of such an approach seems remote. However, such an approach might be successful under the curvature conditions of Frankel's theorem, thereby providing local cross-sections in patches of $\mathcal{M}_{-1}/\mathcal{D}_0$. Indeed, it would be interesting if the curvature conditions of Frankel's theorem were sufficient to provide local cross-sections using a harmonic map approach.

Nevertheless, away from the singular points of $\mathcal{M}_{-1}/\mathcal{D}_0$ we can produce local cross-sections of $\mathcal{M}_{-1} \to \mathcal{M}_{-1}/\mathcal{D}_0$, which for $g \in \mathcal{M}_{-1}$, are modeled on the space $(S_2)^{TT}_g$ of transverse traceless (with respect to g) 2-covariant symmetric tensor fields on M. These local cross-sections yield submanifolds transversal to the orbits of $\mathcal{D}_0$ in $\mathcal{M}_{-1}$.

Thus, in these charts, away from singular points, we can take the formal "L^2-cotangent bundle" $T^*\mathcal{T}$ as representing the reduced phase space on which to write the reduced dynamical equations. This bundle is represented locally as pairs (g, π) where g is a metric lying in some convenient local cross section of $\mathcal{M}_{-1}$ and $\pi \in (S^2_d)^{TT}_g$. Using the standard conformal methods for solving the constraint equations in the constant mean curvature case, it can be shown that such pairs parameterize the solution set of the Einstein constraint equations modulo gauge transformations, as in the $2+1$-dimensional case.

It is now straightforward to show that the (time-dependent) Hamiltonian which generates the induced flow on $T^*\mathcal{T}$ relative to the choice of the time function given by the constant mean curvature slicing is simply the volume functional of the constant mean curvature hypersurfaces. This functional is expressed through the solution of the constraints and the imposition of the gauge condition defining the chosen cross section of $\mathcal{M}_{-1}$ in terms of "coordinates" on $T^*\mathcal{T}$ and the constant mean curvature "time". Thus, modulo the singular points of $\mathcal{M}_{-1}/\mathcal{D}_0$ and the issue of determining a global cross-section of $\mathcal{M}_{-1} \to \mathcal{M}_{-1}/\mathcal{D}_0$, we find that we can extend the analysis for the $2+1$-dimensional to the $3+1$-dimension case.

Further details of the constructions involved here for the $3+1$ formulation will appear elsewhere.

9 Further work

Thus it appears that one can define a naturally reduced Hamiltonian system for $\deg(M) = 0$ manifolds of Yamabe type -1 much as for surfaces with genus $p \geq 2$. But, as we have seen from the considerations regarding linearization stability, if M has $\deg(M) = 0$ and is non-F_6, then $\mathcal{E}^{sc}_{gh}(V)$ is a manifold in a neighborhood of any spatially compact globally hyperbolic Ricci-flat spacetime g_V with a constant mean curvature spacelike Cauchy hypersurface M. Thus we know that for this class of spacetimes, there is some reduced manifold on which Einstein's equations can be written as a Hamiltonian system. However, at present, the conformal method of reduction only appears to work well if we further restrict this class of spacetimes to those spatially compact globally hyperbolic Ricci-flat spacetimes with a constant mean curvature spacelike Cauchy hypersurface M with $\deg(M) = 0$ and with the additional condition that M be of Yamabe type -1. Thus we conjecture that there is some other method of reduction that works for the full class of non-F_6 and $\deg(M) = 0$ manifolds.

Thus the manifolds for which we expect some form of reduction to work but for which the conformal method of reduction at present cannot be fully implemented are the $\deg(M) = 0$ manifolds of Yamabe class $+1$. From Theorems 3.6 and 4.2, assuming the Poincaré and B conjectures, such manifolds are connected sums of orientable spherical types and handles, provided the sum has at least two factors where one of the factors is of spherical type but is not a lens space (see also the remarks following Theorem 4.2).

As a possible approach to dealing with these manifolds, consider Theorem 5.5 which in the Yamabe type -1 case gives an alternate representation of $\mathcal{M}/\mathcal{P} \approx \mathcal{M}_{-1}$ as $\mathcal{M}^1_{\boldsymbol{R}^-}$ rather than as $\mathcal{M}_{-1}$. This alternate representation is introduced as a possible way to generalize these conformal methods to Yamabe type $+1$ manifolds, as the space $\mathcal{M}^1_{\boldsymbol{R}^-}$ naturally generalizes to $\mathcal{M}^1_{\boldsymbol{R}}$.

Unfortunately, essentially the same problems as before arise. Firstly, for Yamabe type $+1$ manifolds, the space $\mathcal{M}^1_{\boldsymbol{R}}$ may have non-manifold points. Secondly, there is not a unique representative pg in each conformal class $\langle g \rangle$ that satisfies $R(pg) = 1$, even with the condition of fixed total volume, as in general there are inequivalent conformal factors which can send a metric of the positive Yamabe class to one of constant scalar curvature and fixed volume. Thus the projection $\pi^1_{\boldsymbol{R}} : \mathcal{M}^1_{\boldsymbol{R}} \to \mathcal{M}/\mathcal{P}$ is not injective, as it is in the Yamabe -1 case (see Section 5), so that $\mathcal{M}^1_{\boldsymbol{R}}$ is not a representation of $\mathcal{M}/\mathcal{P}$. Thus we are looking for a cross-section to $\pi^1_{\boldsymbol{R}}$, whose image will then be in bijective correspondence with $\mathcal{M}/\mathcal{P}$.

To construct such a cross-section, we must make some choice to render a conformal representative in each class unique. One possible approach is to see under what conditions a minimizer of the Yamabe functional (see, for example Lee and Parker [1987]) is unique. In general a minimizer is not unique since, for example, a metric g with a non-isometric conformal isometry $f^*g = pg, p \neq 1$, pulls back under this conformal isometry to the conformally related metric pg with the same value of the Yamabe functional, since the Yamabe functional is invariant with respect to diffeomorphisms of M. Thus in these cases, the minimizer of the Yamabe functional cannot be unique. Moreover, by Theorem 5.3, if $\deg(M) = 0$, then the conformal group of any metric on M must be finite. Thus even if $\deg(M) = 0$, in the presence of non-trivial finite conformal groups, the minimizer of the Yamabe functional will not yield a unique choice of conformal representative within each conformal class. Thus further choices would have to be made in order to get a cross-section χ of $\pi^1_{\boldsymbol{R}}$ and a representation $\chi(\mathcal{M}/\mathcal{P}) \subseteq \mathcal{M}^1_{\boldsymbol{R}}$ of $\mathcal{M}/\mathcal{P}$.

We have noted that $\mathcal{M}^1_{\boldsymbol{R}}$ may have non-manifold points at those g such that $\ker(DR(g))^* = 0$. Another criterion for constructing a cross-section $\chi : \mathcal{M}/\mathcal{P} \to \mathcal{M}^1_{\boldsymbol{R}}$ might be to require that χ be chosen so as to avoid the singularities in $\mathcal{M}^1_{\boldsymbol{R}}$, so that the image $\chi(\mathcal{M}/\mathcal{P}) \subseteq \mathcal{M}^1_{\boldsymbol{R}}$ would be a smooth submanifold of $\mathcal{M}$ (and then would be a submanifold representation of $\mathcal{M}/\mathcal{P}$, analogous to $\mathcal{M}_{-1}$ for the Yamabe -1 case).

A further criteria for χ should be that the resulting space $\chi(\mathcal{M}/\mathcal{P}) \subseteq \mathcal{M}^1_{\boldsymbol{R}}$ must be invariant by $\mathcal{D}_0$, since the next step would involve taking the quotient

$$(\chi(\mathcal{M}/\mathcal{P}))/\mathcal{D}_0$$

for the reduced configuration space. Finally, we remark that the difficulties involving "orbifold" singularities and the existence of a global cross-section, as for $\mathcal{M}_{-1}/\mathcal{D}_0$ in the Yamabe type -1 case, persist in the above construction.

Acknowledgements: Vincent Moncrief would like to acknowledge support for this research by NSF Grants PHY-9201196 and INT-9015153 to Yale University. He would also like to acknowledge the hospitality and support provided by Université Pierre-et-Marie-Curie, the Aspen Institute for Physics, and the Institute for Theoretical Physics at the University of California, Santa Barbara, where part of this research was carried out.

References

[1] Arms, J, Marsden, J, and Moncrief, V (1982), *The structure of the space of solutions of Einstein's equations. II. Many Killing vector fields*, Ann. Phys. **144**, 81-106.

[2] Arnowitt, R, Deser, S, and Misner, C (1962), *The dynamics of general relativity*, in *Gravitation: an introduction to current research*, ed. by L Witten, John Wiley and Sons, Inc., New York.

[3] Aubin, T (1970), *Métrique riemanniennes et courbure*, J. Diff. Geom. **4**, 383-424.

[4] Aubin, T (1982), *Nonlinear Analysis on Manifolds. Monge-Ampère Equations*, Springer-Verlag, New York.

[5] Bochner, S (1946), *Vector fields and Ricci curvature*, Bull. Amer. Math. Soc. **52**, 776-797.

[6] Bröcker, T, and Jänich, (1982), *Introduction to Differential Topology*, Cambridge University Press, Cambridge.

[7] Budic, R, Isenberg, J, Lindblom, L, and Yasskin, P (1978), *On the determinations of Cauchy surfaces from intrinsic properties*, Commun. Math. Phys. **61**, 87-101.

[8] Choquet-Bruhat, Y, Fischer, A, and Marsden, J (1979), *Maximal hypersurfaces and positivity of mass*, in *Proceedings of the International School of Physics "Enrico Fermi"*, Course LXVII, *Isolated Gravitating Systems in General Relativity*, ed. J Ehlers, North Holland Publishing Company, Amsterdam.

[9] Choquet-Bruhat, Y, and York, J (1980), *The Cauchy Problem*, in *General Relativity and Gravitation: Volume 1*, ed. A Held, Plenum Press, New York.

[10] Earle, C, and Eells, J (1969), *A fibre bundle description of Teichmüller theory*, J. Diff. Geom., **3**, 19-43.

[11] Ebin, D (1970), *The space of Riemannian metrics*, Proc. Symp. Pure Math., Amer. Math. Soc. **15**, 11-40.

[12] Edmonds, A (1985), *Transformation groups and low-dimensional manifolds*, in *Group Actions on Manifolds*, Contemporary Mathematics, Volume 36, ed. R Schultz, 339-366.

[13] Ellis, G (1971), *Topology and cosmology*, Gen. Rel. and Gravitation **2**, 7-21.

[14] Fischer, A (1970), *The theory of superspace*, in *Relativity*, eds. M Carmeli, S Fickler, and L Witten, Plenum Press, New York.

[15] Fischer, A, and Marsden, J (1972), *The Einstein equations of evolution - a geometric approach*, J. Math. Phys. **28**, 1-38.

[16] Fischer, A, and Marsden, J (1975a), *Deformations of the scalar curvature*, Duke Mathematical Journal **42**, 519-547.

[17] Fischer, A, and Marsden, J (1975b), *Linearization stability of nonlinear partial differential equations*, Proc. Symp. Pure Math., Amer. Math. Soc. **27**, 219-263.

[18] Fischer, A, and Marsden, J (1977), *The manifold of conformally equivalent metrics*, Canadian Journal of Mathematics, **1**, 193-209.

[19] Fischer, A, and Marsden, J (1979a), *The initial value problem and the dynamical formulation of general relativity*, in *General Relativity, an Einstein Centenary Volume*, eds. S Hawking and W Israel, Cambridge University Press, Cambridge, England.

[20] Fischer, A, and Marsden, J (1979b), *Topics in the dynamics of general relativity*, in *Proceedings of the International School of Physics "Enrico Fermi"*, Course LXVII, *Isolated Gravitating Systems in General Relativity*, ed. J Ehlers, North Holland Publishing Company, Amsterdam.

[21] Fischer, A, Marsden, J, and Moncrief, V, (1980), *The structure of the space of solutions of Einstein's equations I. One Killing field*, Ann. Inst. Henri Poincaré **33**, 147-194.

[22] Fischer, A, and Tromba, A (1984a), *On a purely "Riemannian" proof of the structure and dimension of the unramified moduli space of a compact Riemann surface*, Mathematische Annalen **267**, 311-345.

[23] Fischer, A, and Tromba, A (1984b), *Almost complex principle fiber bundles and the complex structure on Teichmüller space*, J. für die reine und angewandte Mathematik **352**, 151-160.

[24] Fischer, A, and Tromba, A (1984c), *On the Weil-Petersson metric on Teichmüller space*, Trans. Amer. Math. Soc. **284**, 319-335.

[25] Fischer, A, and Tromba, A (1987), *A new proof that Teichmüller space is a cell*, Trans. Amer. Math. Soc. **303**, 257-262.

[26] Fischer, A, and Wolf, J A (1974), *The Calabi construction for compact Ricci flat Riemannian manifolds*, Bull. Amer. Math. Soc. **80**, 92-97.

[27] Fischer, A, and Wolf, J A (1975), *The structure of compact Ricci-flat Riemannian manifolds*, J. Diff. Geom. **10**, 277-288.

[28] Frankel, T (1966), *On a theorem of Hurwitz and Bochner*, J. Math. and Mechanics, **15**, 373-377.

[29] Friedman, J, and Witt, D (1986), *Homotopy is not isotopy for homeomorphisms of 3-manifolds*, Topology **25**, 35-44.

[30] Gromov, M, and Lawson, Jr., H B (1980a), *Spin and scalar curvature in the presence of a fundamental group. I*, Annals of Mathematics, **111**, 209-230.

[31] Gromov, M, and Lawson, Jr., H B (1980b), *The classification of simply connected manifolds of positive scalar curvature*, Annals of Mathematics, **111**, 423-434.

[32] Gromov, M, and Lawson, Jr., H B (1983), *Positive scalar curvature and the Dirac operator on complete Riemannian manifolds*, Institut des Hautes Études Scientifiques, Publications Mathématiques, Number 58, 83-196.

[33] Hawking, S, and Ellis, G (1973), *The Large Scale Structure of Space-Time*, Cambridge University Press, Cambridge, England.

[34] Hsiang, W (1967a), *The natural metric on $\boldsymbol{SO}(n)/\boldsymbol{SO}(n-1)$ is the most symmetric metric*, Bull. Amer. Math. Soc. **73**, 55-58.

[35] Hsiang, W (1967b), *On the bounds on the dimensions of the isometry groups of all possible riemannian metrics on an exotic sphere*, Ann. of Math. **85**, 351-358.

[36] Hsiang, W (1971), *On the degree of symmetry and the structure of highly symmetric manifolds*, Tamkang Journal of Mathematics, Tamkang College of Arts and Sciences, Taipei, **73**, 1-22.

[37] Hu, S-T (1959), *Homotopy Theory*, Academic Press, New York.

[38] Kazdan, J, and Warner, F (1975a), *Scalar curvature and conformal deformation of Riemannian structure*, J. Diff. Geom. **10**, 113-134.

[39] Kazdan, J, and Warner, F (1975b), *Existence and conformal deformation of metrics with prescribed Gaussian and scalar curvatures*, Annals of Mathematics, (2) **101**, 317-331.

[40] Kazdan, J (1985), *Prescribing the curvature of a Riemannian manifold*, Conference Board of the Mathematical Sciences, Regional Conference Series in Mathematics, Number 57, Amer. Math. Soc., Providence, Rhode Island.

[41] Kneser, H (1929), *Geschlossen Flächen in dreidimensionalen Mannigfaltigkeiten*,Jber. Deutsch. Math.-Verein **38**, 248-260.

[42] Kobayashi, S, and Nomizu, K (1963), *Foundations of Differential Geometry*, vol 1, Interscience, John Wiley and Sons, New York.

[43] Lee, J, and Parker, T (1987), *The Yamabe problem*, Bull. Amer. Math. Soc. **17**, 37-91.

[44] Lelong-Ferrand, J (1969), *Transformations conformes et quasiconformes des variétés riemanniennes; application à la démonstration d'une conjecture de A. Lichnerowicz*, C. R. Acad. Sci. Paris **269**, 583-586.

[45] Lelong-Ferrand, J (1971), *Transformations conformes et quasi-conformes des variétés riemanniennes compacts (démonstration de la conjecture de A. Lichnerowicz)*, Acad. Roy. Belg. Cl. Sci. Mem. Coll. 8°(2) **39**, no.5.

[46] Milnor, J (1961), *A unique decomposition theorem for 3-manifolds*, American Journal of Mathematics, 1-7.

[47] Moncrief, V (1975a), *Spacetime symmetries and linearization stability of the Einstein equations. I*, J. Math. Phys. **16**, 493-498.

[48] Moncrief, V (1975b), *Decompositions of gravitational perturbations*, J. Math. Phys. **16**, 1556-1560.

[49] Moncrief, V (1976), *Space-time symmetries and linearization stability of the Einstein equations. II*, J. Math. Phys. **17**, 1893-1902.

[50] Moncrief, V (1989), *Reduction of the Einstein equations in* $2+1$ *dimensions to a Hamiltonian system over Teichmüller space*, J. Math. Phys. **30** (12), 2907-2914.

[51] Moncrief, V (1990), *How solvable is* $(2+1)$*-dimensional Einstein gravity?*, J. Math. Phys. **31** (12), 2978-2982.

[52] Obata, M (1971), *The conjectures on conformal transformations of Riemannian manifolds*, J. Diff. Geom. **6**, 247-258.

[53] Omori, H (1970), *On the group of diffeomorphisms of a compact manifold*, Proc. Symp. Pure Math., Amer. Math. Soc. **15**, 167-183.

[54] Orlik, P, and Raymond, F (1968), *Actions of* $\boldsymbol{SO(2)}$ *on 3-manifolds*, in Proceedings of the Conference on Transformation Groups, New Orleans, 1967 Springer-Verlag, New York.

[55] Raymond, F (1968), *Classification of the action of the circle on 3-manifolds*, Trans. Amer. Math. Soc., **131**, 51-78.

[56] Schoen, R, and Yau, S T (1979a), *Existence of incompressible minimal surfaces and the topology of three dimensional manifolds with non-negative scalar curvature*, Annals of Mathematics, **110**, 127-142.

[57] Schoen, R, and Yau, S T (1979b), *The structure of manifolds with positive scalar curvature*, Manuscripta Math., **28**, 159-183.

[58] Spanier, E (1966), *Algebraic Topology*, McGraw-Hill Book Company, New York.

[59] Thurston, W (unpublished notes), *The geometry and topology of 3-manifolds*, preprint, Princeton University, Princeton, New Jersey.

[60] Tromba, A (1992), *Teichmüller Theory in Riemannian Geometry*, Birkhäuser Verlag, Basel.

[61] Witt, D (1986), *Symmetry groups of state vectors in quantum gravity*, J. Math. Phys. **27** (2),573-592.

[62] Wolf, J A (1972), *Spaces of Constant Curvature*, second edition, Publish or Perish Press, Berkeley, California.

Darboux Transformations for a Class of Integrable Systems in n Variables

GU CHAOHAO (C. H. GU)

University of Science and Technology
of China, Hefei, Anhui, China
and
Institute of Mathematics of Fudan
University, Shanghai, China

Dedicated to Prof. Yvonne Choquet-Bruhat

ABSTRACT. A class of integrable systems in R^n is introduced. It is shown that the Darboux transformation method is valid for obtaining new explicit solutions. A class of solitons in R^n is constructed and some further results are announced.

§ 1. INTRODUCTION

The theory of completely integrable systems which are closely related to the soliton phenomena has been studied extensively. One of the crucial problems is to find exact solutions. Among many beautiful approaches the Darboux transformations are very attractive [1,2]. The advantage is that the algorithm of Darboux transformation is universal and purely algebraic. Moreover, it can be used successively to obtain infinite sequences of exact solutions. In the case of lower dimensions (i.e. R^{1+1} or R^{1+2}) a plenty of exact solutions, in particular, multi-soliton solutions, were obtained[e.g. 2, 3,4,5]. In the case of R^{1+1} there is a unified explicit formula of Darboux matrices for AKNS systems[6,7]. The results were extended to the case of R^{1+2} [8].

For the case of higher dimensions, R. Beals and K. Tenenblat [9] studied two set of integrable systems which are called the intrinsic generalized wave equation (IGWE) and the intrinsic generalized sine-Gordon equation (IGSGE). They gave Bäcklund transformations together with the inverse scattering approach to these equations. However, there are no explicit formulas of exact solutions except the special case $n = 3$[10]. Inspired by their work, in the present paper, we consider a more general system and prove that the formula of Darboux matrix for the R^{1+1} AKNS systems is also valid for the system of higher dimensions. Moreover, a class of interesting special solutions is found explicitly. Such a solution consists of 3-waves propagated in R^{1+2} which behave as soliton-like solution in R^{1+2}.

The work is supported by the China National Program for fundamental research "nonlinear science" .

M. Flato et al. (eds.), Physics on Manifolds, 153–160.

In § 2, we formulate the completely integrable system in R^n which is the generalization of IGWE. § 3 is devoted to proving that the Darboux transformation approach is valid. In § 4 a reduction to the Lie algebra $su(m)$ is obtained. The 3-wave solutions are obtained in § 5. The soliton-like behaviour is elucidated. In § 6 we point out that for the integrable system which generalizes IGSGE the Darboux matrix method is valid as well. In § 7, we announce briefly some results for a class of integrable system in the space-time R^{1+n}, in particular, the interaction of solitons.

§ 2. The system of equations

Let

$$\frac{\partial \Psi}{\partial x_j} = (\lambda J_j + P_j)\Psi \qquad (i = 1, 2, \cdots, n) \tag{2.1}$$

be a system of equations on R^n, where J_j and P_j are $m \times m$ matrices and λ is a parameter. We assume that J_j are constant matrices.

The integrability conditions of system (2.1) are

$$[J_j, J_l] = 0 \tag{2.2}$$

$$[J_j, P_i] + [P_j, J_i] = 0 \tag{2.3}$$

and

$$\frac{\partial P_j}{\partial x_i} - \frac{\partial P_i}{\partial x_j} + [P_j, P_i] = 0 \tag{2.4}$$

Owing to (2.2), without loss of generalities, we may assume that J_j are diagonal matrices. We assume that J_j are linearly independent and the equalities $[J_j, A] = 0$, $(j = 1, 2, \cdots, n)$ imply that A is diagonal. From (2.3) it is seen that there exists an off-diagonal $m \times m$ matrix P uniquely such that

$$P_i = [P, J_i] \tag{2.5}$$

In fact, let

$$P_j = (\underset{j}{p^\alpha_\beta}), \qquad J_j = \mathrm{diag}(\underset{j}{\lambda^1}, \cdots \underset{j}{\lambda^m}) \tag{2.6}$$

(2.3) has the form

$$(\underset{i}{\lambda^\alpha} - \underset{i}{\lambda^\beta})\underset{j}{p^\alpha_\beta} = (\underset{j}{\lambda^\alpha} - \underset{j}{\lambda^\beta})\underset{i}{p^\alpha_\beta} \tag{2.7}$$

and we may take $P = (p^\alpha_\beta)$ with

$$p^\alpha_\beta = \frac{1}{\underset{i}{\lambda^\alpha} - \underset{i}{\lambda^\beta}}\underset{i}{p^\alpha_\beta}, \qquad (\alpha \neq \beta, \underset{i}{\lambda^\alpha} - \underset{i}{\lambda^\beta} \neq 0) \tag{2.8}$$

System (2.4) becomes

$$\frac{\partial [P, J_j]}{\partial x_i} - \frac{\partial [P, J_i]}{\partial x_j} + [[P, J_j], [P, J_i]] = 0 \tag{2.9}$$

The system is an overdetermined one. We will see later that it admits an infinite number of nontrivial solutions, besides the trivial solution $P = 0$.

§ 3. The Darboux transformations

Let P and $\Psi(\lambda)$ be a solution to (2.1). We want to construct matrix α such that

$$\Psi_1(\lambda) = (\lambda I + \alpha)\Psi(\lambda) \tag{3.1}$$

together with some matrix function P_1 satisfies (2.1)

From the requirement

$$\frac{\partial \Psi_1(\lambda)}{\partial x_j} = (\lambda J_j + P_j^1)\Psi_1(\lambda) \tag{3.2}$$

we obtain

$$P_j^1 = P_j + [\alpha, J_j] \tag{3.3}$$

and

$$\frac{\partial \alpha}{\partial x_j} = P_j^1 \alpha - \alpha P_j = [P_j, \alpha] + [\alpha, J_j]\alpha \tag{3.4}$$

For the AKNS systems there is an explicit formula for the matrix α obtained by Sattinger and the present author independently[6,7]. We are going to show that this formula is valid for the system (2.1) in R^n, too.

Let $\underset{\alpha}{h}$ be a column solution of the system (2.1) for some parameter $\lambda = \lambda_\alpha$ ($\alpha = 1, 2, \cdots, m$) and H is the $m \times m$ matrix

$$H = [\underset{1}{h}, \underset{2}{h}, \cdots, \underset{m}{h}] \tag{3.5}$$

Here, we require that $\det H \neq 0$ at least in some region and at least two of λ_α's are unequal. We have

$$\frac{\partial H}{\partial x_j} = [\lambda_1 J_j \underset{1}{h} + P_j \underset{1}{h}, \cdots, \lambda_m J_j \underset{m}{h} + P_j \underset{m}{h}] = J_j H \Lambda + P_j H \tag{3.6}$$

with

$$\Lambda = \begin{pmatrix} \lambda_1 & & & \\ & \lambda_2 & & \\ & & \ddots & \\ & & & \lambda_m \end{pmatrix} \tag{3.7}$$

Let

$$\alpha = -H\Lambda H^{-1} \tag{3.8}$$

Then,

$$\begin{aligned} \frac{\partial \alpha}{\partial x_j} &= -(J_j H\Lambda + P_j H)\Lambda H^{-1} + H\Lambda H^{-1}(J_j H\Lambda + P_j H)H^{-1} \\ &= -J_j\alpha^2 + P_j\alpha + \alpha J_j \alpha - \alpha P_j = [P_j, \alpha] + [\alpha, J_j]\alpha \end{aligned}$$

Hence (3.4) is satisfied.

From (3.3) we have

$$P^1 = P + \alpha^{\text{off}} \tag{3.9}$$

Here α^{off} is the off-diagonal part of α. (3.1) and (3.9) is called a Darboux transformation from (P, Ψ) to (P', Ψ_1). We have:

Theorem. *If (P,Ψ) is a solution to (2.1), then its Darboux transformation (P',Ψ_1) is a solution to (1) too.*

As in the case of R^{1+1} the Darboux transformations can be done successively and the algorithm is purely algebraic.

The construction of α depends on the parameters $\lambda_1,\cdots,\lambda_m$ and the column solutions $\underset{\alpha}{h}$. We may write

$$\underset{\alpha}{h} = \Psi(\lambda_\alpha)\underset{\alpha}{l}, \qquad (\alpha = 1,2,\cdots,m) \tag{3.10}$$

Here $\underset{\alpha}{l}$ are constant columns and we write

$$L = [\underset{1}{l},\underset{2}{l},\cdots,\underset{m}{l}] \tag{3.11}$$

We have the following diagram of the permutability theorem

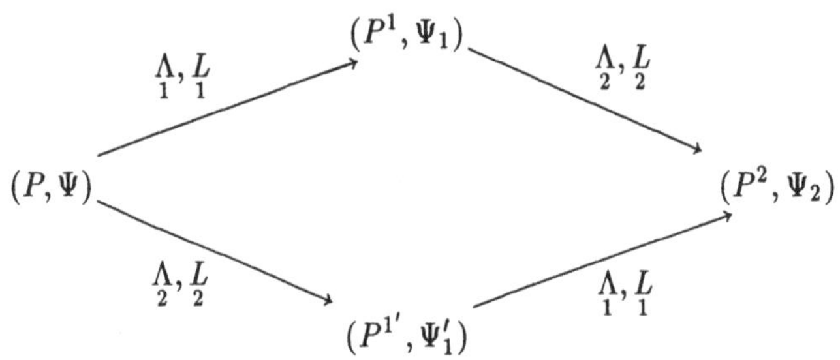

The proof of the permutability theorem is just as the case of AKNS systems in R^{1+1}[7].

§ 4. Reduction to $su(m)$ case.

We suppose that J_i are purely imaginary and P satisfies

$$P^+ + P = 0 \tag{4.1}$$

i.e. P is valued in the Lie algebra $su(m)$. As usual, the symbol $+$ means the conjugation and transpose of a matrix.

The reduction approach is similar to that for the principal chiral field [4].

Let μ be a non real complex number. We take

$$\lambda_\alpha = \mu \text{ or } \overline{\mu}, \qquad (\alpha = 1,2,\cdots,m) \tag{4.2}$$

We may take the column vectors $\underset{\alpha}{h}$ such that the set of columns $\underset{\alpha}{h}$'s with $\lambda_\alpha = \mu$ and the set of column $\underset{\beta}{h}$'s with $\lambda_\beta = \overline{\mu}$ are both linearly independent. Moreover, we choose $\underset{\alpha}{h}$'s such that

$$\underset{\alpha}{h}^+ \underset{\beta}{h} = 0 \qquad (\lambda_\alpha \neq \lambda_\beta) \tag{4.3}$$

at a given point. Then (4.3) holds everywhere since

$$\frac{\partial}{\partial x_j}(\underset{\alpha}{h}^+ \underset{\beta}{h}) = \underset{\alpha}{h}^+(-\overline{\lambda}_\alpha J_j + P_j^+)\underset{\beta}{h} + \underset{\alpha}{h}^+(\lambda_\beta J_j + P_j)\underset{\beta}{h} = 0 \tag{4.4}$$

Hence $\underset{1}{h}, \cdots \underset{m}{h}$ are linearly independent i.e. $\det H \neq 0$ everywhere. From $\alpha H = -H\Lambda$, we have

$$\alpha[\underset{1}{h}, \cdots, \underset{m}{h}] = -[\lambda_1 \underset{1}{h}, \cdots \lambda_m \underset{m}{h}]$$

and hence

$$\alpha \underset{\alpha}{h} = -\lambda_\alpha \underset{\alpha}{h}, \qquad \underset{\alpha}{h}^+ \alpha^+ = -\overline{\lambda}_\alpha \underset{\alpha}{h}^+ \tag{4.5}$$

Consequently,

$$\underset{\alpha}{h}^+ (\alpha^+ + \alpha) \underset{\beta}{h} = -(\lambda_\beta + \overline{\lambda}_\alpha) \underset{\alpha}{h}^+ \underset{\beta}{h} = -(\mu + \overline{\mu}) \underset{\alpha}{h}^+ \underset{\beta}{h} \tag{4.6}$$

since $\underset{\alpha}{h}^+ \underset{\beta}{h} = 0$ for the case $\lambda_\alpha \neq \lambda_\beta$. Hence,

$$\alpha^+ + \alpha = -(\mu + \overline{\mu}) I \tag{4.7}$$

in particular, we have

$$(\alpha^{\text{off}})^+ + \alpha^{\text{off}} = 0 \tag{4.8}$$

From (3.9) it is seen that $P^1 \in su(m)$ if $P \in su(m)$.

§ 5. An example and the soliton solution

Let $m = n = 3$ and

$$J_1 = \begin{bmatrix} i & & \\ & 0 & \\ & & 0 \end{bmatrix}, \qquad J_2 = \begin{bmatrix} 0 & & \\ & i & \\ & & 0 \end{bmatrix}, \qquad J_3 = \begin{bmatrix} 0 & & \\ & 0 & \\ & & i \end{bmatrix} \tag{5.1}$$

The trivial solution is

$$P = 0 \tag{5.2}$$

and

$$\Psi = \begin{bmatrix} e^{i\lambda x_1} & 0 & 0 \\ 0 & e^{i\lambda x_2} & 0 \\ 0 & 0 & e^{i\lambda x_3} \end{bmatrix} \tag{5.3}$$

Take $\lambda_1 = \mu$, $\lambda_2 = \lambda_3 = \overline{\mu}$ $(\mu \neq \overline{\mu})$. We choose $\underset{1}{h}, \underset{2}{h}, \underset{3}{h}$ such that

$$H = \begin{bmatrix} e^{i\mu x_1} & -\overline{a} e^{i\overline{\mu} x_1} & -\overline{b} e^{i\overline{\mu} x_1} \\ a e^{i\mu x_2} & e^{i\overline{\mu} x_2} & 0 \\ b e^{i\mu x_3} & 0 & e^{i\overline{\mu} x_3} \end{bmatrix} \qquad (a \neq 0, b \neq 0) \tag{5.4}$$

It is seen that (4.3) holds and

$$\det H = e^{i(\mu x_1 + \overline{\mu} x_2 + \overline{\mu} x_3)} \Delta \neq 0 \tag{5.5}$$

where $\Delta = 1 + |a|^2 e^{i(\mu - \overline{\mu})(x_2 - x_1)} + |b|^2 e^{i(\mu - \overline{\mu})(x_3 - x_1)}$. By calculations, we obtain

$$\begin{aligned} p = \alpha_3^2 &= a\overline{b}(\mu - \overline{\mu}) e^{i(\mu(x_2 - x_1) - \overline{\mu}(x_3 - x_1))} / \Delta \\ q = \alpha_3^1 &= \overline{b}(\mu - \overline{\mu}) e^{i(\overline{\mu}(x_1 - x_3))} / \Delta \\ r = \alpha_2^1 &= \overline{a}(\mu - \overline{\mu}) e^{i(\overline{\mu}(x_1 - x_2))} / \Delta \end{aligned} \tag{5.6}$$

Consequently, the Darboux transformation gives a non trivial solution

$$P^1 = \begin{bmatrix} 0 & r & q \\ -\bar{r} & 0 & p \\ -\bar{q} & -\bar{p} & 0 \end{bmatrix} \tag{5.7}$$

If we consider the variable x_1 as time t, and let $\xi = x_2 - x_1$, $\eta = x_3 - x_1$. The solution consists of three travelling waves with the velocity (1,1) in the (x_2, x_3)-plane.

Let $\mu = l + ik$. For simplicity, we assume $k > 0$. The amplitudes of the p, q, r waves are

$$\begin{aligned} |p| &= 2|ab|ke^{-k\xi-k\eta}/[1+|a|^2e^{-2k\eta}+|b|^2e^{-2k\xi}] \\ |q| &= 2|b|ke^{-k\eta}/[1+|a|^2e^{-2k\eta}+|b|^2e^{-2k\xi}] \\ |r| &= 2|a|ke^{-k\xi}/[1+|a|^2e^{-2k\eta}+|b|^2e^{-2k\xi}] \end{aligned} \tag{5.8}$$

The solution has the following soliton-like behaviour: As $|\xi|^2 + |\eta|^2 \to \infty$, $|p|$, $|q|$, $|r| \to 0$ rapidly, except the following cases.

(1) When $\xi \to +\infty$ and η keeps bounded, then $|q|$ keeps bounded.
(2) When $\eta \to +\infty$ and ξ keeps bounded, then $|r|$ keeps bounded.
(3) When $\xi + \eta \to -\infty$, $k(\xi - \eta) + \ln|b|/|a|$ keeps finite, then $|p|$ keeps bounded.

In particular, for fixed x_2 and x_3, $|p|$, $|q|$, $|r| \to 0$ rapidly as $t \to \pm\infty$.

Remark. In the present case the system of equations (2.4) is reduced to

$$\begin{aligned} &\frac{\partial p}{\partial x_1} + \frac{\partial p}{\partial x_2} + \frac{\partial p}{\partial x_3} = 0 \qquad \frac{\partial q}{\partial x_1} + \frac{\partial q}{\partial x_2} + \frac{\partial q}{\partial x_3} = 0, \qquad \frac{\partial r}{\partial x_1} + \frac{\partial r}{\partial x_2} + \frac{\partial r}{\partial x_3} = 0 \\ &\frac{\partial p}{\partial x_1} = i\bar{q}r \qquad \frac{\partial q}{\partial x_2} = -ipr \qquad \frac{\partial r}{\partial x_3} = -iq\bar{p} \end{aligned} \tag{5.9}$$

Hence, p, q, r are functions of ξ and η and the equation are reduced to

$$\frac{\partial p}{\partial \xi} + \frac{\partial p}{\partial \eta} = -i\bar{q}r, \qquad \frac{\partial q}{\partial \xi} = -ipr, \qquad \frac{\partial r}{\partial \eta} = -iq\bar{p} \tag{5.10}$$

Hence every solution is a travelling wave solution (consists of 3 waves) with the same velocity (1,1). There are no solutions with different velocities of propagation. Hence, we cannot say anything on the interaction of solitons. However, the successive Darboux transformations give a sequences of global solutions to the nonlinear wave equations (5.10). The case $n = m > 3$ can be treated as well.

§ 6. The generalization of IGSGE

We consider the more complicated system

$$\frac{\partial \Psi}{\partial x_j} - (\lambda J_j + \frac{1}{\lambda} L_j + P_j)\Psi \tag{6.1}$$

Besides P_j, the matrix function L_j are functions of $x_1, \cdots, x_n$ too. The conditions of integrability are

$$\begin{gathered}[J_j, P_i] + [P_j, J_i] = 0.\\ \frac{\partial P_j}{\partial x_i} - \frac{\partial P_i}{\partial x_j} + [P_j, P_i] + [J_j, L_i] + [L_j, J_i] = 0.\\ \frac{\partial L_j}{\partial x_i} - \frac{\partial L_i}{\partial x_j} + [L_j, P_i] + [P_j, L_i] = 0.\\ [L_j, L_i] = 0.\end{gathered} \tag{6.2}$$

which can be considered as equations for P_i and L_i. We still have

$$P_i = [P, J_i] \tag{6.3}$$

As before, let

$$\Psi_1(\lambda) = (\lambda I + \alpha)\Psi(\lambda) \tag{6.4}$$

Ψ_1 satisfies (6.3) with some P^1 and L^1 if and only if

$$P^1 = P + \alpha^{\text{off}} \tag{6.5}$$

$$L_j^1 = \alpha L_j \alpha^{-1} \tag{6.6}$$

and

$$\frac{\partial \alpha}{\partial x_j} = \alpha L_j \alpha^{-1} - L_j + [P_j, \alpha] - [J_j, \alpha]\alpha \tag{6.7}$$

We constructed the matrices H and α as before, i.e. via formula (3.5) and (3.8). However, in the present case, we should have $\lambda_\alpha \neq 0$. We have

$$\frac{\partial H}{\partial x_j} = J_j H \Lambda + L_j H \Lambda^{-1} + P_j H \tag{6.8}$$

By calculation, we see that the matrix α defined by (3.8) satisfies the equations (6.5). Thus we obtain:

Theorem. *If (P, L, Ψ) is a solution to (6.1), then its Darboux transformation (P^1, L^1, Ψ^1) defined by (6.4), (6.5), (6.6) satisfies (6.1) too.*

When $L_i = 0$, this theorem reduced to the theorem proved in § 3 as a special case.

We can use $P = 0$, $L_j =$ a system of constant matrices which are permutative each other as a seed solution and a series of new solution can obtained successively. The method of reduction to $su(n)$ is valid, provided that $L_j^+ = L_j$.

§ 7. A CLASS OF INTEGRABLE SYSTEM IN R^{n+1}

We may consider a system of equations in R^{n+1} by adding a t equation to system (2.1) i.e.

$$\frac{\partial \Psi}{\partial x_i} = (\lambda J_j + P_j)\Psi, \qquad \frac{\partial \Psi}{\partial t} = V(t)\Psi \tag{7.1}$$

where $V(\lambda)$ is a polynomial of λ, e.g.

$$V(\lambda) = V_0 + V_1\lambda + V_2\lambda^2 \tag{7.2}$$

We assume

$$J_1 = \begin{bmatrix} 1 & & & \\ & 0 & & \\ & & \ddots & \\ & & & 0 \end{bmatrix}, \cdots, J_n = \begin{bmatrix} 0 & & & \\ & 0 & & \\ & & \ddots & \\ & & & 1 \end{bmatrix} \tag{7.3}$$

V_2, V_1 and V_0 as polynomials of P and $\frac{\partial P}{\partial x_i}$ and obtain a system of evolution equations

$$\frac{\partial P_i}{\partial t} - \frac{\partial V_0^{\text{off}}}{\partial x_i} + [P_i, V_0]^{\text{off}} = 0 \tag{7.4}$$

It has been shown that the Darboux matrix method is valid as well. For the $su(n)$ case, the single soliton solution may have different velocity of propagations. Moreover, we constructed multi-soliton solutions and proved that the interaction of solitons is elastic if we consider their magnitudes only.

The details of these results will be published elsewhere [11]

References

1. Darboux, G., Comp. Rend. **94** (1882), 1456-1459.
2. Matveev, V. B., & Salle, M. A., *Darboux transformations and solitons*, Springer-Verlag, 1990.
3. Gu, C. H. & Hu, H. S., LMP **11** (1986), 325-335.
4. Gu, C. H. & Zhou, Z., Nonlinear evolution equations: integrability and spectral methods, Manchester Univ. Press, 1990, pp. 115-123.
5. Gu C. H., Zhou, Z., Lett. Math. Phys. **13** (1987), 179-187.
6. Sattinger, D. H. and Zurkowski, V. D., Physics **26D** (1987), 225-250.
7. Gu, C. H., Nankai Lectures on Mathematical Physics (1987), World Scientific Publishing Company, Singapore, 1989, pp. 162-168.
8. Zhou, Z., Lett. Math. Phys. **16** (1988), 9-17.
9. Beals R. & Tenenblat, K., *An intrinsic generalization for the wave and sine-Gordon equation*, Differential Geometry, Putman Monographs # 52, 1991, pp. 25-46.
10. Tenenblat, K, *A Note on Solutions for the Intrinsic Generalized Wave and Sine-Gordon Equations*, (a paper published by Academic Press Inc.).
11. Gu, C. H., *On the Interaction of Solitons for a Class of Integrable systems in the space-time* R^{n+1} *(preprint)*.

GROUP THEORETICAL TREATMENT OF FUNDAMENTAL SOLUTIONS

N. H. IBRAGIMOV
Institute of Mathematical Modeling
Russian Academy of Sciences
Miusskaya Sq. 4
Moscow 125047, Russia

Abstract

The purpose of the paper is to present a new adaptation of Lie's theory to differential equations in the space of distributions. The group theoretic approach leads, e. g. , to a simplification of the construction of the fundamental solutions for many equations of mathematical physics. It is also applicable to linear equations with variable coefficients and is independent on a choice of coordinates.

Le but de ce papier est de présenter une adaptation nouvelle de la théorie de Lie aux équations différentielles dans l'espace des distributions. En particulier, on montre que cette approche simplifie la construction des solutions fondamentales pour nombre d'équations de physique mathématique. La méthode est indépendante des systèmes de coordonnées.

M. Flato et al. (eds.), Physics on Manifolds, 161–175.

Introduction

Lie's theory of differential equations is based on an existance of symmetry groups, and provides a unified group theoretic method instead of ad hoc integration methods used to solve ordinary differential equations (see, e.g. [1]). It also provides, in the case of partial differential equations, simple methods for finding certain classes of solutions, e.g., invariant solutions. But, the group theory was considered, until recently, to be practically useless in boundary value problems for partial differential equations. This point of view is based on the quite reasonable argumentation that even if a given differential equation has a symmetry group, boundary and/or initial conditions, in general, break this symmetry. Exeptions could be for a restricted number of boundary value problems with special datas.

It was observed recently that the last possibility (i.e., problems with special, but property choosen boundary or initial conditions) opens a way for applications of the group theory to general boundary value problems. Namely, it was demonstrated in [2] - [3] that equations defining fundamental solutions and the Goursat problem for Riemann's function heritage symmetry properties of the differential equations under consideration. It follows that one can obtain the fundamental solution (or Riemann's function) as an invariant solution if a given linear differential equation has a sufficiently large symmetry group. Hence a Cauchy problem with an arbitrary data is reduced to the problem of constructing invariant solutions.

The group theoretic approach to fundamental solutions suggeste in [2] and first illustrated by the Laplace and heat equations has been applied in [4] to the wave equations in Riemannian spaces with non-trivial conformal group. The classical wave equation is of this type. But Riemannian spaces with non-trivial conformal group provide a very large variety of hyperbolic equations possessing the Huygens principle (see [5]). These are equations with intrinsically variable coefficients solvable by our approach.

Here, the group theoretic approach to fundamental solutions is discussed in details. It is based on extensions of point transformations to test functions and to distributions. We start with an extension discussed in [6]. In our presentation this extension is slightly changed so that to make it compatible with the group analysis of differential equations involving distributions. Lie's infinitesimal technique is also developed.

1. Transformation groups extended to distributions

Let $x = (x^1, \ldots, x^n)$ and $\varphi(x)$ denote independent variables and test functions. An action of a distribution f is denoted by $\langle f, \varphi \rangle$.

Consider transformations in $\mathbb{R}^n$,

$$\bar{x} = \Phi(x) \tag{1}$$

which convert points $x \in \mathbb{R}^n$ into new positions $\bar{x} \in \mathbb{R}^n$ into new positions $\bar{x} \in \mathbb{R}^n$.

Here $\Phi : \mathbb{R}^n \to \mathbb{R}^n$ is a C^∞-diffeomorphism with the Jacobian

$$\mathcal{J} = \det\left(\frac{\partial \bar{x}^i}{\partial x^j}\right) \neq 0 .$$

Test functions behave as a scalar with respect to point transformations (1). Namely, the transformation $\varphi \to \bar{\varphi}$ of scalar functions is defined by the equation

$$\bar{\varphi}(\bar{x}) = \varphi(x). \tag{2}$$

This equation written in terms of one of the variables x or $\bar{x}$, say $\bar{x}$,

$$\bar{\varphi}(\bar{x}) = \varphi \circ \Phi^{-1}(\bar{x})$$

yields the following transformation law for scalar functions

$$\bar{\varphi} = \varphi \circ \Phi^{-1}. \tag{3}$$

Eq. (2), or the equivalent eq. (3), means that scalar functions are form-invariant. The change in their argument is given by Φ^{-1} in eq. (3).

We define the transformation $f \to \bar{f}$ of distributions by the equation (cf. (2))

$$\langle \bar{f} \circ \Phi^{-1}, \varphi \circ \Phi^{-1} \rangle = \langle f, \varphi \rangle. \tag{4}$$

The extension of the transformation (1), given by eqs. (2) and (4), fits point transformations to invariance properties of differential equations in the space of distributions.

Now we rewrite the right-hand side of eq. (4) in terms of new test functions $\varphi \circ \Phi^{-1}$ by regarding the transformation (1) as a change of variables. Then the following formula applies (see [6], Ch. VI, B4):

$$\langle f, \varphi \rangle = \langle |\mathcal{J}|^{-1} f \circ \Phi^{-1}, \varphi \circ \Phi^{-1} \rangle \tag{5}$$

which is in agreement with the usual rule for the change of variables in an integral,

$$<\bar{f}\circ\Phi^{-1} \ , \ \varphi\circ\Phi^{-1}> = <|g|^{-1} f\circ\Phi^{-1} \ , \ \varphi\circ\Phi^{-1}>$$

and yields the following transformation law for distributions :

$$\bar{f} = |g|^{-1} f \ . \tag{6}$$

In particular, thre transformation of the Dirac measure is given by

$$\bar{\delta} = |g|^{-1}_{x=0} \, \delta \ . \tag{7}$$

Remark

Let the distribution f be identified with a function. Then eq. (6) gives the form-transformation of the function f , while the composition with Φ^{-1} in eq. (4) indicates the change of arguments in both fonctions f and φ .

Now we consider a one parameter local group G of transformations

$$\bar{x} = \Phi(x, a) \ . \tag{8}$$

We assume that

$$\Phi(x, 0) = x \tag{9}$$

and that

$$\Phi(\bar{x}, b) = \Phi(x, a + b)$$

when a and b are sufficiently small. The group G is generated by its infinitesimal transformation

$$\bar{x} \simeq x + a\xi(x) , \tag{10}$$

where the vector field $\xi = (\xi^1, \dots, \xi^n)$ is given by

$$\xi(x) = \frac{\partial\Phi(x,a)}{\partial a}\Big|_{a=0} .$$

It follows from eq. (9) that the Jacobian of transformation (8) is positive in a small neighbourhood of $a = 0$.

$$\bar{f} = g^{-1} f \tag{11}$$

and

$$\bar{\delta} = (g^{-1})_{x=0}\,\delta . \tag{12}$$

For the infinitesimal transformation (10), eqs. and (12) yield:

$$\bar{f} \simeq \left(1 - a\frac{\partial\xi^i(x)}{\partial x^i}\right) f \quad , \tag{13}$$

2. An auxiliary differential equation

Let $\tau(x)$ be a continuously differentiable function such that grad $\tau \neq 0$. Let S be the hypersurface given by

$$\tau(x) = 0 .$$

Denote by ω the Leray form of S ([7], see also [6], p. 439), i.e., an $(n-1)$ – differential form such that

$$d\tau \wedge \omega = dx^1 \wedge \ldots \wedge dx^n \; . \tag{14}$$

The Dirac measure $\delta(\tau)$ on the surface S is defined by

$$< \delta(\tau) \, , \varphi > = \int_{\tau=0} \varphi \, \omega \; .$$

We have from this definition

$$\tau\delta(\tau) = 0 \; .$$

The successive differentiation yields

$$\tau\delta'(\tau) + \delta(\tau) = 0 \quad , \quad \tau\delta''(\tau) + 2\delta'(\tau) = 0 \quad , \quad \ldots \; .$$

Hence

$$\tau\delta^{(m)}(\tau) + m\delta^{(m-1)}(\tau) = 0 \quad , \quad m = 0, 1, 2, \ldots \; . \tag{15}$$

The following first order ordinary differential equation will be necessary in section 5 :

$$\tau f'(\tau) + mf(\tau) = 0 \quad , \quad m = 1, 2, \ldots \; . \tag{16}$$

Its classical solution is $f = \tau^{-m}$. After (15) we know its solution in the space of distributions, namely, $f = \delta^{(m-1)}(\tau)$. These two linearly independent solutions yield the general solution of eq; (16) :

$$f = C_1 \, \delta^{(m-1)}(\tau) + C_2 \, \tau^{-m} \quad ,$$

where C_1 and C_2 are arbitrary constants.

3. Laplace equation

The Laplace equation

$$\Delta u = 0 \tag{17}$$

with $n \geqslant 3$ variables is invariant with respect to the conformal group. The generators of this group are

$$X_i = \frac{\partial}{\partial x^i}, \quad X_{ij} = x^j \frac{\partial}{\partial x^i} - x^i \frac{\partial}{\partial x^j},$$

$$Y_i = (2x^i x^j - r^2 \delta^{ij}) \frac{\partial}{\partial x^j} + (2-n)\, x^i u \frac{\partial}{\partial u}, \quad r = |x| \tag{18}$$

$$Z_i = x^i \frac{\partial}{\partial x^i}, \quad Z_2 = u \frac{\partial}{\partial u} \quad ; \ i, j = 1, \dots, n.$$

Let's find the subgroup of the conformal group which leaves invariant the equation for fundamental solutions:

$$\Delta u = \delta . \tag{19}$$

A necessary condition for this invariance is the immobility of the point $x = 0$. It is evident that $x = 0$ is fixed under the action of the subgroup generated by X_{ij}, Y_i, Z_1 and Z_2. Is eq. (19) invariant with respect to this subgroup ?

Consider first rotations and dilatations generated by X_{ij} and Z_1, Z_2 respectively. Operators Y_i will be discussed later.

Both the Laplace operator Δ and the Dirac measure δ are invariant under rotations (see (12)). Hence eq. (19) is invariant.

Eq. (19) is not invariant with respect to the one parameter group of dilatations generated by Z_1 as well as with respect to dilatations generated by Z_2. Therefore we consider a linear combination of Z_1, Z_2 :

$$Z = x^i \frac{\partial}{\partial x^i} + k\,u \frac{\partial}{\partial u} \tag{20}$$

and find an unknown coefficient $k = \text{const}$ from the invariance condition of eq. (19). Let's write the infinitesimal criterium for this invariance. The infinitesimal transformation (10) corresponding to operator (20) is

$$\bar{x}^i \simeq x^i + a x^i \quad , \quad \bar{u} \simeq u + k a u\,. \tag{21}$$

The differentiation gives the following infinitesimal transformations for derivatives :

$$\bar{u}_i \simeq u_i + (k-1)\,a\,u_i \quad , \quad \bar{u}_{ij} \simeq u_{ij} + (k-2)\,a\,u_{ij}\,.$$

It follows that the infinitesimal transformation of the left-hand side of eq. (19) is

$$\bar{\Delta}\bar{u} \simeq \Delta u + (k-2)\,a\,\Delta u\,. \tag{22}$$

We have, for transformations (21), $\partial \xi^i / \partial x^i = n$. Therefore, according to formula (14), the infinitesimal transformation of the right-hand side of eq. (19) is

$$\bar{\delta} \simeq \delta - n a \delta\,. \tag{23}$$

The infinitesimal criterium for the invariance of eq. (19),

$$(\overline{\Delta}\overline{u} - \overline{\delta})\Big|_{\Delta u = \delta} = 0 \ ,$$

becomes after (22) - (23) $k - 2 + n = 0$. Hence, $k = 2 - n$ in operator (20).

Thus, eq. (189) is invariant with respect to rotations and dilatations generated by

$$X_{ij} = x^j \frac{\partial}{\partial x^i} - x^i \frac{\partial}{\partial x^j} \quad , \quad Z = x^i \frac{\partial}{\partial x^i} + (2 - n)\ u \frac{\partial}{\partial u} \ . \tag{24}$$

Now we find solutions of eq. (19) which are invariant with respect to the group with the generators (24). For the rotations we have two functionally independent invariants, e.g., r and u. The remaining operator, Z from (24), is written in the variables r and u as follows:

$$Z = x^i \frac{\partial}{\partial r} + (2 - n)\ u \frac{\partial}{\partial u} \ .$$

The equation $Z F = 0$ gives one independent invariant

$$F = u\, r^{n-2} \ .$$

It follows that invariant solutions of eq. (19) are given by $F =$ const., i.e.,

$$u = C\, r^{2-n} \quad , \quad C = \text{const.} \tag{25}$$

The substitution of the function (25) in eq. (19) yields

$$C = [\,(2-n)\,\Omega_n\,]^{-1} \quad ,$$

where Ω_n is the area of an n-dimensional unit sphere. Thus, we have obtained the fundamental solution

$$u = \frac{r^{2-n}}{(2-n)\Omega_n} \tag{26}$$

for the Laplace equation from the invariance condition up to a constant factor. We used the differential equation (19) as a scaling condition only.

Remark

The function $F = u\, r^{n-2}$ satisfies the equations $Y_i F = 0$ and, hence, is an invariant for conformal transformations generated by the operators Y_i from (18). Therefore, the fundamental solution (26) has an extra-symmetry given by Y_i.

4. Heat equation

Consider the equation

$$u_t - \Delta u = \delta(t, x) \tag{27}$$

with the Laplace operator Δ in n variables x^i. The generators of rotations, Galilean transformations and dilatations :

$$X_{ij} = x^j \frac{\partial}{\partial x^i} - x^i \frac{\partial}{\partial x^j} \quad , \quad Y_i = 2t \frac{\partial}{\partial x^i} - x^i\, u \frac{\partial}{\partial u} \;,$$
$$Z = 2t \frac{\partial}{\partial x^i} + x^i \frac{\partial}{\partial x^i} - n\, u \frac{\partial}{\partial u} \tag{28}$$

replace here the operators (24) used for eq. (19).

We have the following functionally independent invariants of rotations: t, r, u. The operators Y_i are written in the space of these invariants as follows:

$$Y_i = x^i\left(2\frac{t}{r}\frac{\partial}{\partial x^i} - u\frac{\partial}{\partial u}\right).$$

The equations $Y_i F = 0$ give two independent invariants:

$$t \text{ and } p = u\, e^{r^2/4tb}.$$

The remaining operator, Z from (28), is written in the variables t and p as follows:

$$Z = 2t\frac{\partial}{\partial t} - np\frac{\partial}{\partial p}.$$

The equation $Z F = 0$ gives one invariant

$$F = u\, t^{n/2}\, e^{r^2/4t}.$$

It follows that invariant solutions of eq. (27) are given by

$$u = C\, t^{-n/2}\, e^{-r^2/4t}, \quad C = \text{const.} \tag{29}$$

The substitution of the function (29) in eq. (27) yields $C = (2\sqrt{\pi})^{-n}$. Hence, the fundamental solution is

$$u = (2\sqrt{\pi t})^{-n}\, e^{-r^2/4t}, \quad t > 0. \tag{30}$$

We again used the differential equation (27) as a scaling condition only.

5. Wave equation

Here, the beginning is just as in the proceeding cases. The equation for fundamental solutions,

$$u_{tt} - \Delta u = \delta(t, x) \tag{31}$$

is invariant with respect to the group of rotations, Lorentz transformations and dilatations generated by

$$\begin{aligned} X_{ij} &= x^j \frac{\partial}{\partial x^i} - x^i \frac{\partial}{\partial x^j} \quad , \quad Y_i = t \frac{\partial}{\partial x^i} + x^i \frac{\partial}{\partial t} \, , \\ Z &= t \frac{\partial}{\partial t} + x^i \frac{\partial}{\partial x^i} + (1-n) u \frac{\partial}{\partial u} \; ; \; i, j = 1, \dots, n . \end{aligned} \tag{32}$$

The rotations and Lorentz transformations have two functionally independent invariants :

$$u \text{ and } \tau = t^2 - r^2 .$$

Operator Z from (32) is written in the variables τ and u as follows :

$$Z = 2\tau \frac{\partial}{\partial \tau} + (1-n) u \frac{\partial}{\partial u} . \tag{33}$$

Its invariant is $F = u\,\tau^{(n-1)/2}$, so that an invariant solution could be taken in the form

$$u = C\,\tau^{(1-n)/2} . \tag{34}$$

But, eq. (31) has no classical solutions. Therefore we take invariant solutions in the form

$$u = f(\tau) ,$$

and write the invariance condition under the operator (33). This condition is

$$Z(f(\tau) - u|_{u=f} = 0 ,$$

or

$$2\tau f'(\tau) + (n-1) f(\tau) = 0 . \tag{35}$$

The classical solution of eq. (35) is given by (34). Now we find its solution in the space of distributions, when $n \geqslant 3$ is odd. Fundamental solutions for even n can be constructed by Hadamard's method of descent.

Let $n = 2m + 1$, $m = 1, 2, \ldots$. Then eq. (35) has the form of eq. (16). Hence its solution is given by (17), i.e.

$$f(\tau) = C_1\, \delta^{(\frac{n-3}{2})}(\tau) + C_2\, \tau^{(\frac{1-n}{2})} . \tag{36}$$

The substitution in eq. (31) yields $C_1 = \frac{1}{2}\pi^{-m}$, $C_2 = 0$.

Thus, the fundamental solution of the wave equation with an odd number $n \geqslant 3$ of spacial variables is

$$u = \frac{1}{2}\pi^{(1-n)/2}\, \delta^{(\frac{n-3}{2})}(t^2 - |x|^2) . \tag{37}$$

References

1. Lie, Sophus (1891) Vorlesungen über Differentialgleichungen mit bekannten infinitesimalen Transformationen, B.G. Teubner, Leipzig.

2. Ibragimov, Nail H. (1989) Primer on the group analysis, Znanie, Moscow.

3. Ibragimov, Nail H. (1991) Essays in the group analysis of ordinary differential equations, Znanie, Moscow.

4. Berest, Yurii Yu. (1991) "Construction of fundamental solutions for Huygens equations as invariant solutions", Dokl. Akad. Nauk SSSR 317, 786-789.

5. Ibragimov, Nail H. (1983) Transformations groups applied to mathematical physics, Nauka, Moscow (English transl. D. Reidel, Dordrecht).

6. Choquet-Bruhat, Yvonne, DeWitt-Morette, Cecile, with Dillard-Bleick, Margaret (1982) Analysis, Manifolds and Physics, North-Holland, Amsterdam.

7. Leray, Jean (1953) Hyperbolic differential equations, Institute for Advanced Study, Princeton.

ON THE REGULARITY PROPERTIES OF THE WAVE EQUATION

S. KLAINERMAN[1] and M. MACHEDON[2]

The research we want to report about has been motivated by the study of the regularity properties of solutions to the equations of "wave maps[3] " defined from the Minkowski space-time $\mathbf{M}^{n+1}$ to a Riemannian manifold $\mathcal{N}$. Our interest in these equations is due to the fact that they are the simplest system of wave equations, derived from a relativistic Lagrangean, which contain nonlinear terms quadratic in the first derivatives.

The equations are obtained by virtue of the Action Principle from the Lagrangean density $L = -\frac{1}{2}Tr_{\mathbf{m}}(\phi^* h)$ where $Tr_{\mathbf{m}}(\phi^* h)$ denotes the trace relative to $\mathbf{m}$ of the pull–back $\phi^* h$ of the Riemannian metric h of $\mathcal{N}$. With respect to a system of coordinates y^a, $a = 1, \ldots m$, they have the form,

$$\Box\phi^a + \Gamma^a_{bc}(\phi)\mathbf{m}^{\mu\nu}\partial_\mu\phi^b\partial_\nu\phi^c = 0 \qquad (W.M.)$$

where Γ^a_{bc} are the Christoffel symbols of $(\mathcal{N}, h)$. We are interested of studying the evolution of such wave maps which verify, at time $t = 0$, the initial value problem,

$$\phi(0, x) = f_0(x), \quad \partial_t\phi(0, x) = f_1(x)$$

where f_0, f_1 are compactly supported smooth maps from $\mathbb{R}^n$ to $\mathcal{N}$.

The energy-momentum tensor of (W.M.) is given by, $\mathbf{T}_{\alpha\beta} = \frac{1}{2}\Big(\langle\phi_{,\alpha}, \phi_{,\beta}\rangle - \frac{1}{2}\mathbf{m}_{\alpha\beta}(\mathbf{m})^{\mu\nu}\langle\phi_{,\mu}\phi_{,\nu}\rangle\Big)$, where $<,>$ denotes the scalar product in $\mathcal{N}$. It verifies the conservation law,

$$D^\beta\mathbf{T}_{\alpha\beta} = 0.$$

[1] Supported by the N.S.F. grant DMS-9103613

[2] Supported by a PYI award

[3] These are sometimes referred to as sigma models, or improperly, harmonic maps

M. Flato et al. (eds.), Physics on Manifolds, 177–191.

The total energy enclosed on the hyperplane $\mathcal{H}_{t_0}$, given by $t = t_0$, corresponding to the standard killing vectorfield $T_0 = \partial_t$, is defined by,

$$E(t) = \int_{\mathcal{H}_t} \frac{1}{2}\Big(\sum_{\alpha=0}^{n} |\partial_\alpha \Phi|^2\Big) dx$$

where $|\partial_\alpha \Phi|^2 = h_{ab}\partial_\alpha \Phi^a \partial_\alpha \Phi^b$. In view of the law of conservation of energy we have,

$$E(t) = E(0).$$

Due to the strength of the nonlinear terms, the apriori bound provided by the energy identity is too weak to allow us, by itself, to derive interesting conclusions concerning the behaviuor of solutions to (W.M.).

The central unsolved question in the theory of wave maps is that of the regularity of solutions starting with smooth initial conditions. The present state of the art is as follows: We have a general global regularity result in $n = 1$. For $n = 2$ we have quite general results on the regularity of wave maps which satisfies additional symmetry assumptions. These are the assumption of equivariance, in the case of the work of [Sh–Za], [Gr1], and of spherical symmetry in the case of [Ch-Za]. In dimension $n = 3$ we have specific examples of wave maps[4], defined from $\mathbb{M}^{3+1}$ to S^2, which break-down in finite time. To understand what might happen in the general case it is worthwhile to reexamine, in the context of the equations of the type W.M., the classical result of local in time existence, and uniqueness.

Let us first broadened our discussion and consider all systems of equations of the form,

(1) $$\Box \phi^I = F^I(\phi, D\phi)$$

[4] see [Sh]

with initial values,

$$\phi(0,x) = f_0(x), \quad \partial_t\phi(0,x) = f_1(x) \tag{1a}$$

We shall assume that F^I are smooth functions of $\phi, D\phi$, quadratic relative to $D\phi$.

The classical proof of local existence and uniqueness for (1) rests on the standard energy inequality for solutions to the inhomogeneous wave equation,

$$\Box\phi = F. \tag{W.E.}$$

We have,

$$\|D\phi(t,\cdot)\|_{L^2(\mathbb{R}^n)} \le \|D\phi(0,\cdot)\|_{L^2(\mathbb{R}^n)} + \int_0^t \|F(t,\cdot)\|_{L^2(\mathbb{R}^n)} \tag{2}$$

and, more generally for Sobolev norms, $\| \,\|_s = \| \,\|_{H^s(\mathbb{R}^n)}$

$$\|D\phi(t,\cdot)\|_s \le \|D\phi(0,\cdot)\|_s + \int_0^t \|F(t,\cdot)\|_s. \tag{2'}$$

Applying the inequality (2') to the system (1) we infer that, as long as $|\phi|_\infty$ remains bounded, and for any $s \ge 0$,

$$\|D\phi(t,\cdot)\|_s \le C_s\|D\phi(0,\cdot)\|_s \exp\Big(\int_0^t |D\phi(\tau,\cdot)|_\infty d\tau\Big).$$

In the standard proof for local existence, one uses the Sobolev inequality,

$$|D\phi(\tau,\cdot)|_\infty \le C\|D\phi(\tau,\cdot)\|_{\frac{n}{2}+\epsilon}$$

for any $\epsilon > 0$. Thus, very crudely,

$$\int_0^t |D\phi(\tau,\cdot)|_\infty d\tau \le Ct \sup_{0\le\tau\le t} \|D\phi(\tau,\cdot)\|_{\frac{n}{2}+\epsilon}. \tag{3}$$

These simple inequalities allow us, via a standard iteration scheme, to derive the following,

Theorem 1. *Consider equations of the type (1) in $\mathbf{R}^{1+n}$ subject to the initial conditions (1a). Assume that $f_0 \in H_{\frac{n}{2}+1+\epsilon}, f_1 \in H_{\frac{n}{2}+\epsilon}$. There exists a $T_* > 0$ and a unique solution ϕ defined in the slab $\mathcal{D}_* = [0, T_*] \times \mathbf{R}^n$ verifying the following conditions,*

(i)
$$\sup_{\mathcal{D}_*} |\phi(t,x)| + |D\phi(t,x)| < \infty$$

and,

(ii)
$$\sup_{[0,T_*]} \left(\|\phi(t,\cdot)\|_{\frac{n}{2}+1+\epsilon} + \|\partial_t \phi(t,\cdot)\|_{\frac{n}{2}+\epsilon} \right) < \infty.$$

The question we want to adress here is that of the optimality of the result of Theorem 1. To start with, we first recall the fact[5] that the energy inequality (2) is the only known inequality for (W.E.), for $n \geq 2$, in which the first derivatives of ϕ are estimated in terms of F alone, without derivatives. This fact is crucial in applications to (W.M.) since the nonlinear terms are quadratic in the first derivatives. Therefore (2') seems necessary, the only improvement could come from (3). It is quite obvious that the estimate (3) is too crude. We lose a lot of information by giving up the time integration on the left hand side. To do better we shall make use the following estimate, for solutions of W.E.

Proposition 1. *Consider the inhomogeneous wave equation W.E. in $\mathbf{M}^{n+1}$. Consider first the case $n \geq 3$. For every $\epsilon > 0$ there exists a constant C for which we have the estimate,*

(i)
$$\left(\int_0^t |\phi(\tau,\cdot)|_\infty^2 d\tau\right)^{1/2} \leq C\left(\|D\phi(0,\cdot)\|_{\frac{n-3}{2}+\epsilon} + \int_0^t \|F(\tau,\cdot)\|_{\frac{n-3}{2}+\epsilon} d\tau\right).$$

For $n = 2$ we have on the other hand[6],

[5] see [Li], [Pe] etc.

[6] For a proof of the inequality below see [P] and [Ka]

$$\text{(ii)} \qquad \Big(\int_0^t |D^{1/4}\phi(\tau,\cdot)|_\infty^4 d\tau\Big)^{1/4} \leq C\Big(\|D\phi(0,\cdot)\|_\epsilon + \int_0^t \|F(\tau,\cdot)\|_\epsilon d\tau\Big).$$

The proof of Prop. 1, for $n \geq 3$ is an immediate consequence of the Sobolev inequalities and the following form of the Strichartz inequality (see [P]),

Proposition 2.

Consider the inhomogeneous wave equation W.E. in $\mathbf{M}^{n+1}$, $n \geq 3$. For every $\epsilon > 0$ there exists a constant C for which we have the estimate,

$$\Big(\int_0^t \|D^{\frac{n-3}{2(n-1)}}\phi(\tau,\cdot)\|^2_{\frac{2(n-1)}{n-3}} d\tau\Big)^{1/2} \leq c\|D\phi(0,\cdot)\|_\epsilon$$

Remark 1.

We remark that for $n = 3$ the sharp form of the inequality stated in Proposition 1 is valid provided that ϕ is spherically symmetric. Indeed in that case we have,

$$\Big(\int_0^t |\phi(\tau,\cdot)|^2_\infty d\tau\Big)^{1/2} \leq c\Big(\|D\phi(0,\cdot)\| + \int_0^t \|F(\tau,\cdot)\| d\tau\Big)$$

This is a straightforward application of the Hardy -Littlewood maximal function inequality applied to the specific form of the solution. This result is probably true also for any $n \geq 3$. Unfortunately the inequality fails in general. This has been shown in [Kl-Ma] in the case of dimension $n = 3$. The argument given there can be extended to show that the estimate (i) fails,in general, for any $n \geq 3$.

Proposition 1 allows us to prove the following sharper version of the local existence theorem (see [Po-Si]).

Theorem 2. *Consider equations of the type (1) in $\mathbf{M}^{1+n}$, $n \geq 3$, subject to the initial conditions (1a). Assume that $f_0 \in H_{\frac{n-1}{2}+1+\epsilon}$, $f_1 \in H_{\frac{n-1}{2}+\epsilon}$. There exists a*

$T_* > 0$ and a unique solution ϕ defined in the slab $\mathcal{D}_* = [0, T_*] \times \mathbf{R}^n$ verifying the following conditions,

(i)
$$\sup_{\mathcal{D}_*} |\phi(t,x)| < \infty$$

(ii)
$$\int_0^{T_*} |D\phi(\tau,\cdot)|_\infty^2 d\tau < \infty$$

and,

(iii)
$$\sup_{[0,T_*]} \left(\|\phi(t,\cdot)\|_{\frac{n-1}{2}+1+\epsilon} + \|\partial_t\phi(t,\cdot)\|_{\frac{n-1}{2}+\epsilon} \right) < \infty.$$

For $n = 2$ the same result holds true for data $f_0 \in H_{\frac{3}{4}+1+\epsilon}, f_1 \in H_{\frac{3}{4}+\epsilon}$.

Is the result of Theorem 2 optimal ? Obviously, if the sharp form of Proposition 1 was true, in other words if the inequalities (i), (ii), were true with $\epsilon = 0$, then Theorem 2 could be improved to hold for data $f_0 \in H_{\frac{n-1}{2}+1}, f_1 \in H_{\frac{n-1}{2}}$. In view of Remark 1 this is the case for $n \geq 3$ and sperically symmetric solutions of (1), if they exist. We conjecture however that, for general nonspherical symmetric solutions the result of Theorem 2 is optimal in the following sense;

Conjecture 1. *The initial value problem for general systems of equations[7] (1), is not well posed for general initial data $f_0 \in H_{\frac{n-1}{2}+1}, f_1 \in H_{\frac{n-1}{2}}$.*

Recently H. Lindblad has obtained a result in this direction for equations of the form $\Box\phi = \phi^2$ in $\mathbf{R}^{3+1}$. He has been able to show that the initial value problem is not well posed for $f_0 \in L_2(\mathbf{R}^3), f_1 \in H_{-1}(\mathbf{R}^3)$.

[7]with only the assumption that F are smooth functions quadratic in $D\phi$.

Until now we have considered in (1) only general nonlinear terms $F(\phi, D\phi)$ quadratic in $D\phi$. Nevertheless, the type of nonlinear terms appearing on the right hand side of W.M. have special form. Indeed they satisfy the null condition, as defined in [Kl1], [Kl2], [Ch]. For systems of equations of the form (1) this means that the nonlinear terms $F(\phi, D\phi)$ have the form

$$F^I = \sum_{J,K} \Gamma^I_{J,K}(\phi) B^I_{JK}(D\phi^J, D\phi^K) \tag{4}$$

where the B^I_{JK} are any of the null forms,

$$Q_0(\phi, \psi) = \partial_\alpha \phi \cdot \partial^\alpha \psi = -\partial_t \phi \partial_t \psi + \sum_{i=1}^n \partial_i \phi \partial_i \psi \tag{4a}$$

$$Q_{\alpha\beta}(\phi, \psi) = \partial_\alpha \phi \partial_\beta \psi - \partial_\beta \phi \partial_\alpha \psi \quad 0 \le \alpha < \beta \le n. \tag{4b}$$

We remark that the class of systems verifying the null condition include the equations for wave maps (W.M.) defined from the Minkowski space–time $\mathbf{R}^{n+1}$ with values in an arbitrary Riemannian manifold $(\mathcal{N}, h)$.

To see whether the null condition allows us to prove a better local existence and uniqueness result we consider the following case of scalar equations, which can be explicitely solved,

$$\Box \phi = \Gamma(\phi)(\phi_t^2 - |\nabla \phi|^2) \tag{5}$$

$$\phi(0, x) = 0, \quad \partial_t \phi(0, x) = f. \tag{5a}$$

Let F be the solution of the following differential equation

$$F''(\phi) = \Gamma(\phi) F'(\phi)$$

$$F(0) = 0, \quad F'(0) = 1.$$

and F^{-1} be the inverse function to F, well defined in some neighborhood of $\phi = 0$. Now let ψ be the solution to the linear wave equation,

$$\Box\psi = 0 \tag{6}$$

$$\psi(0,x) = 0, \quad \partial_t\psi(0,x) = f. \tag{6a}$$

We observe that $\phi = F^{-1}(\psi)$ defines our desired solution to (5). It makes sense, provided that ψ stays in the region where F^{-1} is well defined. This is certainly the case provided that $|\psi(t,\cdot)|_\infty$ remains small. As an example consider $\Gamma(\phi) = 1$. Then $\phi = \log(1+\psi)$.

The analysis of the previous example tempts us to conjecture that, if the null condition is satisfied, one can improve the necessary condition of well posedness of the initial conditions in Theorem 2 according to,

Conjecture 2. *The initial value problem (1)–(1a) is well posed for*

$$f_0 \in H_{\frac{n-2}{2}+1+\epsilon}(\mathbf{R}^n), \; f_1 \in H_{\frac{n-2}{2}+\epsilon}(\mathbf{R}^n).$$

Another plausible conjecture is the following,

Conjecture 3. *The initial value problem (1)–(1a) is well posed for,*

$$f_0 \in W^{1,n}(\mathbf{R}^n), f_1 \in W^{1,n-1}(\mathbf{R}^n).$$

While we cannot so far prove or disprove either of the above conjectures we want to discuss the recent, significant improvement of Theorem 2, for equations verifying the null condition. The result was obtained in [Kl-Ma], for n=3.

Theorem 3. *Consider equations of the type (1), verifying the null condition (4), in* $\mathbf{R}^{1+n}$, $n \geq 1$, *subject to the initial conditions*

$$f_0 \in H_{\frac{n-2}{2}+1}(\mathbf{R}^n), \ f_1 \in H_{\frac{n-2}{2}}(\mathbf{R}^n).$$

There exists a $T_* > 0$ *and a unique solution* ϕ *of (1.1)–(1.4) defined in the slab* $\mathcal{D}_* = [0, T_*] \times \mathbf{R}^3$ *verifying the following conditions*

(i) $$\sup_{\mathcal{D}_*} |\phi(t, x)| < \infty$$

(ii) $$\int\int_{\mathcal{D}_*} \left(|Q(\phi, \phi)|^2 + |D^{\frac{n-1}{2}} Q(\phi, \phi)|^2 \right) dt dx < \infty$$

for any of the null quadratic forms (1.4a), (1.4b). In addition,

(iii) $$\sup_{[0,T_*]} \left(\|\phi(t, \cdot)\|_{H^{\frac{n-1}{2}+1}(\mathbf{R}^n)} + \|\partial_t \phi(t, \cdot)\|_{H^{\frac{n-1}{2}}(\mathbf{R}^n)} \right) < \infty.$$

The basic new tool which allows us to prove our main theorem stated above is contained in the following

Proposition 3. *Consider the solutions* ϕ, ψ *of homogeneous wave equations in Minkowski space-time* $\mathbf{R}^{n+1}$,

$$\Box \phi = F, \qquad \Box \psi = G$$

Then, for any of the null forms (4a)-(4b) and any $T \geq 0$ *we have,*

$$\int_0^T \int_{\mathbf{R}^n} |Q(\phi, \psi)|^2 \, dx dt \leq c \Big(\|D\phi(0, \cdot)\| + \int_0^T \|F(t, \cdot)\| dt \Big)^2 \cdot \Big(\|D^{1+\frac{n-1}{2}} \psi(0, \cdot)\| + \int_0^T \|D^{\frac{n-1}{2}} G(t, \cdot)\| dt \Big)^2$$

The proof of the Proposition is based on the Parseval identity and a concrete representation of the space–time Fourier transform of $Q(\phi,\psi)$. We note that the result of Theorem 3 is false if Q is replaced by an arbitrary quadratic form in $D\phi, D\psi$, for example $\partial_t\phi\cdot\partial_t\psi$. On the other hand the result is certainly not sharp, it can be improved in many directions. Here are two examples of such results, in $n=3$, which strengthen Proposition 3 considerably.

Proposition 4. *Consider the solution ϕ_i, $i=1,\dots,4$ of the homogeneous wave equation $\Box\phi_i=0$ subject to the initial conditions $\phi_i=0$, $\partial_t\phi_i=f_i$ at $t=0$. Then for all $f_i\in S(\mathbf{R}^3)$, $i=1,\dots,4$*

$$\left|\int_{-\infty}^{\infty}\int_{\mathbf{R}^3}\phi_1\phi_2 Q(\phi_3,\phi_4)dxdt\right|\le C\|f_1\|_{L^2}\cdots\|f_4\|_{L^2}$$

and a constant C independent of the functions $(f_i)_{i=1,\dots,4}$. Here Q denotes any of the null forms $Q_0, Q_{\alpha\beta}$ defined by (4a), (4b).

Proposition 5. *Let ϕ,ψ be the solutions to the homogeneous wave equation subject to the initial conditions $\phi=\psi=0$, $\partial_t\phi=f$, $\partial_t\psi=g$ at $t=0$. Then, for all $f,g\in S(\mathbf{R}^3)$,*

$$\text{(i)}\qquad \int_{-\infty}^{\infty}\int_{\mathbf{R}^3}|\Delta^{-1/2}Q(\phi,\psi)|^2dtdx\le C\|f\|_{L^2}^2\|g\|_{L^2}^2$$

$$\text{(ii)}\qquad \int_{-\infty}^{\infty}\int_{\mathbf{R}^3}|\Delta^{-1/2}Q(\phi,\psi)|^2dtdx\le C\|\Delta^{-1/2}f\|_{L^2}^2\|\Delta^{1/2}g\|_{L^2}^2$$

with C independent of f,g. Here Q denotes any of the null forms Q_{ij}, $1\le i<j\le 3$ defined by (1.4b).

We note that the results of Proposition 5 may be false for the null forms Q_0 and Q_{0i}. This is due to the " low frequency region" in Fourier space.

Finally we remark that, though the invariant scalar Q_0 is the only null form which appear in the expression of the nonlinear terms in (W.M.), the other null forms $Q_{\alpha\beta}$ also appear naturally. To illustrate this we consider the Maxwell equations in $\mathbf{R}^{3+1}$,

$$dF = 0 \quad d^*F = 0 \tag{Ma}$$

with F the electromagnetic field, ${}^*F$ its Hodge dual, i.e. ${}^*F_{\alpha\beta} = \frac{1}{2}\in_{\alpha\beta\gamma\delta} F^{\gamma\delta}$. Therefore if F is a solution of (Ma) so is its dual ${}^*F$.

Consider the initial value problem at $t = 0$,

$$E = E_{(0)}, H = H_{(0)}$$

where $E_i = F_{0i}, H_i = \frac{1}{2}\in_{ijk} F^{jk}$ are, respectively, the electric and magnetic components of F. Let A^μ be a vector potential for F verifiying the Lorentz condition $D_\mu A^\mu = 0$. Then each component A^μ verifies the wave equation $\Box A^\mu = 0$. We can also choose $A^0 = 0$ and A^i such that, at $t = 0$ we have $\partial_t A^i = E^i_{(0)}$, $\partial_i A_j - \partial_j A_i = H_{(0)ij}$, $\partial_i A^i = 0$. Then $E_i = \partial_t A_i$ and $H_i = \in_{ijk} \partial_j A_k$. Consider two solutions F, F' of (Ma) and the invariant $F \cdot F' = E \cdot E' - H \cdot H'$. Let A, A' the vector potentials of F, F', as above. Then, $E \cdot E' - H \cdot H' = \sum_i \big((\partial_t A_i)\cdot(\partial_t A_i') - \sum_j (\partial_j A_i)\cdot(\partial_j A_i')\big) + \sum_{i,j} \partial_i A_j \cdot \partial_j A_i'$. Hence $E^2 - H^2 = \sum_i Q_0(A_i, A_i') + \sum_{i,j} Q_{ij}(A_i, A_j')$. We can therefore make use of the result of Prop. 3 to prove the following,

Proposition 6. *Consider two solutions of the Maxwell equations (Ma) subject to the initial conditions $E = E_{(0)}, H = H_{(0)}, E' = E'_{(0)}, H' = H'_{(0)}$. Then,*

$$\int_{\mathbf{R}^{3+1}} |F \cdot F'|^2 \, dx dt \le c\Big(\|E_{(0)}\|^2 + \|H_{(0)}\|^2\Big) \cdot \Big(\|E'_{(0)}\|_1^2 + \|H'_{(0)}\|_1^2\Big)$$

$$\int_{\mathbf{R}^{3+1}} |F \cdot^* F'|^2 \, dx dt \le c\Big(\|E_{(0)}\|^2 + \|H_{(0)}\|^2\Big) \cdot \Big(\|E'_{(0)}\|_1^2 + \|H'_{(0)}\|_1^2\Big)$$

A similar remark applies to the Dirac equation,

$$i\gamma^{\mu}\partial_{\mu}\Psi = 0 \tag{D}$$

and the invariant forms $\Psi^*\gamma^0\Psi$ and $\Psi^*\gamma^1\gamma^2\gamma^3\Psi$. Here $\gamma^0 = \begin{pmatrix} I & 0 \\ 0 & -I \end{pmatrix}$, $\gamma^i = \begin{pmatrix} 0 & \sigma^i \\ -\sigma^i & 0 \end{pmatrix}$ where σ^i are the Pauli matrices $\sigma^1 = \begin{pmatrix} 0 & 1 \\ 1 & 0 \end{pmatrix}$, $\sigma^2 = \begin{pmatrix} 0 & -i \\ i & 0 \end{pmatrix}$, $\sigma^3 = \begin{pmatrix} 1 & 0 \\ 0 & -1 \end{pmatrix}$. Recall that $\sigma^a\sigma^b = \delta^{ab} + i\epsilon^{abc}\sigma^c$ and $\gamma^\alpha\gamma^\beta + \gamma^\beta\gamma^\alpha = -2\mathbf{m}^{\alpha\beta}$. We also remark that $\gamma^1\gamma^2\gamma^3 = i\begin{pmatrix} 0 & -I \\ I & 0 \end{pmatrix}$.

The Dirac spinor Ψ can be written in terms of 2-spinors u, v. More precisely, $\Psi = \begin{pmatrix} u+v \\ u-v \end{pmatrix}$ where $u = \begin{pmatrix} u_1 \\ u_2 \end{pmatrix}$ and $v = \begin{pmatrix} v_1 \\ v_2 \end{pmatrix}$ verify the equations

$$(\partial_t - A)u = 0$$

and,

$$(\partial_t + A)v = 0$$

with $A = \sum_{i=1}^{3} \sigma^i\partial_i$. Clearly, $A^2 = \Delta$ and hence $(\partial_t - A)(\partial_t + A) = (\partial_t + A)(\partial_t - A) = -\Box$. Therefore $u = (\partial_t + A)U$, $v = (\partial_t - A)V$ where U, V are solutions of the wave equation, $\Box U = \Box V = 0$, subject to the initial conditions $U(0,x) = 0, \partial_t U = u(0,x)$ and $V(0,x) = 0, \partial_t V(0,x) = v(0,x)$.

Now, if Ψ and Ψ' are two solutions to (D) with initial conditions $\Psi_{(0)}$ and $\Psi'_{(0)}$, we have

$$\Psi^*\gamma^0\Psi' = 2(\bar{u}\cdot v + \bar{v}\cdot u)$$

and

$$\Psi^*\gamma^1\gamma^2\gamma^3\Psi' = 2i(\bar{u}\cdot v - \bar{v}\cdot u)$$

We claim that both invariants can be written as linear combinations of null forms in U, V. This means that for any U, V verifying $\square U = \square V = 0$, the expression $\overline{(\partial_t U + AU)} \cdot (\partial_t V - AV)$ is a linear combination of such null expressions. It clearly suffices to prove the statement for real U, V. Now, since we have $(\sigma^a)^t = \overline{\sigma^a}$ and $\sigma^a \sigma^b = \delta^{ab} + i\epsilon^{abc}\sigma^c$ we write $(\partial_t U + \overline{\sigma}^a \partial_a U) \cdot (\partial_t V - \sigma^b \partial_b V) = \partial_t U \cdot \partial_t V - \sum_a \partial_a U \cdot \partial_a V + (\overline{\sigma^a} \partial_a U) \cdot \partial_t V - (\overline{\sigma^a} \partial_t U) \cdot \partial_a V - i \in^{cab} \overline{\sigma}^c \partial_a U \partial_b V$. Hence our expression can be written as a linear combination of $Q_0(U, V)$ and $Q_{\alpha\beta}(U, V)$ which proves the assertion.

We therefore infer that,

Proposition 7. *Consider two solutions of the Dirac equation* (D) *subject to the initial conditions* $\Psi = \Psi_{(0)}$, $\Psi' = \Psi'_{(0)}$. *Then*

$$\int_{\mathbb{R}^{3+1}} |\Psi^* \gamma^0 \Psi'|^2 \, dx dt \le c \|\Psi_{(0)}\|^2_{L^2} \|\Psi'_{(0)}\|^2_{H^1}$$

and

$$\int_{\mathbb{R}^{3+1}} |\Psi^* \gamma^1 \gamma^2 \gamma^3 \Psi'|^2 \, dx dt \le c \|\Psi_{(0)}\|^2_{L^2} \|\Psi'_{(0)}\|^2_{H^1}$$

The estimates in Proposition 3 and Proposition 7 were generalized to the case of the Klein -Gordon equation and Dirac equation with mass in [B-B] and [Gr2].

Finally we remark that the result of Proposition 6 applies not only to the scalar invariant $F \cdot F'$ but also to antisymmetric expressions of the type $D_{\alpha\beta}(F, F') = F_{\alpha\mu} F_\beta'^{\ \mu} - F'_{\alpha\mu} F_\beta^{\ \mu}$. Let A, A' be the vector potentials of F, F' defined as in the proof of Proposition 6. For $\alpha = 0, \beta = i$ we can write, $D_{0i} = \partial_t A_j (\partial_i A'_j - \partial_j A'_i) - \partial_t A'_j (\partial_i A_j - \partial_j A_i) = Q_{0i}(A_j, A'_j) - \partial_t A_j \partial_j A'_i + \partial_t A'_j \partial_j A_i$. Now, since $\partial^i A_i = 0$ and $\square A_i = 0$ we can write $A_i = \in_{iab} \partial_a B_b$ with $\partial^a B_a = 0$ and $\square B_a = 0$. Therefore $\partial_t A_j \partial_j A'_i = \in_{jab} \partial_a \partial_t B_b \partial_j A'_i = \frac{1}{2} \in_{jab} Q_{aj}(\partial_t B_b, A'_i)$. Therefore,

$$D_{0i} = Q_{0i}(A_j, A'_j) + \frac{1}{2} \in_{ajb} Q_{aj}(\partial_t B_b, A'_i) - \frac{1}{2} \in_{ajb} Q_{aj}(\partial_t B'_b, A_i)$$

with B' the corresponding potential for A'. Proceeding in the same manner we find,

$$D_{ij}(F,F') = +Q_0(A_i, A'_j) - Q_0(A'_i, A_j) + \frac{1}{2}\,\epsilon_{akb}\left(Q_{ak}(\partial_i B_b, A'_j) + Q_{ak}(\partial_i B'_b, A_j)\right)$$
$$+\frac{1}{2}\,\epsilon_{akb}\left(Q_{ak}(A_i, \partial_j B'_b) - Q_{ak}(A'_i, \partial_j B_b)\right) + Q_{ij}(A_k, A'_k).$$

∎

Thus, under the same assumptions as those of Proposition 6 we have,

$$\int_{\mathbb{R}^{3+1}} \sum_{0\leq\alpha<\beta\leq 3} |D_{\alpha\beta}(F,F')|^2\,dxdt \leq c\left(\|E_{(0)}\|^2 + \|H_{(0)}\|^2\right)$$
$$\cdot\left(\|E'_{(0)}\|_1^2 + \|H'_{(0)}\|_1^2\right).$$

Simmilar antisymmetric expressions can be written for the Dirac equation.

(1) [B-B] M. Beals - M. Bezard, Personal communication.

(2) [Ch] D. Christodoulou, "Global solutions for nonlinear hyperbolic equations for small data," *Comm. Pure Appl. Math.* **39** 1986, 267–282.

(3) [Ch-Za] D. Christodoulou - A. Tahvildar-Zadeh, " Regularity of Spherically symmetric Harmonic Maps of the 2+1 Minkowski Space", Preprint.

(4) [Gr1] M. Grillakis " Classical solutions for the Equivariant Wave Maps in 1+2 dimensions" preprint.

(5) [Gr2] M. Grillakis, Personal communication.

(6) [Ka] L. Kapitanski *J.Soviet Math.*,**vol.56**, no.2, 1991, pp.2348-2389

(7) [Kl1] S. Klainerman, "Long time behaviour of solutions to nonlinear wave eqts." *Proc. of the I.C.M. Warszawa 1983* pp. 1209–1215

(8) [Kl2] S. Klainerman, "The Null Condition and the global existence for non-linear waves", *Lectures in Appl. Math.*, **23**, pp. 293-325, (1986).

(9) [Kl-Ma] S. Klainerman - M. Machedon, "Space-Time Estimates for null Forms and the Local Existence Theorem", Preprint.

(10) [Li] W. Littman " The wave operator and L^p norms." *Journal of Math. and Mech.* **12** pp. 55-68(1963)

(11) [P] H. Pecher, "Nonlinear Small Data Scattering for the Wave and Klein-Gordon Equation", *Math. Z.*, **185**, pp. 261-270, (1984).

(12) [Pe] J.C.Peral "L^p estimates for the Wave Equation", *Jour. of Funct. Analysis* **36** pp. 114-145.

(13) [Po-Si] G. Ponce - T. Sideris, "Local Regularity of Nonlinear Wave Equations in Three Space Dimensions", Preprint.

(14) [Sh] J. Shatah, "Weak Solutions and the Development of Singularities of the SU(2) σ-Model", *Comm. Pure Appl. Math.*, **41**, pp459-469, (1988).

(15) [Sh-Za] J. Shatah - A. Tahvildar-Zadeh, "Reg. of Harmonic Maps From Minkowski Space-Time into Rot. Symmetric Manifolds", preprint.

Le problème de Cauchy linéaire et analytique pour un opérateur holomorphe et un second membre ramifié

The linear Cauchy analytic problem with holomorphic operator and ramified second member

Jean LERAY

Abstract :

Asymptotic and exact solutions of a Cauchy problem have been often obtained by means of functions U (x, $k_1(x)$, ..., $k_n(x)$) where the k_j are characteristic functions of the differential operator. In this way Y. Hamada constructed ramified solutions of Cauchy problems with ramified data.

Recently he paved the way to new results by choosing U more claborately ; namely he defines U by a suitable Goursat problem, whose solutions have a remarquable analytic continuation ; hence analytic continuations of solutions of Cauchy problems.

I sketched them last October at the franco-japanese symposium held in Marseille-Luminy. But then only preliminaries were known about the case of a ramified second member. Those preliminaries have been now completed.

M. Flato et al. (eds.), Physics on Manifolds, 193.

ON BOLTZMANN EQUATION

P.L. LIONS
Ceremade
Université de Paris-Dauphine
Place de Lattre de Tassigny
75775 Paris Cedex 16

dédié à Y. Choquet-Bruhat

ABSTRACT: We present a global existence result of weak solutions for Boltzmann equation obtained by R.J. DiPerna and the author. This result is based upon convergence properties of solutions of Boltzmann equation that we recall and discuss.

RÉSUMÉ: Nous présentons un résultat d'existence globale de solutions faibles de l'équation de Boltzmann obtenu par R.J. DiPerna et l'auteur. Ce résultat est basé sur des propriétés de convergence des solutions de l'équation de Boltzmann, propriétés que nous rappelons et discutons.

SUMMARY:

1. Introduction

The well-known Boltzmann equation ([**1**]) is one of the basic models in statistical physics or mechanics that describe the evolution of a large number of particles, say gas molecules, in a rarefied gas. If f denotes the density of particles at position $x \in \mathbb{R}^N$, with velocity $v \in \mathbb{R}^N$, at time $t \geq 0$, the Boltzmann equation can be written as follows

$$\frac{\partial f}{\partial t} + v \cdot \nabla_x f = Q(f,f) \quad \text{in} \quad \mathbb{R}_x^N \times \mathbb{R}_v^N \times [0,\infty) . \tag{1}$$

Here and everywhere below, we denote indifferently by $a \cdot b$ or (a,b) the scalar product in $\mathbb{R}^N$ of a and b. The unknown function f being a density is a nonnegative function.

M. Flato et al. (eds.), Physics on Manifolds, 195–201.

The right-hand side of the equation (the so-called collision term) describes the rate of changes in the velocities distribution due to collisions. In particular, if there were no collisions (i.e. $\varphi \equiv 0$), (1) would simply mean that f is constant along particle paths ($\dot{x} = v$, $\dot{v} = 0$). In the presence of collisions, the term Q is given by

$$Q(f,f) = \int_{\mathbb{R}^N} dv_* \int_{S^{N-1}} d\omega\, B(v - v_*, \omega)(f' f'_* - f f_*) \tag{2}$$

where $f_* = f(x, v_*, t)$, $f' = f(x, v', t)$, $f'_* = f(x, v'_*, t)$ and where

$$v' = v - (v - v_*, \omega)\omega \quad , \quad v'_* = v_* + (v - v_*, \omega)\omega \,. \tag{3}$$

Here v, v_* represent the ingoing velocities of two particles colliding at (x,t) and v', v'_* are the outgoing velocities : (3) is then nothing else than a representation, parametrized by $\omega \in S^{N-1}$ (an angle), of all possible elastic collisions (i.e. preserving momentum and kinetic energy).

The function $B = B(z, \omega)$, called the collision kernel, takes into account the interaction between particles and is a given nonnegative function depending only on $|z|$ and (z, ω). A model example is given by the hard-spheres collision kernel $B = |(z, \omega)|$. We shall assume that B satisfies (in addition to the conditions we already mentioned)

$$B \in L^1_{\text{loc}}(\mathbb{R}^N \times S^{N-1}) \quad ; \quad \text{for all} \quad R \in (0, \infty)$$
$$\frac{1}{1+|z|^2} \int_{|v-z| \le R} \int_{S^{N-1}} dv\, d\omega\, B(v, \omega) \to 0 \quad \text{as} \quad |z| \to +\infty \,.$$

Even if this condition looks rather general, it is far from being optimal since some physically realistic interactions (like inverse-power potentials) lead to singular kernels B that present singularities at $\theta = \pm\pi/2$, where $\cos\theta = |(z, \omega)|$, which are not integrable. This classical difficulty is circumvented by a truncation of this singularity, called angular cut-off, which is thus implicit in the condition (2). The angular cut-off assumption amounts to neglect grazing collisions. For more details on this and on the derivation of Boltzmann equation, we refer the reader to C. Cercignani's book [**3**].

We are concerned here with global existence results in the large for the Cauchy problem associated with (1) i.e. when we complement (1) with an initial condition

$$f|_{t=0} = f_0 \quad \text{in} \quad \mathbb{R}^{2N}_{x,v}$$

where $f_0 \ge 0$ is given on $\mathbb{R}^{2N}_{x,v}$. It was shown in R.J. DiPerna and P.L. Lions [**4**],[**5**], that one can construct a global weak solution of (1),(3) (in a sense to be recalled below) provided that f^0 satisfies the natural conditions

$$\iint_{\mathbb{R}^{2N}} f_0 \left(1 + |x|^2 + |v|^2 + |\log f_0|\right) dx\, dv < \infty \,. \tag{4}$$

As we shall see, the regularity and the uniqueness of such solutions are outstanding open problems. Let us also mention that other existence results have been shown by various authors (see the references in [**3**] and [**4**]) and concern the global existence of smooth solutions in various regimes : close to an equilibrium (a Maxwellian) or to the vacuum, space homogeneous... Let us recall for instance the classical study of T. Carleman [**2**] in the space homogeneous case.

2. A priori estimates and the existence result

The only known a priori estimates for Boltzmann equation are consequences of various conservations and identities that we briefly recall now.

First of all, the "symmetric" nature of $Q(f,f)$ yields easily the following formal identities for solution of (1)

$$\frac{d}{dt}\iint_{\mathbb{R}^{2N}} f\,\psi\,dx\,dv = 0 \quad , \quad \text{for} \quad \psi \equiv 1\,,\ v_k\ (1 \leq k \leq N)\,,\ |v|^2\,,\ |x-vt|^2\,. \tag{5}$$

Next, solutions of (1) should, at least formally, satisfy the following "entropy identity"

$$\begin{aligned}&\frac{d}{dt}\iint_{\mathbb{R}^{2N}} f\log f\,dx\,dv +\\ &\quad + \frac{1}{4}\int_{\mathbb{R}^N} dx \int_{S^{N-1}} B\,d\omega \iint_{\mathbb{R}^{2N}} dv\,dv_*\,[f'f'_* - ff_*]\,\log\frac{f'f'_*}{ff_*} = 0\,.\end{aligned} \tag{6}$$

One observes that the second term is clearly nonnegative and thus (6) yields, in particular, that $\iint_{\mathbb{R}^{2N}} f\log f\,dx\,dv$ should be nonincreasing with respect to $t \geq 0$.

From these identities, one easily deduces the following a priori bounds : let $R \in (0,\infty)$, there exists $C > 0$ depending only on R such that if the initial condition f_0 satisfies

$$\iint_{\mathbb{R}^{2N}} f_0\,(1+|x|^2+|v|^2+|\log f_0|)\,dx\,dv \leq R\,, \tag{7}$$

then the corresponding solution of (1) should satisfy

$$\sup_{t\geq 0}\iint_{\mathbb{R}^{2N}} f(t)\,(1+|x-vt|^2+|v|^2+|\log f(t)|)\,dx\,dv \leq C\,. \tag{8}$$

At this stage, it is quite clear that the bounds which are available do not even allow to define quadratic terms like those which are present in the collision term $Q(f,f)$. This is why a new formulation of (1) was proposed in [**4**], based upon some further a priori estimates. In order to introduce this notion of solutions of (1) - called renormalized solutions in [4] -, we first recall the observation made in the Introduction namely that, if there are no collisions and $Q \equiv 0$, (1) means that f is constant along particles paths. But this statement is clearly equivalent to the same statement with f replaced by $\beta(f)$ where β is any one to one application from $[0,\infty)$ into $[0,\infty)$. This remark leads to the investigation of the equation satisfied by $\beta(f) = \log(1+f)$ (for instance) whenever f satisfies (1) : we immediately find of course :

$$\left(\frac{\partial}{\partial t} + v \cdot \nabla_x\right) \beta(f) = \beta'(f)\, Q(f,f) \quad \text{in } \mathbb{R}^{2N}_{x,v} \times (0,\infty)\,. \tag{9}$$

Next, we observe that denoting by

$$\begin{aligned} Q^+(f,f) &= \int_{\mathbb{R}^N} dv_* \int_{S^{N-1}} d\omega\, B(v-v_*,\omega)\, f' f'_*\,, \\ Q^-(f,f) &= \int_{\mathbb{R}^N} f_* \left(\int_{S^{N-1}} B(v-v_*,\omega)\, d\omega\right) dv_*\,, \end{aligned} \tag{10}$$

we have $Q = Q^+ - Q^-$ and one shows easily that (8) implies

$$\begin{aligned} &Q^-(f,f)(1+f)^{-1} \in L^\infty(0,\infty; L^1(\mathbb{R}^N_x \times K)) \\ &\qquad \text{for any compact set } K \text{ of } \mathbb{R}^N_v\,. \end{aligned} \tag{11}$$

And, using (9), one can show that the following holds for all $T \in (0,\infty)$

$$\begin{aligned} &Q^+(f,f)(1+f)^{-1} \in L^1(0,T; L^1(\mathbb{R}^N_x \times K)) \\ &\qquad \text{for any compact set } K \text{ of } \mathbb{R}^N_v\,. \end{aligned} \tag{12}$$

These observations lead to the following

DEFINITION. *$f \in C([0,\infty); L^1(\mathbb{R}^{2N}_{x,v}))$ is said to be s renormalized solution of (1) if $\sup_{t\geq 0}\left\{\iint_{\mathbb{R}^{2N}} f(t)\,(1+|x-vt|^2+|v|^2+|\log f(t)|)\, dx\, dv\right\} < \infty$; (10), (11) and (9) holds (with $\beta(t) = \log(1+t)$) and if we have*

$$\begin{aligned} &\left(\iint_{\mathbb{R}^{2N}} f \log f\, dx\, dv\right)(t) + \\ &+ \frac{1}{4}\int_0^t ds \int_{\mathbb{R}^N} dx \int_{S^{N-1}} B\, d\omega \iint_{\mathbb{R}^{2N}} dv\, dv_* (f'f'_* - ff_*) \log \frac{f'f'_*}{ff_*} \\ &\qquad \leq \left(\iint_{\mathbb{R}^{2N}} f \log f\, dx\, dv\right)(0) \end{aligned} \tag{13}$$

for all $t \geq 0$.

REMARKS: i) The equation (9) holds in distributions sense for example.

ii) One can show that one obtains equivalent definitions by requesting that (9) holds for all $\beta \in C^1([0,\infty),\mathbb{R})$ such that $|\beta'(t)|(1+t)^{-1}$ is bounded on $[0,\infty)$. It is also shown in [**4**] that functions with the bounds stated in the definition satisfy (9) if and only if (1) holds in integrated form over almost all particle paths (a more standard formulation sometimes called mild solutions).

iii) If a renormalized solution f satisfies : $Q^{\pm}(f,f) \in L^1_{\text{loc}}$, then (1) holds in distributions sense. □

THEOREM. *Let f_0 satisfy (4). Then, there exists a renormalized solution f which satisfies the initial condition (3).*

REMARK: As we already mentioned in the Introduction, the regularity and the uniqueness of these solutions are completely open.

3. Convergence properties

As it is often the case when trying to build global weak solution of nonlinear evolution problems, the existence is obtained if we are able to pass to the limit in the equation for sequences of solutions satisfying, uniformly, the a priori bounds which are available. A complete existence proof is then obtained by applying this "stability" property on sequences of approximated solutions (as, for instance, solutions of approximated, simpler problems having the same general structure and yielding the same a priori estimates). This is precisely the strategy followed in [**4**],[**5**] in order to prove the theorem stated in the preceding section. This is why the theorem we give in this section is probably the central result that allows to obtain the existence result stated above.

We thus consider a sequence $(f^n)_n$ of renormalized solutions of (1) that correspond to initial conditions $(f_0^n)_n$. We assume that f_0^n, f^n satisfy respectively (7),(8) uniformly in n. Then, without loss of generality, we may assume that f_0^n, f^n converge weakly respectively in $L^1(\mathbb{R}^{2N}_{x,v})$, $L^1(\mathbb{R}^{2N}_{x,v} \times [0,T])$ $(\forall\, T \in (0,\infty))$ to some f_0, f.

THEOREM. *i) f is a renormalized solution satisfying (3).*

ii)

$$\int_{\mathbb{R}^N} f^n \psi \, dv \underset{n}{\rightarrow} \int_{\mathbb{R}^N} f \psi \, dv \quad in \quad L^p(\mathbb{R}^N_x \times [0,T])$$

for all $\psi \in L^\infty(\mathbb{R}^N_v)$, $T \in (0,\infty)$, $1 \le p < \infty$.

iii)

$$\int_{\mathbb{R}^N} Q^+(f^n, f^n)\psi\, dv \underset{n}{\to} \int_{\mathbb{R}^N} Q^+(f,f)\psi\, dv\,,$$

$$\int_{\mathbb{R}^N} Q^-(f^n, f^n)\psi\, dv \underset{n}{\to} \int_{\mathbb{R}^N} Q^-(f,f)\psi\, dv$$

in measure locally on $\mathbb{R}^N \times [0,\infty)$, *for all* $\psi \in L^\infty(\mathbb{R}^N_v)$ *with compact support.*

REMARKS: i) To be really precise, part i) requires to assume $\underline{\lim}_n \int_{\mathbb{R}^{2N}} f_0^n \log f_0^n \, dx\, dv = \iint_{\mathbb{R}^{2N}} f_0 \log f_0 \, dx\, dv$. Otherwise, f satisfies (13) with $\iint_{\mathbb{R}^{2N}} f_0 \log f_0 \, dx\, dv$ replaced by $\underline{\lim}_n \iint_{\mathbb{R}^{2N}} f_0^n \log f_0^n \, dx\, dv$.

ii) Part ii) of the above result shows that velocity averages (called macroscopic quantities in Physics) are more regular (in the sense that they enjoy compactness properties) than the original densities. □

We would like to conclude with a few brief remarks. First of all, we recall that we are using in the results above the angular cut-off assumption, neglecting thus grazing collisions. On the contrary, it was shown in P.L. Lions [**6**] that, for models which concentrate on grazing collisions like for instance the Landau model, a stronger compactness phenomenon than ii), iii) above occurs and one obtains in this case that f^n converges (strongly) in $L^p(0,T;L^1(\mathbb{R}^{2N}_{x,v}))$ ($\forall\, p \in [1,\infty)$, $\forall\, T \in (0,\infty)$). And it was also shown in [**6**] that if this holds for the Boltzmann models we are considering here (hard spheres, or angular cut-off), then f_0^n converges in $L^1(\mathbb{R}^{2N}_{x,v})$ to f_0. It is then tempting to raise the following issues : i) for models allowing grazing collisions (at least certain classes), does one have the same compactification in all variables property ?, ii) for the models we are considering here, assuming that f_0^n converges in $L^1(\mathbb{R}^{2N}_{x,v})$ to f_0, is it true that f^n converges to f in $C([0,T];L^1(\mathbb{R}^{2N}_{x,v}))$ ($\forall\, T \in (0,\infty)$)? We certainly hope to come back on these questions in future publications.

References

[1] L. Boltzmann, Weitere studien über das wärmegleichgenicht unfer gasmoleküler, Sitzungsberichte der Akademie der Wissenschaften, Wien, 66 (1978), 275-370. Translation : Further studies on the thermal equilibrium of gaz molecules, 88-174 ; In *Kinetic Theory 2*, ed. S.G. Brush, Pergamon, Oxford, 1966.

[2] T. Carleman, *Problèmes mathématiques dans la théorie cinétique des gaz.* Notes rédigées par L. Carleson et O. Frostman, Publications mathématiques de l'Institut Mittag-Leffler, Almquist and Wikselles, Uppsala, 1957.

[3] C. Cercignani, *The Boltzmann equation and its applications.* Springer, Berlin, 1988.

[4] R.J. DiPerna and P.L. Lions, On the Cauchy problem for Boltzmann equations : Global existence and weak stability, Ann. Math., 130 (1989), 321-366.
[5] R.J. DiPerna and P.L. Lions, Global solutions of Boltzmann's equation and the entropy inequality, Arch. Rat. Mech. Anal., 114 (1991), p. 47-55.
[6] P.L. Lions, On Boltzmann and Landau equations, preprint.

STAR PRODUCTS AND QUANTUM GROUPS

Carlos MORENO

LPTM, Université Paris VII, Tour Centrale, 2 Place Jussieu, F-75251 Paris Cedex 05
Departamento de Física Teórica, Universidad Complutense, E-28040 Madrid

and

Luis VALERO

Instituto "Giner de los Rios", E-28100 Madrid
Departamento de Física Teórica, Universidad Complutense, E-28040 Madrid

For Yvonne Choquet-Bruhat

Lecture to the International Colloquium in Honour of Yvonne Choquet-Bruhat, *Analysis, Manifolds and Physics*, Collège de France, Paris, June 3-5, 1992

M. Flato et al. (eds.), Physics on Manifolds, 203–234.

ABSTRACT

We provide proofs of Drinfeld's theorems about the construction of all invariant Star Products on a simply connected Lie group G, endowed with an invariant symplectic structure β_1. Solutions of the Triangular Yang-Baxter equation follow, as well as quantizations of the underlying Triangular Poisson-Lie structure on G, i.e. the corresponding Triangular Quantum Groups.

RÉSUMÉ

A partir des travaux de Berezin sur la quantification nous avons construit précédemment des Produits Star Invariants sur les espaces symétriques kähleriens et sur les espaces homogènes compacts quantifiables. Nous prouvions que la fonction de Bochner-Calabi de l'espace détermine la classe des diviseurs du fibré positif de la quantificatión, sa classe de Chern, le noyau de la quantificatión, etc., tandis que le développement de Taylor de cette fonction détermine les 2-cochaines invariantes du Produit Star. De manière analogue, la théorie de Drinfeld des Groupes Quantiques Triangulaires se développe à partir de la théorie des produits star invariants sur un groupe de Lie G muni d'une structure symplectique invariante β_1. Nous prouvons ici quelques théorèmes énoncés par Drinfeld sur la construction et l'équivalence des tels produits star. La série formelle de Campbell - Hausdorff des extensions de G par de cocycles β_h déformés de β_1, joue dans, ce cas, le rôle de la fonction de Bochner-Calabi, et la quantification de Weyl remplace celle de Berezin. Ces théorèmes s'appuient de manière essentielle sur l'interprétation cohomologique de l'équatión Quantique de Yang-Baxter que nous avions dégagé dans un travail précédent à partir des résultâts de Drinfeld.

1. INTRODUCTION

A.- The notions of *Poisson-Lie Group*, *Quantized Poisson-Lie Group*, *Quantum Group*, *Quasi-Quantum Group* etc. have been introduced and developed in [1-3] as a general mathematical framework within which it is possible to understand the results on infinite (and finite) dimensional Hamiltonian Integrable Systems obtained by the *Inverse Scattering Method* applied to many specific cases of partial differential equations. The classical part of these results are in [4], and one can adventageously follow the historical development of the subject from the Notes and References to each one of its chapters. In particular, the key notion of *classical R-matrix* is developed in the now classical article [5], and references [1], [6-10] contain the main results on Poisson-Lie Groups. Reference [12] contains recent developments and shows the profound importance of the notions involved.

B.- The starting point of Drinfeld's theory of Triangular Quantum Groups [2], [11], [12] is the theory of Star Products [13], [14], [15] on a simply connected Lie Group endowed with an invariant Poisson structure. The construction of these star products [11], [16], [17], [46] amounts to solving the equation

$$\begin{aligned} F(x+y;z)\,F(x;y) &= F(x;y+z)\,F(y;z) \\ F(0;x) = F(x;0) &= 1\,, \end{aligned} \tag{1}$$

where $F(x;y) \in \mathfrak{A}(\mathfrak{g})^{\otimes 2}[[h]]$ and $+$ stand for the coproduct in the enveloping algebra $\mathfrak{A}(\mathfrak{g})$ of the Lie algebra $\mathfrak{g}$ of $\boldsymbol{G}$ [11], [18], [19]. Equation (1) is equivalent to the nullity of an infinity sequence of 3-cocycles in the cohomology of the Hochschild complex $(\mathcal{T}\mathfrak{A}(\mathfrak{g});\boldsymbol{d})$ and another reduced one [20-22] as described in [3], [20-23]. The main result concerning this complex is the classical result on formal groups [3], [24]: 1) the skewsymmetrized $\boldsymbol{A}\alpha$ of a k-cocycle α is a k-cocycle and an element of $\bigwedge^k(\mathfrak{g})$, 2) $\alpha = \boldsymbol{A}\alpha + \boldsymbol{d}E$, where E is a (k 1)-cochain, 3) the cohomology class of α is zero if and only if $\boldsymbol{A}\alpha = 0$; that is $H^k(\mathcal{T}\mathfrak{A}(\mathfrak{g});\boldsymbol{d}) \approx \bigwedge^k(\mathfrak{g})$. Lichnerowicz-Vey theorems in [14], [25], referring to manifolds, can be seen as a generalization of the foregoing results. For the theory of star products, only the results for $k = 2,3$ are needed, so that we initially used the results from Lichnerowicz [14] page 199 in preparing [18], [19].

C.- The relation between equation (1) and the Quantum Yang-Baxter Equation (Q.Y.B.E.), as discovered by Drinfeld [11] is developed in [18], [19], [26], [27], [46]. Related questions in an abstract setting are in [28], [29].

The generalized notion of equation (1) in the context of Quasi-Quantum-Groups [3] is completely new, but still related to star products, at least on Quasi-Triangular Quantum Groups [3], [27].

D.- The way in which invariant star products (1) on a Lie group endowed with an invariant symplectic structure $\beta_1 \in \mathfrak{g}^* \wedge \mathfrak{g}^*$ are constructed in [11] is a direct generalization of the way in which the usual Moyal star product on the abelian group $(\mathbb{R}^{2n}; \beta_1)$ can be obtained [30], [31] from the Heisenberg group $\boldsymbol{H}_n$, viewed as the central extension of $\mathbb{R}^{2n}$ by a factor corresponding to the cocycle β_1. In this way $(\mathbb{R}^{2n}; \beta_1)$ can be identified with an orbit of the coadjoint representation of $\boldsymbol{H}_n$, and the exponential in the integral form of the Moyal star product can essentially be viewed as determined by the Campbell-Hausdorff formal group $\bar{\gamma}(\bar{x}; \bar{y})$ of $\boldsymbol{H}_n$, which by the commutativity of $\mathbb{R}^{2n}$ consist of only two terms. The important thing in [11] is that this approach to the usual Moyal star product has a meaning for $(\boldsymbol{G}; \beta_1)$. The associativity of the star product on the orbit comes essentially from the associativity of the formal law group $\bar{\gamma}(\bar{x}; \bar{y})$. To obtain the invariant star product on $\boldsymbol{G}$, Drinfeld remarks that a certain mapping to the orbit is a local diffeomorphism at the point $e \in \boldsymbol{G}$. The invariance of the star product on $\boldsymbol{G}$, is due to the invariance of the star product on the orbit under the coadjoint action and to the fact that this action and left-action of the group on itself are equivaried by the local diffeomorphism.

To obtain other invariant star products on $\boldsymbol{G}$, [11] takes the liberty of replacing the cocycle β_1, in the foregoing construction, by any cocycle:

$$\beta_h = \beta_1 + \beta_2 \, h + \cdots + \beta_R \, h^{R-1} + \cdots ,$$

where h is the parameter of deformation. The development of this idea leads to the result that any invariant star product on $(\boldsymbol{G}; \beta_1)$ is equivalent to one so obtained from one of this cocycles.

E.- No proofs are given in [11] of the theorems stated therein and we have no knowledge of any proof of them in the literature. Given to the obvious importance [3], [11], [26], [27], of results in [11], their implications for further developments concerning Quasi-Quantum-Groups [3], [32], [33], [27], and their relation to the theory of Star Products, we belive it worthwhile that the detailed proofs be made available.

In the proof given here of those results, a basic result implicit in reference [11] plays a major role. The reader is refered to [18], [19] for a statement and proofs of that basic result.

All constructions here are in the framework of Star Products [13-15], [16], [25]. For algebraic and analytic techniques for use in the program of *Strict Quantization* [34], see references [35-38].

2. The Geometric Framework

A.- Let $(G;\Lambda)$ be a simply connected Lie group G, endowed with a left-invariant symplectic structure whose Poisson tensor is Λ, [39-42]. Let $\wedge_r^I(G) \equiv \mathfrak{g}\wedge \overset{r}{\cdots} \wedge\mathfrak{g}$ be the space of skew-symmetric invariant r-tensors on G (r-tensors for short) and $\wedge^{I,r}(G) \equiv \mathfrak{g}^*\wedge \overset{r}{\cdots} \wedge\mathfrak{g}^*$, the space of invariant r-differential forms on G. We know that

$$[\wedge_r^I(G);\wedge_s^I(G)] \subset \wedge_{r+s-1}^I(G)$$

where the bracketing denotes the Schouten bracket [39]. If we define

$$\partial : \wedge_r^I(G) \longrightarrow \wedge_{r+1}^I(G)$$

by $\partial A = -[\Lambda; A]$ we have *the invariant* Λ-complex [39] $\left(\oplus \wedge_r^I(G);\partial\right)$ and the Λ-cohomology $H_S^*(G)$ on G. If $\mu : \wedge_r^I(G) \rightarrow \wedge^{I,r}(G)$ is the usual isomorphism [39], it induces an isomorphism $\bar{\mu} : H_S^r(G) \rightarrow H_R^r(G)$ ($R \equiv$ de Rahm).

B.- $\{e_i; i = 1,\ldots,n\}$ will always be a basis of $\mathfrak{g}$, $\{e^i\}$ its dual basis and $\{\bar{e}^i\}$ the corresponding left-invariant vector fields on G. $\mathfrak{A}(\mathfrak{g})$ is the enveloping algebra of $\mathfrak{g}$, allways identified with the algebra of the invariant differential operators on G, and in general $\mathfrak{A}(\mathfrak{g})^{\otimes k}$ is identified with the algebra of invariant k-multidifferential operators on G.

In normal coordinates on a neighborhood $U(e) \subset G$ we obtain:

$$\bar{e}_j(x) = B(x)_j^k \frac{\partial}{\partial x^k}$$

where:

$$B(x)_j^k = \delta_j^k + \frac{1}{2}x^i C_{ij}^k + \sum_{n\geq 1} \frac{1}{(2n)!} b_{2n} \left(\sum_{i_1,\ldots,i_{2n}=1}^{\dim \mathfrak{g}} N_{i_1,\ldots,i_{2n};j}^k x^{i_1}\cdots x^{i_{2n}} \right)$$

$$N_{i_1,\ldots,i_{2n};j}^k = C_{i_1 j_1}^k \cdot C_{i_2 j_2}^{j_1} \cdots C_{i_{2n-1} j_{2n-1}}^{j_{2n-2}} \cdot C_{i_{2n} j}^{j_{2n-1}}$$

b_{2n} are the Bernouilli numbers [43] and $[e_i; e_j] = C_{ij}^k e_k$.

C.- Let

$$\overline{M}(x) = \frac{1}{2!}\overline{M}^{ij}(x)\frac{\partial}{\partial x^i}\wedge\frac{\partial}{\partial x^j}$$

be one invariant 2-tensor on G, where $\overline{M}^{ij}(x) = M^{ab}\cdot B(x)_a^i \cdot B(x)_b^j$.

Let $\overline{N}(x)$ be another one. Then:

$$\left[\overline{M};\overline{N}\right]^{a\,b\,c}(x)=\frac{1}{3!}\left[\overline{M};\overline{N}\right]^{i\,j\,k}(x)\,\frac{\partial}{\partial x^i}\wedge\frac{\partial}{\partial x^j}\wedge\frac{\partial}{\partial x^k}\,.$$

We then have:

$$\left[\overline{M};\overline{N}\right]^{a\,b\,c}(x)=[M;N]^{i\,j\,k}\cdot B_i^a(x)\cdot B_j^b(x)\cdot B_k^c(x)\,.$$

If as usual in a Yang-Baxter context, we define:

$$M^{1\,2}=M\otimes 1\;;\quad M^{2\,3}=1\otimes M\;;\quad M^{1\,3}=P^{2\,3}\,M^{1\,2}\,P^{2\,3}\,,$$

we can write:

$$\begin{aligned}[M;N]=-\,\{&[M^{1\,2};N^{1\,3}]+[M^{1\,2};N^{2\,3}]+[M^{1\,3};N^{2\,3}]+\\&+[N^{1\,2};M^{1\,3}]+[N^{1\,2};M^{2\,3}]+[N^{1\,3};M^{2\,3}]\}\,.\end{aligned}$$

Therefore:

Proposition 1.- *With the foregoing notations:*

a) $$[\bar{M};\bar{N}]=0\iff[M;N]=0$$

b) *Let $r\in g\wedge g$ (or s_1 or Λ_1) be the value of Λ at $e\in \boldsymbol{G}$; then:*

$$[\Lambda;\Lambda]=0\iff\left[r^{1\,2};r^{1\,3}\right]+\left[r^{1\,2};r^{2\,3}\right]+\left[r^{1\,3};r^{2\,3}\right]=0\;;\quad\text{(C.Y.B.E.)}$$

c) *$h_R\in\mathfrak{g}\wedge\mathfrak{g}$ is a Schouten cocycle if and only if*

$$\left[h_R^{1\,2};S_1^{1\,3}\right]+\left[h_R^{1\,2};S_1^{2\,3}\right]+\left[h_R^{1\,3};S_1^{2\,3}\right]+\left[S_1^{1\,2};h_R^{1\,3}\right]+\left[S_1^{1\,2};h_R^{2\,3}\right]+\left[S_1^{1\,3};h_R^{2\,3}\right]=0\,,$$

where, as usual, $S_1^{1\,2}=S_1\otimes 1$, $S_1^{1\,3}=P^{2\,3}\,S_1^{1\,2}\,P^{2\,3}$, $S_1^{2\,3}=1\otimes S_1$. ■

3. The Central Extension $\bar{\mathfrak{g}}_t=\mathfrak{g}\times_{\beta_t}\mathbb{R}E$. Adjoint and Coadjoint Representations of $\boldsymbol{G}$ and $\overline{\boldsymbol{G}}_t$

A.- Let $(\boldsymbol{G};\Lambda)$ be as in §1. Let $\beta_1\in\mathfrak{g}^*\wedge\mathfrak{g}^*$ be the 2-cocycle $\beta_1=\mu(r)$ and β_i, $i\in\mathbb{N}$, any 2-cocycles. Consider:

$$\beta_t=\beta_1+\beta_2+\cdots+\beta_R\,t^{R-1}$$

where t is a real parameter. Also consider the isomorphims:

$$\tilde{\beta}_1:\mathfrak{g}\longrightarrow\mathfrak{g}^*\;;\quad\tilde{\Lambda}_1:\mathfrak{g}^*\longrightarrow\mathfrak{g}\,,$$

$\tilde{\beta}_1\circ\tilde{\Lambda}_1=1_{\mathfrak{g}^*}$; $\tilde{\Lambda}_1\circ\tilde{\beta}_1=1_{\mathfrak{g}}$. Or in components: $\left(\tilde{\beta}\right)_{i\,j}\left(\tilde{\Lambda}_1\right)^{k\,j}=\delta_i^k$.

For small t, similar isomorphisms $\tilde{\beta}_t$ and $\tilde{\Lambda}_t$ are defined. Thus,

$$\tilde{\Lambda}_t = \sum_{j\geq 1} \tilde{\Lambda}_j \cdot t^{j-1}$$

and

$$\tilde{\Lambda}_{l+1} = -\tilde{\Lambda}_1 \cdot \tilde{\beta}_{l+1} \cdot \tilde{\Lambda}_1 - \tilde{\Lambda}_1 \cdot \sum_{i=2}^{l} \tilde{\beta}_i \cdot \tilde{\Lambda}_{l+2-i} \,.$$

At this point we may remark that the first term in this sum is the Schouten cocycle corresponding to the de Rahm cocycle $\tilde{\beta}_{l+1}$.

B.- Let $\overline{G}_t$ be the simply-connected Lie group with Lie algebra $\bar{\mathfrak{g}}_t = \mathfrak{g} \oplus \mathbb{R}\, E$,

$$[x + aE; y + bE] = [x; y] + \beta_t(x; y)E \,.$$

With the usual notations, adjoint and coadjoint actions of $\bar{G}_t$ are given by:

$$\overline{Ad} \cdot \overline{exp}(x + aE) \cdot (y + bE) = Ad\, exp\, x \cdot y + (b + \beta_t(x; A(x)\, y)\, E \,,$$

where

$$A(x) = \frac{exp\, ad\, x - I}{ad\, x} \in L(\mathfrak{g}; \mathfrak{g}) \,,$$

The coadjoint action defined by:

$$\overline{Ad}^* \cdot \overline{exp}(x + aE) = {}^t\left[\overline{Ad}\, \overline{exp}\, (-(x + aE))\right]$$

is given by:

$$\overline{Ad}^*\, \overline{exp}(x + a\, E) \cdot (f_1 + f_2) = f_1' + f_2' \,,$$

where:

$$f_1' = Ad^* exp\, x \cdot f_1 + f_2(E) f_{\beta_t}(-x)$$

$$f_2' = f_2$$

and

$$f_{\beta_t}(x) = \tilde{\beta}_t(x) \circ A(x) \in \mathfrak{g}^* \,.$$

In particular, if $f_1 = 0$ and $f_2 = E^*$, we have:

$$f_1' = f_{\beta_t}(-x)$$

$$f_2' = E^* \,.$$

C.- The following properties will be essential.

Proposition 2.- *For small t, the mapping*

$$x \in \mathfrak{g} \xrightarrow{\Phi_t} f_{\beta_t}(-x) \in \mathfrak{g}^* ,$$

is a local diffeomorphism at $x = 0$.

Proof.-

$$d_0\Phi_t \cdot y = \lim_{u\to 0} \frac{\Phi_t(u\,y) - \Phi_t(0)}{u} =$$

$$= \lim_{u\to 0} \frac{f_{\beta_t}(-u\,y) - f_{\beta_t}(0)}{u} = \tilde{\beta}_t(-y) \cdot \lim_{u\to 0} A(-u\,y) = -\tilde{\beta}_t(y) .$$

Therefore: $d_0\Phi_t = -\tilde{\beta}_t$, and this map has an inverse. ∎

Corollary 3.- *The mapping:*

$$exp\, W(0) = W(e) \subset \boldsymbol{G} \xrightarrow{K_t} \mathfrak{g}^* + E^*$$

$$exp\, x \longrightarrow \overline{Ad}^*\, \overline{exp}\, x \cdot E^*$$

is a diffeomorphism if $W(0) \subset \mathfrak{g}$ is small, t being small. ∎

Proposition 4.- *Let $W(e) \subset U(e)$ symmetric such that $W(e) \cdot W(e) \subset U(e)$ and $W(0) \subset U(0)$ such that $exp\, W(0) = W(e)$. Then $\forall z \in W(0)$ the following diagram is commutative:*

$$\begin{array}{ccc} W(e) = exp\, W(0) \subset U(e) & \xrightarrow{K_t} & K_t\,(W(e)) \subset V(E^*) \subset \mathcal{O}_{E^*} \\ L_{exp\, z} \downarrow & & \downarrow \overline{Ad}^* \cdot \overline{exp}\, z \\ U(e) = exp\, U(0) & \xrightarrow{K_t} & \mathcal{O}_{E^*} \end{array}$$

Proof.- Let x be an element in $W(0)$. From the hypothesis there is $\gamma(z;x) \in U(0)$ such that $\gamma(z;x) = lg(exp\, z \cdot exp\, x)$. From Corollary 7 below we have:

$$\overline{exp}\, z \cdot \overline{exp}\, x = \overline{exp}\, \overline{\gamma}_t(z;x) ,$$

where

$$\overline{\gamma}_t(z;x) = \gamma(z;x) + \hat{\gamma}_t(z;x)E ; \qquad \hat{\gamma}_t(z;x) \in \mathbb{R} .$$

We then obtain

$$\left(\overline{Ad}^* \cdot \overline{exp}\, z \circ K_t\right) exp\, x = \overline{Ad}^*\, \overline{exp}\, z \circ \overline{Ad}^*\, \overline{exp}\, x \cdot E^* =$$

$$= \overline{Ad}^*\, \overline{exp}\, \overline{\gamma}_t(z;x) \cdot E^* = \overline{Ad}^* \cdot \overline{exp}\, (\gamma(z;x) + \hat{\gamma}(z;x)\, E) \cdot E^* =$$

$$= \overline{Ad}^*\, \overline{exp}\, \gamma(z;x) \cdot E^* ,$$

where the last equality follows from B. In the same way:

$$(K_t \circ L_{exp\, z})\, exp\, x = K_t(exp\, z \cdot exp\, x) = K_t\,(exp\, \gamma(z;x)) = \overline{Ad}^* \cdot \overline{exp}\, \gamma(z;x) \cdot E^* ,$$

which ends the proof. ∎

4. Hausdorff Series of Algebras $\mathfrak{g}$ and $\overline{\mathfrak{g}}_t$

With the usual notations [43,44],

$$\overline{\gamma}_t(\overline{x};\overline{y}) = \sum_{n=1}^{\infty} \overline{z}_n(\overline{x};\overline{y})$$

$$\overline{z}_n(\overline{x};\overline{y}) = \frac{1}{n} \sum_{p+q=n} \left(\overline{z}'_{p,q}(x;y) + \overline{z}''_{p,q}(x;y)\right)$$

$$\overline{z}'_{p,q}(\overline{x};\overline{y}) = \sum \frac{(-1)^{m+1}}{m} \frac{\left(\overline{ad\,x}\right)^{p_1} \left(\overline{ad\,y}\right)^{q_1} \cdots \left(\overline{ad\,x}\right)^{p_m} \overline{y}}{(p_1!\,q_1!) \cdots (p_{m-1}!\,q_{m-1}!)\; p_m!}, \tag{2}$$

where the sum is extended to $p_1+\cdots+p_m=p,\ q_1+\cdots+q_{m-1}=q-1,\ p_i+q_i \geq 1,\ i=1,\ldots,m-1.$

$$\overline{z}''_{p,q}(\overline{x};\overline{y}) = \sum \frac{(-1)^{m+1}}{m} \frac{\left(\overline{ad\,x}\right)^{p_1} \left(\overline{ad y}\right)^{q_1} \cdots \left(\overline{ad\,y}\right)^{q_{m-1}} \overline{x}}{(p_1!\,q_1!) \cdots (p_{m-1}!\,q_{m-1}!)}$$

where now $p_1+\cdots+p_{m-1}=p-1,\ q_1+\cdots+q_{m-1}=q,\ p_i+q_i\geq 1,\ i=1,\ldots,m-1,$

$$\overline{\gamma}_t\left(\overline{\gamma}_t\left(\overline{x};\overline{y}\right);\overline{z}\right) = \overline{\gamma}_t\left(\overline{x};\overline{\gamma}_t\left(\overline{y};\overline{z}\right)\right).$$

And similar expressions for $\mathfrak{g}$.

Proposition 5.- *Let $\overline{z} = x + aE;\ \overline{y} = y + bE;\ \overline{z} = z + cE$. Then:*

$$\left(\overline{ad\,x}\right)^r \cdot \left(\overline{ad\,y}\right)^s \cdot \overline{z} = (ad\,x)^r \cdot (ad\,y)^s \cdot z + \beta_t\left(x; (ad\,x)^{r-1}(ad\,y)^s\, z\right) E,$$

for $r \geq 1;\quad s \geq 0$.

Proof.- Straightforward. ∎

We therefore obtain,

Proposition 6.- *If $p + q \neq 1$*

$$\overline{z}'_{p,q}\left(\overline{x};\overline{y}\right) = z'_{p,q}(x;y) + \hat{z}'_{p,q}(x;y)E$$

$$\overline{z}''_{p,q}\left(\overline{x};\overline{y}\right) = z''_{p,q}(x;y) + \hat{z}''_{p,q}(x;y)E$$

$$\overline{\gamma}_t\left(\overline{x};\overline{y}\right) = \gamma(x;y) + \left(\hat{\gamma}_t(x;y) + a + b\right)E,$$

where $\hat{z}'_{p,q}(x;y)$ is obtained by substituting the operators $\tilde{\beta}_t(x)$ or $\tilde{\beta}_t(y)$ for the first operator $ad\,x$ or $ad\,y$ on the left of the expression (2) of $z'_{p,q}(x;y)$. Similary for $\hat{z}''_{p,q}(x;y)$.

Corollary 7.- *Suppose $\overline{x} = x;\ \overline{y} = y;\ \overline{z} = z$. Then:*

1) $$\overline{\gamma}_t(x;y) = \gamma(x;y) + \hat{\gamma}(x;y)E$$

2) $$\hat{\gamma}_t\left(\gamma(x;y);z\right) + \hat{\gamma}_t(x;y) = \hat{\gamma}_t\left(x;\gamma(y;z)\right) + \hat{\gamma}_t(y;z)$$ ■

If $R \in \mathbb{N}$, we can define

$$\overline{\gamma}_{t;\,R+1}\left(\overline{x};\overline{y}\right) = \sum_{n=1}^{R+1} \overline{z}_n\left(\overline{x};\overline{y}\right),$$

and give similar definitions for $\gamma_{t;\,R+1}(x;y)$ and $\hat{\gamma}_{t;\,R+1}(x;y)$.

Proposition 8.- *Let $\overline{x} = x$; $\overline{y} = y$; $\overline{z} = z$. Then*

1) $$\overline{\gamma}_{t;\,R+1}(x;y) = \gamma_{R+1}(x;y) + \hat{\gamma}_{t;\,R+1}(x;y)E$$

2) *The term*

$$\gamma_{R+1}\left(\gamma_{R+1}(x;y);z\right) + \left[\hat{\gamma}_{t;\,R+1}\left(\gamma_{R+1}(x;y);z\right) + \hat{\gamma}_{t;\,R+1}(x;y)\right]E$$

contains all terms of total degree $\leq R+1$ in the formal series $\overline{\gamma}_t\left(\overline{\gamma}_t(x;y);z\right)$.

3) *The term*

$$\gamma_{R+1}\left(x;\gamma_{R+1}(y;z)\right) + \left[\hat{\gamma}_{t;\,R+1}\left(x;\gamma_{R+1}(y;z)\right) + \hat{\gamma}_{t;\,R+1}(y;z)\right]E$$

contains all terms of total degree $\leq R+1$ in the formal series $\overline{\gamma}_t\left(x;\overline{\gamma}_t(y;z)\right)$. In particular, polynomials 2) and 3) coincide to the degree $(R+1)$. ■

5. Power Expansion of $exp - \frac{2\pi i}{\varepsilon}(\xi + E^*) \cdot \overline{\gamma}_t(\varepsilon\, x;\varepsilon\, y)$

ε is a real parameter; $\xi \in \mathfrak{g}^*$.

We obtain:

$$\begin{aligned} &exp - \frac{2\pi i}{\varepsilon}(\xi + E^*) \cdot \overline{\gamma}_t(\varepsilon\, x;\varepsilon\, y) = \\ &\quad = exp\left(-2\pi i \xi(x+y)\right) \cdot exp\left(-2\pi i \sum_{k=1}^{\infty} J_k(x;y;\xi)\,\varepsilon^k\right) = \\ &\quad = exp\left(-2\pi i \xi(x+y)\right) \cdot \left[1 + \sum_{R=1}^{\infty} B_R(x;y;\xi)\,\varepsilon^R\right], \qquad (3) \end{aligned}$$

where:

$$J_k(x;y;\xi) = \xi \cdot z_{k+1}(x;y) + \hat{z}_{k+1}(x;y)$$

$$B_R(x;y;\xi) = \sum_{l=1}^{R} \left[\frac{1}{l!}(-2\pi i)^l \sum J_{k_1}(x;y;\xi) \cdots J_{k_l}(x;y;\xi)\right],$$

where the 2nd sum is extended to $k_1+\cdots k_l=R\,;\ k_1,\ldots,k_l\geq 1$. $B_R(x;y;\xi)$ is a polynomial of maximal degree $2R$ and minimal degree $(R+1)$ on the components of x, y. In the component $\xi \in \mathfrak{g}^*$ this polynomial is of degree R. (To be used in Proposition 10).

Also:

$$
\begin{aligned}
& exp - \frac{2\pi i}{\varepsilon}(\xi + E^*)\,\overline{\gamma}_{t;\,R+1}(\varepsilon\, x; \varepsilon\, y) = \\
& \qquad = exp\,(-2\pi i\,\xi(x+y)) \cdot exp\left(-2\pi i \sum_{n=1}^{R} J_n(x;y;\xi)\,\varepsilon^n\right) = \\
& \qquad = exp\,(-2\pi i\xi(x+y))\left[1 + \sum_{s=1}^{\infty} \hat{B}_s(x;y;\xi)\,\varepsilon^s\right]
\end{aligned} \tag{4}
$$

It is easy to prove:

Proposition 9.- *Expressions* (3) *and* (4) *coincide to the order* ε^R. *That is* $B_1 = \hat{B}_1\,; \ldots; B_R = \hat{B}_R$. ■

6. An Invariant Star Product on $\mathfrak{g}^* + E^*$

A.- Let $\mathcal{F}(\overline{\mathcal{F}})$ be the Fourier transform [45]. Let us define bidifferential operators on $\mathfrak{g}^* + E^*$ by the expession:

$$
D_R\,(\varphi_1;\varphi_2)\,(\xi) = \int_{\mathbb{R}^{2n}\times\mathbb{R}^{2n}} e^{-2\pi i\xi(x+y)} \cdot B_R(x;y)\,\left(\overline{\mathcal{F}}\varphi_1\right)(x)\,\left(\overline{\mathcal{F}}\varphi_2\right)(y)dx\,dy\,,
$$

where φ_1, φ_2 are *convenable* functions or distributions.

Proposition 10.- *If* φ_1, φ_2 *are polynomials on the components of* $\xi \in \mathfrak{g}^*$, $D_R\,(\varphi_1;\varphi_2)$ *is also a polynomial, and if R is large enought* $D_R\,(\varphi_1;\varphi_2) = 0$. ■

With φ_1, φ_2 being polynomials, let us define:

$$
(\varphi_1 \triangleright \varphi_2)\,(\xi) = \varphi_1 \cdot \varphi_2 + \sum_{R\geq 1} D_R\,(\varphi_1;\varphi_2)\,\varepsilon^R \tag{5}
$$

or in the obvious formal sense,

$$
\begin{aligned}
(\varphi_1 \triangleright \varphi_2)\,(\xi) = & \int_{\mathbb{R}^{2n}\times\mathbb{R}^{2n}} exp\,(-2\pi i\xi(x+y)) \cdot \\
& \cdot exp\left(-2\pi i \sum_{k=1}^{\infty} J_k(x;y;\xi)\,\varepsilon^k\right) \cdot \left(\overline{\mathcal{F}}\varphi_1\right)(x) \cdot \left(\overline{\mathcal{F}}\varphi_2\right)(y)\,dx\,dy\,,
\end{aligned}
$$

or

$$(\varphi_1 \triangleright \varphi_2)(\xi) = \int_{\mathbb{R}^{2n}\times\mathbb{R}^{2n}} exp\left(-\frac{2\pi i}{\varepsilon}(\xi + E^*)\,\overline{\gamma}_t(\varepsilon\, x; \varepsilon\, y)\right)\cdot(\overline{\mathcal{F}}\varphi_1)(x)\cdot(\overline{\mathcal{F}}\varphi_2)(y)\,dx\,dy\,, \tag{6}$$

as well.

In the case $\mathfrak{g} \equiv \mathbb{R}^{2n}$; $G \equiv \mathbb{R}^{2n}$, the above integral coincides with the usual Moyal $*$-product related with Weyl's quantization.

It is clear that for the polynomials φ_1, φ_2 we can choose R large enough so that

$$(\varphi_1 \triangleright \varphi_2)(\xi) = \int_{\mathbb{R}^{2n}\times\mathbb{R}^{2n}} exp\left(-\frac{2\pi i}{\varepsilon}(\xi + E_*)\cdot\overline{\gamma}_{t;\,R+1}(\varepsilon\, x; \varepsilon\, y)\right)(\overline{\mathcal{F}}\varphi_1)(x)\,(\overline{\mathcal{F}}\varphi_2)(y)\,dx\,dy\,, \tag{7}$$

where the integral is now considered in the usual distributional sense. From the usual calculations with distributions with compact support (punctual support!) and the Fourier transform, we can prove:

Proposition 11.- *Composition $\triangleright$ defined by (5), (6) or (7), endows the space of polynomials on $\mathfrak{g}^*$ with an associative unital structure:*

$$\varphi_1 \triangleright 1 = 1 = 1 \triangleright \varphi\,; \qquad (\varphi_1 \triangleright \varphi_2) \triangleright \varphi_3 = \varphi_1 \triangleright (\varphi_2 \triangleright \varphi_3)\,. \qquad \blacksquare$$

B.- Suppose now that $\varphi_1, \varphi_2, \varphi_3 \in \mathcal{C}^\infty(\mathfrak{g}^*)$. Let $\xi_0 \in \mathfrak{g}^*$. Consider Taylor expansions to the relevant orders:

$$\varphi_i(\xi) = P_i(\xi - \xi_0) + R_i(\xi; \xi_0)\ ;\ \ i = 1, 2, 3\,.$$

P_i are polynomials. Define as before (D_k are bidiferential operators):

$$(\varphi_1 \triangleright \varphi_2)(\xi) = \sum_{K\geq 0} D_k(\varphi_1; \varphi_2)(\xi)\cdot\varepsilon^K\,,$$

and now also:

$$(\varphi_1 \triangleright \varphi_2) \triangleright \varphi_3 = \sum_{K \geq 0} (D_k(\varphi_1; \varphi_2) \triangleright \varphi_3) \varepsilon^K =$$

$$= \sum_{K \geq 0} (D_L(D_K(\varphi_1; \varphi_2); \varphi_3) \varepsilon^L) \varepsilon^K =$$

$$= \sum_{S \geq 0} \left[\sum_{K+L=S} D_L(\varphi_1; D_K(\varphi_2; \varphi_3)) \right] \varepsilon^S .$$

For any fixed S,

$$\sum_{K+L=S} D_L(D_K(\varphi_1; \varphi_2); \varphi_3)(\xi_0) ,$$

is clearly defined by means of the polynomials P_i and is therefore equal to

$$\sum_{K+L=S} D_L(D_K(P_1; P_2); P_3)(\xi_0) , \tag{8}$$

In the same way:

$$\sum_{K+L=S} D_L(\varphi_1; D_K(\varphi_2; \varphi_3))(\xi_0)$$

is equal to

$$\sum_{K+L=S} D_L(P_1; D_K(P_2; P_3))(\xi_0) . \tag{9}$$

From Proposition 11, expressions (8) and (9) coincide.

We have thus proved:

Proposition 12.- *$(\mathcal{C}^\infty(\mathfrak{g}^*)[[\varepsilon]]; \triangleright)$ is an associative unital algebra. That is, if $\varphi_1, \varphi_2, \varphi_3 \in \mathcal{C}^\infty(\mathfrak{g}^*)$, we have $\forall S \in \mathbb{N}$,*

$$\sum_{L+K=S} D_L(D_K(\varphi_1; \varphi_2); \varphi_3) = \sum_{L+K=S} D_L(\varphi_1; D_K(\varphi_2; \varphi_3)) ,$$

$$1 \triangleright \varphi_1 = \varphi_1 \triangleright 1 = \varphi_1 .$$

■

C.- If h is a real parameter, the following formal expansion is obviously true:

$$exp\left[\frac{1}{h}(\xi + E^*)(\overline{\gamma}_t(-2\pi ihx; -2\pi ihy) + 2\pi ihx + 2\pi ihy)\right] =$$

$$= 1 + \sum_{R=1}^{\infty} (-2\pi i)^R B_R(x; y; \xi) h^R .$$

From the homogeneity of order $(k_a + 1)$ in x, y of polynomial $J_a(x; y; \xi)$ we obtain:

$$\begin{aligned}
&(-2\pi i)^R B_R(x; y; \xi) = \\
&\quad = \sum_{l=1}^{\infty} \frac{1}{l!} \left(\sum J_{k_1}(-2\pi i x; -2\pi i y; \xi) \cdots J_{k_l}(-2\pi i x; -2\pi i y; \xi) \right) = \\
&\quad \equiv P_R(-2\pi i x; -2\pi i y; \xi),
\end{aligned}$$

where the sum is extended to $k_1 + \cdots + k_l = R;\ k_1, \ldots, k_l \geq 1$. P_R is now a real polynomial and:

$$\begin{aligned}
D_R(\varphi_1; \varphi_2)(\xi) = B_R\left(-\frac{1}{2\pi i}\frac{\partial}{\partial \xi}; -\frac{1}{2\pi i}\frac{\partial}{\partial \xi} \right) (\varphi_1 \otimes \varphi_2)(\xi) = \\
= \frac{1}{(-2\pi i)^R} P_R\left(\frac{\partial}{\partial \xi}; \frac{\partial}{\partial \xi}; \xi \right) (\varphi_1 \otimes \varphi_2)(\xi).
\end{aligned}$$

We have then proved (see Theorem 6 in [11]):

Theorem 13.- *With the foregoing notations:*

$$exp\,(\xi + E^*) \left[h^{-1}\, \overline{\gamma}_t(h\, x; h\, y) - x - y \right] = 1 + \sum_{R=1}^{\infty} P_R(x; y; \xi)\, h^R .$$

Operators $P_R(\partial/\partial\xi; \partial/\partial\xi; \xi)$ satisfie the associativity conditions $\forall S \in \mathbb{N}$:

$$\sum_{L+K=S} P_L\left(P_K(\varphi_1; \varphi_2); \varphi_3\right) = \sum_{L+K=S} P_L\left(\varphi_1; P_K(\varphi_2; \varphi_3)\right),$$

$\varphi_1, \varphi_2, \varphi_3 \in C^\infty(\mathfrak{g}^*)$. Therefore:

$$\varphi_1 * \varphi_2 = \sum_{R \geq 0} P_R(\varphi_1; \varphi_2)\, h^R, \tag{10}$$

is a star product on $\mathfrak{g}^* + E^*$. (Recall that β_t is present in P_R). ■

D.- Let us write:

$$\overline{Ad}^*\, \overline{exp}(z + aE) \cdot (\xi + E^*) = \sigma(z)\xi + E^*,$$

$$\sigma(z)\xi = Ad^*\, exp\, z \cdot \xi + f_{\beta_t}(-z).$$

A straighforward calculation gives:

Propsition 14.-

a) $J_n(x; y; \sigma(z)\xi) = J_n\left(Ad\, exp(-z) \cdot x; Ad\, exp(-z) \cdot y; \xi\right)$

b) $\quad B_R(x;y;\sigma(z)\xi) = B_R\left(Ad\,exp(-z)\cdot x; Ad\,exp(-z)\cdot y;\xi\right)$

c) $\quad \hat{z}_{n+1}\left(Ad\,exp\,z\cdot x; Ad\,exp\,z\cdot y\right) = \hat{z}_{n+1}(x;y) + \beta_t\left(z; A(z)\cdot z_{n+1}(x;y)\right)$

■

Now:

$$D_R\left(\varphi_1\circ\sigma(z);\varphi_2\circ\sigma(z)\right)(\xi) = \int_{\mathbb{R}^{2n}\times\mathbb{R}^{2n}} exp\left(-2\pi i\xi(x+y)\right)\cdot$$
$$\cdot B_R(x;y;\xi)\cdot\left(\overline{\mathcal{F}}\left(\varphi_1\circ\sigma(z)\right)(x)\right)\cdot\left(\overline{\mathcal{F}}\left(\varphi_2\circ\sigma(z)\right)\right)(y)\,dx\,dy\,.$$

And

$$\overline{\mathcal{F}}\left(\varphi_1\circ\sigma(z)\right)(x) =$$
$$= |det\,Ad^*\,exp(-z)|\cdot$$
$$\cdot\int_{\mathbb{R}^{2n}} exp\,2\pi ix\left(Ad^*\,exp(-z)\left(\tau - f_{\beta_t}(-z)\right)\right)\cdot\varphi_1(\tau)\,d\tau =$$
$$= |det\,Ad^*\,exp(-z)|\cdot$$
$$\cdot\,exp\left(-2\pi i\left(Ad\,exp\,z\cdot x; f_{\beta_t}(-z)\right)\right)\cdot\left(\overline{\mathcal{F}}\varphi_1\right)\left(Ad\,exp\,z\cdot x\right)\,,$$

(φ_1,φ_2 are polynomials, are in $C_0^\infty(\mathfrak{g}^*)$, are in $S(\mathfrak{g}^*)$, or in $S'(\mathfrak{g}^*),\ldots$).

With $x' = Ad\,exp\,z\cdot x$ and $y' = Ad\,exp\,z\cdot y$ we get:

$$D_R\left(\varphi_1\circ\sigma(z);\varphi_2\circ\sigma(z)\right)(\xi) = \cdots = D_R\left(\varphi_1;\varphi_2\right)\left(\sigma(z)\cdot\xi\right).$$

Theorem 15.-

a) *Operators* $J_n\left(\partial/\partial\xi;\partial/\partial\xi;\xi\right)$ *and* $B_R\left(\partial/\partial\xi;\partial/\partial\xi;\xi\right)$ *are invariants under* $\overline{Ad}^*\,\overline{G}_t$.

b) *The star product* (10) *is invariant under* $\overline{Ad}^*\,\overline{G}_t$. *That is to say:*

$$\left(\varphi_1\circ\overline{Ad}^*\,\overline{g}\right)*\left(\varphi_2\circ\overline{Ad}^*\,\overline{g}\right) = \left(\varphi_1*\varphi_2\right)\circ\overline{Ad}^*\,\overline{g}\,,$$
$$\varphi_1,\varphi_2\in C^\infty(\mathfrak{g}^*)\,;\qquad\forall\overline{g}\in\overline{G}_t\,.$$

■

7. An Invariant Star Product on the Group $(\boldsymbol{G};\beta_t)$

If $U(e)\subset\boldsymbol{G}$ is a neigbourhood as in Proposition 4., let us define bidifferential operators by:

$$F_R^t\left(\psi_1;\psi_2\right)(exp\,x) = P_R\left(\psi_1\circ K_t^{-1};\psi_2\circ K_t^{-1}\right)(\xi)\,;\quad\xi = K_t(exp\,x)\,,$$

$\psi_1, \psi_2 \in \mathcal{C}^\infty(U(e))$. F_R^t depends on t through K_t^{-1} and β_t.

Let us now define:

$$(\psi_1 *_t \psi_2)(exp\, x) = \left((\psi_1 \circ K_t^{-1}) * (\psi_2 \circ K_t^{-1})\right)(\xi). \tag{11}$$

We therefore have a star product on $(U(e); \beta_t)$.

If $z \in W(0)$, we have (Propositions 4, 15):

$$\begin{aligned}\left(F_R^t(\psi_1;\psi_2) \circ L_{exp\, z}\right)(exp\, x) &= F_R^t(\psi_1;\psi_2)(L_{exp\, z} \circ exp\, x) = \\ &= P_R\left(\psi_1 \circ K_t^{-1}; \psi_2 \circ K_t^{-1}\right) \cdot \left(\overline{Ad}^*\, \overline{exp}\, z \circ K_t\right)(exp\, x) = \\ &= P_R\left(\psi_1 \circ K_t^{-1} \circ \overline{Ad}^*\, \overline{exp}\, z; \psi_2 \circ K_t^{-1} \circ \overline{Ad}^*\, \overline{exp}\, z\right) K_t(exp\, x).\end{aligned}$$

From Proposition 4,

$$K_t^{-1} \circ \overline{Ad}\, \overline{exp}\, z = L_{exp\, z} \circ K_t^{-1},$$

on $K_t(exp\, x)$. Therefore:

$$\left(F_R^t(\psi_1;\psi_2) \circ L_{exp\, z}\right)(exp\, x) = F_R^t(\psi_1 \circ L_{exp\, z}; \psi_2 \circ L_{exp\, z})(exp\, x),$$

$\forall x, z \in W(0)$. This relation defines F_R^t in a unique way as an invariant bidifferential operator on $\boldsymbol{G}$, because $U(e)$ generates $\boldsymbol{G}$. (See Theorem 6 in [11].)

Theorem 16.- *Expression* (11) *defines an invariant star product on* $(\boldsymbol{G}; \beta_t)$. *Identifying* F_R^t *with the corresponding element in* $\mathfrak{A}(\mathfrak{g}) \otimes \mathfrak{A}(\mathfrak{g})$ *and writing:*

$$F^t(x;y) = 1 + \sum F_R^t(x;y)\, h^R,$$

we can now write the associativity condition $(\psi_1 *_t \psi_2) *_t \psi_3 = \psi_1 *_t (\psi_2 *_t \psi_3)$, *as:*

$$F^t(x+y;z) \cdot F^t(x;y) = F^t(x;y+z)\, F^t(y;z).$$

And $1 * \psi = \psi = \psi * 1$ *as*

$$F^t(0;y) = F^t(x;0) = 1.$$

See [11], [18], [19]. ■

8. A New Star Product on $(\boldsymbol{G}; \beta_1)$

So far, the presence of parameter t, has not been sufficiently justified. It is an underlying idea of Theorem 7 in [11].

The bidifferential operator $F_R^t \in \mathfrak{A}(\mathfrak{g}) \otimes \mathfrak{A}(\mathfrak{g})$ in the star product of Theorem 16 is, with obvious notations,

$$F_R^t(x;y) = \sum_{l=1}^{R} \frac{1}{l!} \sum \left[\hat{z}_{k_1+1} \left((K_t^{-1})_* \frac{\partial}{\partial \xi}; (K_t^{-1})_* \frac{\partial}{\partial \xi} \right) \cdots \right.$$

$$\left. \cdots \hat{z}_{k_l+1} \left((K_t^{-1})_* \frac{\partial}{\partial \xi}; (K_t^{-1})_* \frac{\partial}{\partial \xi} \right) \right]_{e \in G},$$

where $(K_t^{-1})_*$ is the tangent map to the map K_t^{-1}, and the second sum is extended to $k_1 + \cdots k_l = R,\ k_1, \ldots, k_l \geq 1$.

Term $\hat{z}_{k_i+1}(x;y)$ has the form:

$$(\beta_t)_{a\,b}\, x^a \cdot L^b(x;y) \quad \text{or} \quad (\beta_t)_{a\,b}\, y^a \cdot L^b(x;y),$$

where $L^b(x;y)$ is a homogeneus polynomial of degree k_i with coefficients determined by the structure constants of $\mathfrak{g}$. Hence:

$$\hat{z}_{k_i+1} \left(\frac{\partial}{\partial \xi}; \frac{\partial}{\partial \xi} \right) = (\beta_t)_{a\,b} \left(\frac{\partial}{\partial \xi_a} \otimes 1 \right) \cdot L^b \left(\frac{\partial}{\partial \xi}; \frac{\partial}{\partial \xi} \right),$$

or ....

The corresponding operator on G is defined by:

$$\left[\hat{z}_{k_i+1} \left((K_t^{-1})_* \frac{\partial}{\partial \xi}; (K_t^{-1})_* \frac{\partial}{\partial \xi} \right) \right]_{e \in G} =$$

$$= (\beta_t)_{a\,b} \left[\left((K_t^{-1})_* \frac{\partial}{\partial \xi_a} \otimes 1 \right) \cdot L^b \left((K_t^{-1})_* \frac{\partial}{\partial \xi}; (K_t^{-1})_* \frac{\partial}{\partial \xi} \right) \right]_{e \in G} \tag{12}$$

which can be evaluated by means of the local expression of K_t^{-1}:

$$x^b = \Lambda_t^{b\,a} \xi_a + (\beta_t)_{c\,k} \cdot \Lambda_t^{b\,a} \cdot x^c(\xi) \cdot \sum_{r \geq 1} N^k_{i_1 \ldots i_r;a}\, x^{i_1}(\xi) \cdots x^{i_r}(\xi), \tag{13}$$

by considering only the terms in the expansion (13) with $\leq (k_i + 1)$ components of ξ (in fact, evenless, but a finite number suffices). Terms with a greater number of components of ξ do not contribute to the operator (12), as it is a linear combination with constant coefficients of operators of the form:

$$\left[(\beta_t)_{a\,b} \cdot (K_t^{-1})_* \left(\frac{\partial^p}{\partial \xi^a\, \partial \xi^{i_1} \ldots \partial \xi^{i_{p-1}}} \otimes \frac{\partial^q}{\partial \xi^{j_1} \ldots \partial \xi^{j_q}} \right)_{p+q=k_i+1} \right]_{e \in G}, \tag{14}$$

or with operator $\frac{\partial}{\partial \xi^a}$ in the term after $\otimes$.

Looking at the expression (13) we remark that this operator depends on t, by way of $(k_i + 1) - 1 = k_i$ components of Λ_t; i.e. one for each opertor $\frac{\partial}{\partial \xi^l}$ in the expression (14), less one which will contract with $(\beta_t)_{a\,b}$.

In the same way to evaluate the operator:

$$\left[\hat{z}_{k_1+1}\left((K_t^{-1})_* \frac{\partial}{\partial\xi}; (K_t^{-1})_* \frac{\partial}{\partial\xi}\right) \cdots \hat{z}_{k_l+1}\left((K_t^{-1})_* \frac{\partial}{\partial\xi}; (K_t^{-1})_* \frac{\partial}{\partial\xi}\right)\right]_{e\in G}, \tag{15}$$

we need only consider in expression (13) the terms with a number of components $\leq (R+l)$ (in fact evenless, but allways a finite number). Also operator (15) depends on t by way of a product of:

$$(k_i+1)-1+\cdots+(k_l+1)-1=R,$$

components of Λ_t.

Thus, to evaluate $F_R^t(x;y)$ we need only consider in expression (13) the terms with a number of components $\leq 2R$ (in fact, evenless) and $F_R^t(x;y)$ depends on t by way of a product of R components of Λ_t. By expanding these products as power series of t (§ 3,A.-), we clearly obtain a formal expression such as:

$$F_R^t(x;y) = \sum_{L\geq 0} F_{RL}(x;y)\, t^L, \tag{16}$$

(of course the operators $\partial^a/(\partial x^{i_1}\cdots\partial x^{i_a})$ are here written through expression in § 2,B.- as linear combinations of invariant operators $\overline{e}_1\cdot\overline{e}_2\cdot\cdots\cdot\overline{e}_r$, but the parameter t does not appear in the change of basis), where $F_{RL}(x;y)\in\mathfrak{A}(\mathfrak{g})^{\otimes 2}$ and:

$$F_R^t(1;\psi)=0 \implies F_{RL}(1;\psi)=0;\quad R>0;\quad L\geq 0. \tag{17}$$

Theorem 17.- *For each $H\in\mathbb{N}$, the following relations holds:*

$$\begin{aligned}\sum_{R+S+T_1+T_2=H} & F_{RT_1}\left(\psi_1; F_{ST_2}\left(\psi_2;\psi_3\right)\right) = \\ &= \sum_{R+S+T_1+T_2=H} F_{RT_1}\left(F_{ST_2}\left(\psi_1;\psi_2\right);\psi_3\right),\end{aligned} \tag{18}$$

$\psi_1,\psi_2,\psi_3\in C^\infty(G)$.

Proof.- From Theorem 16 we have the star product:

$$\psi_1 *_t \psi_2 = \sum_{S\geq 0} F_S^t(\psi_1;\psi_2)\, h^S$$

That is, for each $L\in\mathbb{N}$,

$$\sum_{R+S=L} F_R^t\left(F_S^t\left(\psi_1;\psi_2\right);\psi_3\right) = \sum_{R+S=L} F_R^t\left(\psi_1; F_S^t\left(\psi_2;\psi_3\right)\right)$$

In view of relation (16), we can write:

$$\sum_{R+S=L}\sum_{T_1,T_2\geq 0} F_{R\,T_1}\left(\psi_1; F_{S\,T_2}(\psi_2;\psi_3)\right) t^{T_1+T_2} =$$
$$= \sum_{R+S=L}\sum_{T_1+T_2\geq 0} F_{R\,T_1}\left(F_{S\,T_2}(\psi_1;\psi_2);\psi_3\right) t^{T_1+T_2}\,.$$

That is, for each $L \in \mathbb{N}$ and each $T \in \mathbb{N}$, we have:

$$\sum_{R+S=L}\sum_{T_1+T_2=T} F_{R\,T_1}\left(\psi_1; F_{S\,T_2}(\psi_2;\psi_3)\right) =$$
$$= \sum_{R+S=L}\sum_{T_1+T_2=T} F_{R\,T_1}\left(F_{S\,T_2}(\psi_1;\psi_2);\psi_3\right)\,. \tag{19}$$

Let us now decompose H as $H = L + T$ in all possible forms with $L, T \in \mathbb{N}$. By forming the corresponding sommes in (19), we obtain (18). ∎

Theorem 18.- *Let us take $\psi_1, \psi_2 \in \boldsymbol{G}$. In the expression:*

$$\psi_1 *_t \psi_2 = \sum_{S,T\geq 0} F_{S\,T}(\psi_1;\psi_2)\, h^S t^T\,, \tag{20}$$

let us write $t = h$ and define the operators

$$F_A(x;y) = \sum_{S+T=A} F_{S\,T}(x;y)\,,$$

and the expression:

$$\psi_1 * \psi_2 = \sum_{A\geq 0} F_A(\psi_1;\psi_2)\, h^A\,. \tag{21}$$

This expression is an invariant star product on the group $\boldsymbol{G}$ with invariant symplectic structure β_1. We say it is determined by the cocycle $\beta_h = \beta_1 + \beta_2 h + \cdots + \beta_k h^{k-1} + \cdots$.

Proof.- Already proved: Theorem 17 and equalities (17). ∎

9. The Relation $\overline{F}_R = F_R - \frac{1}{2}\mu^{-1}(\beta_R)$

A.- It is clear enough that the cochains $F_A(x;y)$ in the star product in Theorem 18 can be determined by first putting $t = h$ in the star product (10) on $\mathfrak{g}^* + E^*$ in Theorem 13 and then $t = h$ in the local expression of K_t. Obviously we thus obtain expression (20) where $t = h$. Hence we first spand:

$$exp(\xi + E^*)\left[h^{-1}\overline{\gamma}_h(hx;hy) - x - y\right]$$

as power series of h. We then have:

$$J_n(x;y;\xi) = \xi \cdot z_{n+1}(x;y) + \hat{z}_{n+1}(x;y) = \xi \cdot z_{n+1}(x;y) + \sum_{j\geq 1} \hat{z}_{n+1;j}(x;y)\, h^{j-1}$$

where $\hat{z}_{n+1;j}(x;y)$ is defined as $\hat{z}_{n+1}(x;y)$ with β_j in place of β_t. We thus obtain:

$$exp\,(\xi + E^*)\,[h^{-1}\overline{\gamma}_h(hx;hy) - x - y] = 1 + \sum_{R=1}^{\infty} Q_R(x;y;\xi)\, h^R$$

where:

$$Q_R(x;y;\xi) = \sum_{l=1}^{R} \frac{1}{l!}\left[\sum H_{n_1}(x;y;\xi)\cdot \cdots \cdot H_{n_l}(x;y;\xi)\right],$$

where the second sum is extended to $n_1+\cdots+n_l=R$, $n_1,\ldots,n_l\geq 1$, and

$$H_n(x;y;\xi) = \xi \cdot z_{n+1}(x;y) + \sum \hat{z}_{k+1;j}(x;y),$$

where now the second sum is extended to $k+j-1=n$, $k,j\geq 1$.

Looking for the first term in which cocycle β_R appears in the formal series:

$$1 + \sum_{R+1}^{\infty} Q_R\left(\frac{\partial}{\partial\xi};\frac{\partial}{\partial\xi};\xi\right)\, h^R$$

and again looking for the term in which this cocycle appears in expression (21) (that is, after the change of variables (13) with $t = h$) we arrive at the following:

Theorem 19.- *Let $F(x;y) = 1+\sum_{i\geq 1} F_i(x;y)\, h^i$ be the invariant star product (21) on $(\boldsymbol{G};\beta_1)$ determined by the cocycle:*

$$\beta_h = \beta_1 + \beta_2\, h + \cdots + \beta_{R-1}\, h^{R-2} + \cdots .$$

Let $\overline{F}(x;y) = 1+\sum_{i\geq 1} \overline{F}_i(x;y)\, h^i$ be the invariant star product corresponding to the cocycle:

$$\overline{\beta}_h = \beta_h + \beta_R\, h^{R-1} .$$

Then:

$$\overline{F}_1(x;y) = F_1(x;y);\ldots;\overline{F}_{R-1}(x;y) = F_{R-1}(x;y),$$

and:

$$\overline{F}_R(x;y) = F_R(x;y) - \frac{1}{2}\mu^{-1}\,(\beta_R)\,.$$

In components

$$[\mu^{-1}\,(\beta_R)]^{lm} = -\Lambda_1^{lk}\,(\beta_R)_{kr}\,\Lambda_1^{rm},$$

$\mu^{-1}\,(\beta_R)$ is therefore the Schouten 2-cocycle corresponding to the de Rahm 2-cocycle β_R. ■

Remark.- Of course F and $\overline{F}$ can still be equivalent.

10. A Result about Star Products Equal to Certain Order

The notations are those of [19] in particular §9.

Let $F(x;y), \overline{F}(x;y)$ be two star products on $(\boldsymbol{G};\beta_1)$. Therefore

$$F_1(x;y) - F_1(y;x) = S_1(x;y) \qquad \overline{F}_1(x;y) - \overline{F}_1(y;x) = S_1(x;y)\,,$$

where $S_1(x;y) \in \mathfrak{g}\wedge\mathfrak{g}$ is the Poisson tensor of $(\boldsymbol{G};\beta_1)$, $(S_1(x;y) \equiv \Lambda_1 \equiv r!!)$. Let us recall that they are equivalent to the order k, if there is an element:

$$E(x) = 1 + \sum_{l=1}^{k} E_i(x)\,h^i\,; \qquad E_i(x) \in \mathfrak{A}(\mathfrak{g})\,,$$

such that:

$$\overline{F}_i - F_i + G_i\left(E_1,\ldots,E_{i-1};\overline{F}_1,\ldots,\overline{F}_{i-1};F_1,\ldots,F_{i-1}\right) = \boldsymbol{d}\,E_i\,,$$

$i = 1,2,\ldots,k$. Where $\boldsymbol{d}$ is the differential of the Hochschild complex [3], [18], $G_1(\cdots) = 0$ and $G_i(\cdots)$ is some cochain defined from the arguments between brackets, ([19], expression 21).

From the Gerstenhaber theory [16], [17], [14], the cochain:

$$\overline{F}_{k+1} - F_{k+1} + G_{k+1}\left(E_1,\ldots,E_k;\overline{F}_1,\ldots,\overline{F}_k;F_1,\ldots,F_k\right)\,,$$

is a Hochschild 2-cocycle and from Theorem 13 in [19] or Proposition 2.2 in [3], there is $h_{k+1} \in \bigwedge^2(\mathfrak{g})$ and $E_{k+1} \in \mathfrak{A}(\mathfrak{g})$ such that:

$$\overline{F}_{k+1} - F_{k+1} + G_{k+1}\left(E_1,\ldots,E_k;\overline{F}_1,\ldots,\overline{F}_k;F_1,\ldots,F_k\right) = h_{k+1} + \boldsymbol{d}\,E_{k+1}\,.$$

Now, let us define the Q.Y.B. solutions [11], [19]:

$$S(x;y) = F^{-1}(y;x)\,F(x;y)\,; \qquad \overline{S}(x;y) = \overline{F}^{-1}(y;x)\,\overline{F}(x;y)\,.$$

From Lemma 34 in [19] we have:

$$\overline{S}_{k+1} - S_{k+1} = 2h_{k+1} + A_{k+1}\left(F_1,\ldots,F_k;\overline{F}_1,\ldots,\overline{F}_k;E_1,\ldots,E_k\right)\,,$$

where $A_{k+1}(\cdots)$ is some cochain defined from the arguments between brackets. We now prove:

Proposition 20.- *Let $\overline{F}, F$ be two invariant star products on $(\boldsymbol{G};\beta_1)$ which coincide to the order k. That is to say;*

$$F_1 = \overline{F}_1\,; \qquad F_2 = \overline{F}_2\,;\ldots;\,F_k = \overline{F}_k\,.$$

Then:

a) *There are $h_{k+1} \in \mathfrak{g} \wedge \mathfrak{g}$, $E_{k+1} \in \mathfrak{A}(\mathfrak{g})$ such that:*

$$\overline{F}_{k+1} - F_{k+1} = h_{k+1} + \boldsymbol{d}\, E_{k+1} \,.$$

b) $$S_1 = \overline{S}_1 \,; \quad S_2 = \overline{S}_2 \,; \ldots ; \; S_k = \overline{S}_k \,.$$

c) $$\overline{S}_{k+1} - S_{k+1} = 2h_{k+1} \,.$$

Moreover, h_{k+1} is not only a Hochschild 2-cocycle but also a Schouten 2-cocycle.

Proof.- a) and b) follow easily from the fact that $\overline{F}$ and F are equivalent to the order k, since they coincide to the order k. That h_{k+1} is a Schoutem cocycle follows from Theorem 2, [19], by substracting the Q.Y.B.E. to the order $(k+1)$ (definition 3, [19]) relative to F from the one relative to $\overline{F}$. The result of the substraction is an expression like in Proposition 1, c) for h_{k+1}, which means that h_{k+1} is a Schouten 2-cocycle. ■

11. ALL THE INVARIANT NON EQUIVALENT STAR PRODUCTS ON $(\boldsymbol{G}; \beta_1)$. FIRST PART

Let $F'(x;y)$ be any star product on $(\boldsymbol{G}; \beta_1)$. Therefore $F_1'(x;y) - F_1'(y;x) = S_1(x;y)$.

Let $F(x;y)$ be the star product in Theorem 18 determined by the cocycle:

$$\beta_t = \beta_1 + \beta_2\, h + \cdots + \beta_{R-1}\, h^{R-2} \,.$$

Suppose $F'(x;y)$ and $F(x;y)$ are equivalent to the order $(R-1)$. That is to say, there is:

$$E(x) = 1 + E_1\, h + \cdots + E_{R-1}\, h^{R-1} \,,$$

such that:

$$F_i - F_i' + G_i \left(E_1, \cdots, E_{i-1}; F_1, \cdots, F_{i-1}; F_1', \cdots, F_{i-1}'\right) = \boldsymbol{d}\, E_i \,,$$

$i = 1, 2, \ldots, R-1$, and there are $h_R \in \mathfrak{g} \wedge \mathfrak{g}$, $E_R \in \mathfrak{A}(\mathfrak{g})$ such that:

$$F_R - F_R' + G_R \left(E_1, \ldots, R_{R-1}; F_1, \ldots, F_{R-1}; F_1', \ldots, F_{R-1}'\right) = \frac{1}{2} h_R + \boldsymbol{d}\, E_R \,. \tag{22}$$

Let us define:

$$\overline{E}(x) = E(x) + E_R\, h^R\,,$$

and also the star product to any order ([19], proposition 29)

$$\overline{F}(x;y) = \overline{E}^{-1}(x+y)\, F'(x;y)\, \overline{E}(x)\, \overline{E}(y)\,,$$

In particular we can write

$$\overline{F}_i - F'_i + G_i\left(E_1,\ldots,E_{i-1};\overline{F}_1,\ldots,\overline{F}_{i-1};F'_1,\ldots,F'_{i-1}\right) = \boldsymbol{d}\, E_i\,, \tag{23}$$

$i = 1, 2, \ldots, R.$

F and $\overline{F}$ are abviously two star products which coincide to the order $(R-1)$. Relation (23) for $i = R$ is:

$$\overline{F}_R - F'_R + G_R\left(E_1,\ldots,E_{R-1};F_1,\ldots,F_{R-1};F'_1,\ldots,F'_{R-1}\right) = \boldsymbol{d}\, E_R\,.$$

If we compare this relation with (22), we obtain:

$$F_R - \overline{F}_R = \frac{1}{2} h_R\,.$$

By Proposition 20, h_R is a Schouten 2-cocycle. Let $\beta_R = \mu\,(h_R)$ the corresponding de Rahm cocycle. We have:

$$\overline{F}_R = F_R - \frac{1}{2}\mu^{-1}\,(\beta_R)\,.$$

Going back to Theorem 19, we see that $\overline{F}_R$ is the Rth term, $\hat{F}_R$, of the star product $\hat{F}$ on $(\boldsymbol{G};\beta_1)$ in Theorem 18 determined by the cocycle:

$$\hat{\beta}_H = \beta_h + \beta_R\, h^{R-1}\,.$$

From Theorem 19, $\hat{F}_i = \overline{F}_i (= F_i)$, $i = 1, 2, \ldots, R-1$ and therefore relations (23) can be written:

$$\hat{F}_i - F'_i + G_i\left(E_1,\ldots,E_{i-1};\hat{F}_1,\ldots,\hat{F}_{i-1};F'_1,\ldots,F'_{i-1}\right) = \boldsymbol{d}\, E_i\,,$$

$i = 1, \ldots, R.$

By directly checking the induction hypothesis for $i = 1$ which is obvious, we have obtained, see Theorem 7 in [11],

Theorem 21.- *Let F' any star product on $(\boldsymbol{G};\beta_1)$. There is a cocycle:*

$$\beta_R = \beta_1 + \beta_2\, h + \cdots + \beta_R\, h^{R-1} + \cdots\,,$$

uniquely determined by F', and an element $E(x) \in \mathfrak{A}(\mathfrak{g})\,[[h]]$ such that:

$$F'(x;y) = E(x+y)\, F(x;y)\, E^{-1}(x)\, E^{-1}(y)\,,$$

where $F(x;y)$ is the star product on $(\boldsymbol{G};\beta_1)$ in Theorem 18 determined by the cocycle β_h.

12. ALL THE INVARIANT STAR PRODUCTS ON $(G;\beta_1)$, SECOND PART

A.- To complete the proof of Theorem 7 in [11], we need to consider the relation between the two star products on G obtained in Theorem 18 from cocycles

$$\beta_t = \beta_1 + \beta_2\, t + \cdots + \beta_k\, t^{k-1} + \cdots , \tag{24}$$

and

$$\omega_t = \beta_1 + (\beta_2 + d_R\alpha_2)\, t + \cdots + (\beta_k + d_R\alpha_k)\ , t^{k-1} + \cdots .$$

here $\alpha_R \in \mathfrak{g}^*$ and d_R is the exterior defferential on invariant forms on G. That is, *we need to consider the consequences of the isomorphism* $\overline{\mathfrak{g}}_{\omega_t} \approx \overline{\mathfrak{g}}_{\beta_t}$.

If $x, y \in \mathfrak{g}$,

$$d_R\alpha_k(x;y) = -\alpha_k[x;y] .$$

We will write:

$$\omega_t = \beta_t + d_R\alpha_t \tag{25}$$

where $\alpha_t = \alpha_2\, t + \alpha_3\, t^2 + \cdots + \alpha_k\, t^{k-1} + \cdots \in \mathfrak{g}^*$. *As before, only a finite number of terms are not zero but a number as big as we need.*

Let L_{α_t} be the operator of traslation in $\mathfrak{g}^*$ defined by

$$L_{\alpha_t} : \lambda + E^* \in \mathfrak{g}^* + E^* \longrightarrow \lambda + \alpha_t + E^* \in \mathfrak{g}^* + E^* .$$

We write $\overline{Ad}^* \cdot \overline{exp}_{\beta_t}\, x$ for the coadjoint action of the group $\overline{G}_{\beta_t}$ as in §3. Similary for $\overline{Ad}^* \cdot \overline{exp}_{\omega_t}\, x$, and the cocycle ω_t. These two actions are related.

Proposition 22.- *The following diagram is commutative:*

$$\begin{array}{ccccc} & \mathfrak{g}^* + E^* & \xrightarrow{L_{\alpha_t}} & \mathfrak{g}^* + E^* & \\ \overline{Ad}^* \cdot \overline{exp}_{\beta_t}\, x & \downarrow & & \downarrow & \overline{Ad}^* \cdot \overline{exp}_{\omega_t}\, x \\ & \mathfrak{g}^* + E^* & \xrightarrow{L_{\alpha_t}} & \mathfrak{g}^* + E^* & \end{array}$$

Proof.- We have:

$$\begin{aligned} \left(\overline{Ad}^* \cdot \overline{exp}_{\omega_t} \cdot L_{\alpha_t}\right)(\xi + E^*) &= Ad^*\, exp\,(\xi + \alpha_t) + f_{\omega_t}(-x)\, E^* = \\ &= Ad^*\, exp\, x \cdot \xi + \\ &\quad + Ad^*\, exp\, x \cdot \alpha_t + \tilde{\beta}_t(-x) \cdot A(-x) + \widetilde{d\,\alpha_t}(-x) \circ A(-x) + E^* = \\ &= \overline{Ad}^* \cdot \overline{exp}_{\beta_t}\, x \cdot \xi + Ad^*\, exp\, x \cdot \alpha_t + \widetilde{d\,\alpha_t}(-x) \cdot A(-x) + E^* . \end{aligned}$$

If $y \in \mathfrak{g}$ we obtain:

$$[d\,\tilde{\alpha}_t(-x) \circ A(-x)]\,(y) = d\,\alpha_t\,(-x; A(-x)\,y) = -\alpha_t\,(ad(-x) \circ A(-x)\,y)\ .$$

We know from §3:

$$ad(-x) \cdot A(-x) = Ad\,exp(-x) - I\,.$$

Therefore:

$$d\,\tilde{\alpha}_t(-x) \circ A(-x) + (Ad^*\,exp\,x - I)\,\alpha_t = 0\,.$$

We finally get

$$\left(\overline{Ad}^* \cdot \overline{exp}_{\omega_t}\,x \circ L_{\alpha_t}\right)(\xi + E^*) =$$
$$= \overline{Ad}^* \cdot \overline{exp}_{\beta_t}\,x \cdot \xi + \alpha_t + E^* = \left(L_{\alpha_t} \circ \overline{Ad}^* \cdot \overline{exp}_{\beta_t}\,x\right)(\xi + E^*)\ . \qquad \blacksquare$$

B.- Let φ_1, φ_2 be two functions in $\mathcal{C}^\infty(\mathfrak{g}^*)$. The star product on $\mathfrak{g}^* + E^*$ determined by the cocycle ω_t in Theorem 13 is:

$$(\varphi_1 *_{\omega_t} \varphi_2)\,(\xi) = \sum_{R \geq 0} P_R^{\omega_t}\left(\frac{\partial}{\partial \xi};\frac{\partial}{\partial \xi};\xi\right)(\varphi_1 \otimes \varphi_2)\,(\xi)\,h^R\,,$$

where $P_R^{\omega_t}$ is defined as in §6,c, and we explicitely write down the cocycle ω_t, which determines the group $\overline{\boldsymbol{G}}_{\omega_t}$. Also:

$$(\varphi_1 *_{\beta_t} \varphi_2)\,(\xi) = \sum_{R \geq 0} P_R^{\beta_t}\left(\frac{\partial}{\partial \xi};\frac{\partial}{\partial \xi};\xi\right)(\varphi_1 \otimes \varphi_2)\,(\xi)\,h^R\,,$$

where $P_R^{\beta_t}$ is defined in a similar way.

Proposition 23.-

1) *With the foregoing notations the two star products $*_{\omega_t}$, $*_{\beta_t}$ are related as in the expresion*

$$(\varphi_1 *_{\omega_t} \varphi_2) \circ L_{\alpha_t} = (\varphi_1 \circ L_{\alpha_t}) *_{\beta_t} (\varphi_1 \circ L_{\alpha_t})\ .$$

2) *We can write the following asymptotic expression*

$$(\varphi_1 \circ L_{\alpha_t})\,(\xi) = 1 + \sum_{l \geq 1} H_l(\varphi)\,t^l \equiv H(t)\,\varphi\,,$$

where $\{H_l, l \geq 1\}$ are differential operators on $\mathfrak{g}^* + E^*$. Then, the following relation between formal power series in h holds:

$$\varphi_1 *_{\omega_h} \varphi_2 = H(h)^{-1}\,(H(h)\varphi_1 *_{\beta_h} H(h)\varphi_2)\ ,$$

and $H(h)$ is an equivalence between star products on $\mathfrak{g}^* + E^*$. (Which equivaries the invariance of star products $*_{\beta_t}$, $*_{\omega_t}$).

Proof.-

1) from §5 and §6,c), operators $P_R^{\omega_t}$ and $P_R^{\beta_t}$ are constructed respectively from:

$$J_k^{\omega_t}(x;y;\xi) = \xi \cdot z_{k+1}(x;y) + \hat{z}_{k+1}^{\omega_t}(x;y)\,,$$

and

$$J_k^{\beta_t}(x;y;\xi) = \xi \cdot z_{k+1}(x;y) + \hat{z}_{k+1}^{\beta_t}(x;y)\,.$$

Obviously

$$\hat{z}_{k+1}^{\omega_t}(x;y) = \hat{z}_{k+1}^{\beta_t}(x;y) - \alpha_t\,(z_{k+1}(x;y))\,.$$

Hence

$$J_k^{\omega_t}(x;y;\xi) = J_k^{\beta_t}(x;y;\xi-\alpha_t)\,,$$

and

$$P_R^{\omega_t}\left(\frac{\partial}{\partial\xi};\frac{\partial}{\partial\xi};\xi\right) = P_R^{\beta_t}\left(\frac{\partial}{\partial\xi};\frac{\partial}{\partial\xi};\xi-\alpha_t\right)\,.$$

We can now write:

$$\begin{aligned}
P_R^{\omega_t}&\left(\frac{\partial}{\partial\xi};\frac{\partial}{\partial\xi};\xi\right)(\varphi_1\otimes\varphi_2)(\xi) = \\
&= P_R^{\beta_t}\left(\frac{\partial}{\partial\xi};\frac{\partial}{\partial\xi};\xi-\alpha_t\right)(\varphi_1\otimes\varphi_2)(\xi) = \\
&= P_R^{\beta_t}\left(\frac{\partial}{\partial\xi'};\frac{\partial}{\partial\xi'};\xi'\right)((\varphi_1\circ L_{\alpha_1})\otimes(\varphi_2\circ L_{\alpha_t}))(\xi')\,,
\end{aligned}$$

where $\xi = \xi' + \alpha_t$. And finally we obain:

$$(\varphi_1 *_{\omega_t} \varphi_2)(\xi) = (((\varphi_1\circ L_{\alpha_t}) *_{\beta-t} (\varphi_2\circ L_{\alpha_t}))\circ L_{-\alpha_t})(\xi)\,.$$

2) As a power series in t and h, relation in 1) can be writen

$$\varphi_1 *_{\omega_t} \varphi_2 = H(t)^{-1}\,(H(t)\,\varphi_1 *_{\beta_t} H(t)\,\varphi_2)\,.$$

If we now take $t = h$, the same proof as for Theorem 17 allows us to get the result. ∎

c.- Let us recall that $*_{\omega_t}$ is invariant by $\overline{Ad}^* \cdot \overline{exp}_{\beta_t}\, x$ and $*_{\beta_t}$ is invariant by $\overline{Ad}^* \cdot \overline{exp}_{\omega_t}\, x$. Notations are now as in Proposition 4. Let ψ_1, ψ_2 be two functions in $C^\infty(U(e))$ and let $K_{\beta_t}, K_{\omega_t}$ be the two local diffeomorphisms in Proposition 4 corresponding respectiely to the cocycles β_t and ω_t. We denote also by $*_{\omega_t}$ the invariant star product on $(\boldsymbol{G};\omega_t)$ obtained in §7. ($*_{\beta_t}$ denotes also the similarly determined invariant star product on $(\boldsymbol{G};\beta_t)$).

Proposition 24.- *With the foregoing notations and with t small enough, we have:*

$$\psi_1 *_{\omega_t} \psi_2 = (\psi_1\circ M_t *_{\beta_t} \psi_2\circ M_t)\circ M_t^{-1}\,,$$

where $M_t = K_{\omega_t}^{-1}\circ L_{\alpha_t}\circ K_{\beta_t}$ is a mapping from $W(e)\subset U(e)$ to $U(e)$.

Proof.- From Proposition 23, we can write in a small neighborhood of E^* in $E^* + \mathfrak{g}^*$:

$$\begin{aligned}
&\psi_1 \circ K_{\omega_t}^{-1} *_{\omega_t} \psi_2 \circ K_{\omega_t}^{-1} = \\
&\quad = \left(\psi_1 \circ K_{\omega_t}^{-1} \circ L_{\alpha_t} *_{\beta_t} \psi_2 \circ K_{\omega_t}^{-1} \circ L_{\alpha_t}\right) \circ L_{-\alpha_t} = \\
&\quad = \left(\psi_1 \circ K_{\omega_t}^{-1} \circ L_{\alpha_t} \circ K_{\beta_t} *_{\beta_t} \psi_2 \circ K_{\omega_t}^{-1} \circ L_{\alpha_t} \circ K_{\beta_t}\right) \circ K_{\beta_t}^{-1} \circ L_{-\alpha_t} .
\end{aligned}$$

From the definition (11) we therefore obtain:

$$\begin{aligned}
\psi_1 *_{\omega_t} \psi_2 &= \left(\psi_1 \circ K_{\omega_t}^{-1} *_{\omega_t} \psi_2 \circ K_{\omega_t}^{-1}\right) \circ K_{\omega_t} = \\
&= \left(\psi_1 \circ M_t *_{\omega_t} \psi_2 \circ M_t\right) \circ M_t^{-1} .
\end{aligned}$$

■

We also prove:

Proposition 25.- *If $z \in W(0)$ and h are small, we have:*

$$L_{exp\, z} \circ M_t = M_t \circ L_{exp\, z} .$$

Proof.- From diagrams in Proposition 4, relatif to K_{ω_t} and K_{β_t} and from Proposition 22, we get:

$$\begin{aligned}
&L_{\exp\, z} \circ M_t = L_{\exp\, z} \cdot K_{\omega_t}^{-1} \circ L_{\alpha_t} \circ K_{\beta_t} = \\
&\quad = K_{\omega_t}^{-1} \circ \overline{Ad}^* \, \overline{exp}_{\omega_t} z \circ L_{\alpha_t} \circ K_{\beta_t} = K_{\omega_t}^{-1} \circ L_{\alpha_t} \circ \overline{Ad}^* \, \overline{exp}_{\beta_t} z \cdot K_{\beta_t} = \\
&\quad = K_{\omega_t}^{-1} \circ L_{\alpha_t} \circ K_{\beta_t} \circ \overline{Ad}^* \, \overline{exp}_{\beta_t} z = M_t \circ L_{exp\, z} .
\end{aligned}$$

■

D.- In normal coordinates $\{x^a \,;\, a = 1, \ldots, dim\, \boldsymbol{G} = n\}$, in a small neighborhood of $e \in \boldsymbol{G}$, we write:

$$x' = M_t \cdot x .$$

In components we clearly have:

$$(x')^a = \sum_{k \geq 0} A_k^a(x)\, t^k \,; \qquad A_0^a(x) = x^a ,$$

where $A_k^a(x)$ are analytic functions and the expansion in t comes from the obvious analiticity in t of $M_t = K_{\omega_t}^{-1} \circ L_{\alpha_t} \circ L_{\beta_t}$. If $l \in \mathbb{N}$, we get

$$((x')^a - (x^a))^l = \left(\sum_{k_1 \geq 1} A_{k_1}^a(x)\, t^{k_1}\right) \cdots \left(\sum_{k_l \geq 1} A_{k_l}^a(x)\, t^{k_l}\right) = \sum_{R \geq l} \Omega_R^{a,\, l}(x)\, t^R ,$$

where

$$\Omega_R^{a,l}(x) = \sum_{k_1+\cdots+k_l=R} A_{k_1}^a(x)\cdots A_{k_l}^a(x)\,.$$

From the Taylor expansion of $\psi_l(x')$ at x we get

$$\begin{aligned}\psi_1(x') &= \sum_{l_1,\dots,l_n\geq 0} \frac{1}{L!}\frac{\partial^{l_1+\cdots+l_n}\psi}{(\partial x^1)^{l_1}\cdots(\partial x^n)^{l_n}}(x)\,(x'^1-x^1)^{l_1}\cdots(x'^n-x^n)^{l_n} = \\ &= \sum_{R\geq o}\left[\sum_{R_1+\cdots+R_n=R} \frac{1}{L!}\frac{\partial^{l_1+\cdots+l_n}\psi}{(\partial x^1)^{l_1}\cdots(\partial x^n)^{l_n}}(x)\cdot\Omega_{R_1}^{1,l_1}(x)\cdots\Omega_{R_n}^{n,l_n}(x)\right] t^R \equiv \\ &\equiv \sum_{R\geq 0}(L_R\psi)(x)\,t^R\,,\end{aligned}$$

where, in the 3th sum $R_1\geq l_1\geq 0,\dots,R_n\geq l_n\geq 0$, and L_R is the differential operator defined by the last equality. We have obtained:

$$(\psi_1\circ M_t)(x) = \sum_{R\geq 0}(L_R\,\psi_1)(x)\,t^R\,.$$

Recall that in the theory of $*$-products, functions appears throught its Taylor expansion at each point, equality sign is therefore justified.

If $z\in\mathfrak{g}$ is small:

$$(\psi_1\circ M_t)(L_{exp\,z}\,x) = \sum_{R\geq 0}(L_R\,\psi_1)(L_{exp\,z}\cdot x)\,t^R\,.$$

From Proposition 25 we can also write:

$$(\psi_1\circ M_t)(L_{exp\,z}\cdot x) = (\psi_1\circ L_{exp\,z})(M_t x) = \sum_{R\geq 0} L_R\,(\psi_1\circ L_{exp\,z})(x)\,t^R\,.$$

Hence, we get:

$$(L_R\psi_1)\circ L_{exp\,z} = L_R\,(\psi_1\circ L_{exp\,z})\,,$$

in a small neigborhood of $e\in \boldsymbol{G}$. This relation define L_R as an invariant diferential operator on $\boldsymbol{G}$. With the usual identification: $L_R\in\mathfrak{A}(\mathfrak{g})$.

We can state:

Proposition 26.- *With the foregoing notations we have the expansion:*

$$(\psi_1\circ M_t)(x) = \sum_{R\geq 0}(L_R\psi_1)(x)\,t^R\,, \tag{26}$$

where L_R is an invariant diferential operator on $\boldsymbol{G}$, and $L_0 = 1$. ■

E.- Let us now put $t = h$ in Proposition 24. *As a power series in h*, $\psi_1 *_{\omega_h} \psi_2$ is the star product on $\boldsymbol{G}$ in Theorem 18 determined by the cocycle ω_h. Similarly $\psi_1 *_{\beta_h} \psi_2$ *as a power series in h* is the star product determined by the cocycle β_h. If we now refer to expansion (26), we see that the formal power series in h, $((\psi_1 \circ M_h) *_{\beta_h} (\psi_2 \circ M_h)) \circ M_h^{-1}$ defines an invariant star product on $\boldsymbol{G}$, equivalent to the star product $\psi_1 *_{\beta_h} \psi_2$, which coincides with the star product $\psi_1 *_{\omega_h} \psi_2$. In the usual notations of [11], [18], [19] for invariant star products on $\boldsymbol{G}$, we can therefore state:

Theorem 27.- *Let $F^\omega(x;y) = \sum_{R\geq 0} F_R^\omega(x;y)\, h^R$ be the invariant star product on $\boldsymbol{G}$ in Theorem 18 determined by the cocycle $\omega_h = \beta_h + \boldsymbol{d}_R\, \alpha_h$, where β_h and α_h are defined by (24) and (25). Let $F^\beta(x;y) = \sum_{R\geq 0} F_R^\beta(x;y)\, h^R$ the invariant star product on $\boldsymbol{G}$ in Theorem 18 , determined by the cocycle β_h. Let $L(x) = \sum_{R\geq 0} L_R(x)\, h^R$ the element in $\mathfrak{A}(\mathfrak{g})[[h]]$ which defines the expansion (26). Then:*

$$F^\omega(x;y) = L^{-1}(x+y)\, F^\beta(x;y)\, L(x)\, L(y)\,.$$

That is, $F^\omega(x;y)$ and $F^\beta(x;y)$ are equivalent through the element $L(x) \in \mathfrak{A}(\mathfrak{g})[[h]]$.

Remark.- From Theorem 19 we must have:

$$F_R^\omega(x;y) = F_R^\beta(x;y) + \frac{1}{2}\partial\mu^{-1}(\alpha_R)\,, \qquad R \geq 1\,.$$

In the equivalence of the theorem we must have $L_1(x) \neq 0$ but $\boldsymbol{d}\, L_1(x) = 0$ and then:

$$F_2^\omega - F_2^\beta + G_2\,(L_1; F_1; F_1') = d\, E_2\,,$$

where $F_1 = F_1'$ and $G_2\,(L_1; F_1; F_1') \neq 0$. Similarly, for any R.

From Theorems 21 and 27 we obviously get:

Theorem 28.-(Drinfeld) *Choose a vector subspace $\boldsymbol{V}$ in the space $Z^2(\mathfrak{g})$ of invariant de Rham 2-cocycles on $\mathfrak{g}$, which is a supplementary space of de Rham 2-exact cocycles $B^2(\mathfrak{g})$, i.e. $Z^2(\mathfrak{g}) = \boldsymbol{V} \oplus B^2(\mathfrak{g})$. Let $F'(x;y)$ be any invariant star product on $(\boldsymbol{G};\beta_1)$. Then, $F'(x;y)$ is equivalent to one obtained in Theorem 18 from a cocycle:*

$$\beta_h = \beta_1 + \beta_2\, h + \cdots + \beta_R\, h^{R-1} + \cdots\,,$$

such that $\beta_k \in \boldsymbol{V}$ if $k > 1$. Moreover, $\{\beta_k\}$ is uniquely determined by $F'(x;y)$.■

This is exactly the statement of Theorem 7 in [11].

ACKNOWLEDGEMENTS

The authors wishes to express their gratitude to professors R. F. Alvarez-Estrada, M. Flato, A. Lichnerowicz and D. Sternheimer for iluminating discussions and friendly encouragement.

REFERENCES

1. V. G. Drinfeld, Hamiltonian structures on Lie groups, Lie bialgebras and the geometric meaning of the classical Yang-Baxter equations, Sov. Math. Dokl. **27** (1983) 68–71.
2. V. G. Drinfeld, Quantum groups, in: ICM Proc. , (Berkeley 1986) pp. 798–820.
3. V. G. Drinfeld, Quasi-Hopf algebras, Leningrad Math. J. **1** (1990) n$^{\underline{o}}$6.
4. L. D. Faddeev, L. A. Takhtajan, *Hamiltonian methods in the theory of solitons*, (Springer-Verlag, Berlin, 1987).
5. M. A. Semenov-Tian-Shansky, What is a classical R-matrix?, Funct. Anal. Appl. **17**(4) (1983) 259–272.
6. Y. Kosmann-Schwarzbach, F. Magri, Poisson-Lie groups and complete integrability, Ann. Inst. Henri Poincaré, Vol. 49, n$^{\underline{o}}$4 (1988) 433–460.
7. Y. Kosmann-Schwarzbach, F. Magri, Poisson-Nijenhuis structures, Ann. Inst. Henri Poincaré, Vol. 53, n$^{\underline{o}}$1 (1990) 35–81
8. J. H. Lu, A. Weinstein, Poisson Lie groups, dressing transformations and the Bruhat decomposition, J. Diff. Geom. **31** (1990) 501–526.
9. A. Weinstein, Some remarks on dressing transformations, preprint, (University Tokyo 1987).
10. M. A. Semenov-Tian-Shansky, Dressing transformation, Publ. RIMS Kyoto Univ. **21** (1985) 1237–1260.
11. V. G. Drinfeld, On constant, quasiclassical solutions of the Yang-Baxter quantum equation, Soviet Math. Dokl. **28** (1983) 667–671.
12. Quantum groups, Lectures Notes in Mathematics, Vol. 1510, (Springer, Berlin, 1992).
13. F. Bayen, M. Flato, C. Fronsdal, A. Lichnerowicz, D. Sternheimer, Deformation Theory and Quantization I and II, Ann. of Phys. **111** (1978) 61–151.
14. A. Lichnerowicz, Déformations d'algèbres associées à une variété symplectique (les $*_\nu$-produits), Ann. Inst. Fourier **32** (1982) 157–209.
15. M. De Wilde, P. B. A. Lecompte, in: *Deformation theory of algebras and structures and applications*, Eds. M. Hazewinkel, M. Gerstenhaber, (Kluwer, Dordrecht, 1988).
16. M. Gerstenhaber, Deformation of rings and algebras, Ann. of Math. **79** (1964) 59–103.
17. M. Gerstenhaber, The cohomology structure of an associative ring, Ann. Math. **78** (1963) 267–288.

18. C. Moreno, L. Valero, Produits star invariants et equation de Yang-Baxter quantique constante, in: Proc. of the " Journnées Relativistes", (Grenoble, 1990) pp. 225–256.

19. C. Moreno, L. Valero, Star products and quantization of Poisson-Lie groups, Jour. Geom. Phys. to appear.

20. R. Bott, Notes on Gelfand-Fucks cohomology and characteristics classes, in: Lectures on the Eleventh Holiday Symposium at New Mexico State University, (Las Cruces, 1973).

21. M. Gerstenhaber, A. Giaquinto, S. D. Schack, Quantum symmetry, in: Lecture Notes in Mathematics Vol. 1510 (Spriger, Berlin 1992).

22. M. J. May, *Simplicial Objects in Algebraic Topology*, (Van Nostrand, Princeton, 1967).

23. P. Hilton, S. Wylie, *Homology Theory*, (Cambridge, 1960).

24. Hyperalgèbres et groupes de Lie, in: Séminaire Sophus Lie, 2e. année, (Paris, 1955/56).

25. J. Vey, Déformations du crochet de Poisson sur une variété symplectique, Comm. Math. Helv. **50** (1975) 421–454.

26. L. A. Takhtajan, Lectures on quantum groups, in: Introduction to Quantum Grpoups and Integrable Massive Models of Quantum Field Theory, Eds. Mo-Lin Ge, Bao-Heng Zhao, (World Scientific, Singapore, 1990).

27. L. A. Takhtajan, Introduction to quantum groups, in: Lectures Notes in Physics Vol. 370 (Springer, Berlin, 1990).

28. C. Roger, Algèbres de Lie graduées et quantization, in: *Symplectic Geometry and Mathematical Physics, Actes du Colloque en l'honneur de Jean Marie Souriau*, Eds. P. Donato, C. Duval, J. Elhadad, G. M. Tuynman, (Progress in Math. 99, Birkhäuser, 1991) pp. 374–421.

29. J. Grabowski, Abstract Jacobi and Poisson structures. Quantization and star products, Journ. Geom. Phys. **9** (1992) 45–73.

30. F. A. Berenzin, Quantization, Izv. Akad. Nauk. SSSR **38** (1974) 1109–1165.

31. C. Moreno, Geodesic symmetries an invariant star products on Kahler symmetric spaces, Lett. Math. Phys. **13** (1987) 245–257.

32. N. Reshetikhin, Quasitriangularity of quantum groups and Poisson-quasitriangular Hopf-Poisson algebras, preprint, Berkeley (1992).

33. N. Reshetikhin, Quantization of Lie-bialgebras, preprint, Berkeley (1992).

34. M. A. Rieffel, Deformation quantization for actions of R^d, preprint, Berkeley september (1991).

35. M. A. Rieffel, Deformation quantization of Heisenberg manifolds, Comm. Math. Phys. **122** (1989) 531–562.

36. M. A. Rieffel, Lie groups convolution algebras as deformation quantization of linear-Poisson structures, Amer. J. Math. **112** (1990) 657–686.

37. S. L. Woronowicz, Twisted $SU(2)$. An example of a noncommutative differetial calculus, Publ. Res. Inst. Math. Sci. Kyoto **23** (1987) 117–181.

38. S. L. Woronowicz, Compact matrix pseudogroups, Comm. Math. Phys **111** (1987) 613–665.

39. A. Lichnerowicz, Les variétés de Poisson et leurs algèbres de Lie associées, J. Diff. Geom. **18** (1983) 523.

40. A. Weinstein, The local structure of Poisson manifolds, J. Diff. Geom. **18** (1983) 523–557.

41. P. Liberman, Ch. Marle, *Symplectic Geometry and Analytical Mechanics* (Reidel, Dordrech, 1987).

42. Y. Choquet-Bruhat and C. De Witt-Morette, *Analysis, Manifolds and Physics. Part II: 92 Applications* (North-Holland, Amsterdam, 1989).

43. N. Bourbaki, *Groupes et algèbres de Lie*, Chapitres 2 et 3 (Diffusion C.C.L.S. Paris, 1972).

44. J. P. Serre *Lie algebras and Lie groups. Lectures given at Harvard University* (W. A. Benjamin, Inc. New York-Amsterdam 1965).

45. F. Treves, *Topological Vector Spaces, Distributions and Kernels* (Academic Press, New York-San Francisco-London 1967).

46. C. Moreno, L. Valero, Produits star sur certains G/K Kälériens. Equation de Yang-Baxter quantique constante, in: Lecture Notes in Mathematics, Vol. 1416 (Springer, Berlin, 1990).

e-mail: W535emducm11.bitnet

O.A. Oleinik

On asymptotic of solutions of a nonlinear elliptic equation in a cylindrical domain

Dedicated to Yvonne Choquet-Bruhat

In this lecture an asymptotic behaviour at infinity of solutions of a nonlinear second order elliptic equation in cylindrical domains is considered. This question arises in many problems of mathematical physics and was studied in many papers (see, for examples, [1] - [6], and references there).

Let us introduce some notations. We set $x = (x_1, \dots, x_n), x' = (x_1, \dots, x_{n-1}), S(a, b) = \{x : x' \in \omega, a < x_n < b\}, \sigma(a, b) = \{x : x' \in \partial\omega, a < x_n < b\}; \omega$ is a bounded domain in $\mathbf{R}^{n-1} = (x_1, \dots, x_{n-1})$ with a smooth boundary $\partial\omega$.

Here we study the asymptotic behaviour as $x_n \to \infty$ of solutions of the equation

$$\Delta u - |u|^{p-1}u = 0 \qquad \text{in } S(0, \infty) \tag{1}$$

with the boundary condition

$$\frac{\partial u}{\partial \nu} = 0 \text{ on } \sigma(0, \infty), \tag{2}$$

where ν is the exterior unit normal to $\sigma(0, \infty), p > 1$. These results are obtained jointly with V.A. Kondratiev. In paper [4] the problem (1), (2) is also considered and the following theorem is proved.

Theorem 1. *Let $u(x)$ be a solution of the equation (1) in $S(0, \infty)$ with the boundary condition (2) on $\sigma(0, \infty), p > 1$. Assume that $u(x)$ attains positive and negative values (oscillates) in any domain $S(k, \infty), k > 0$.*

Then

$$|u(x)| \le Ce^{-\alpha x_n}, \alpha, C = \text{const} > 0. \tag{3}$$

In paper [5] the following theorem is contained.

Theorem 2. *Let $u(x)$ be a positive solutions of the problem (1), (2) in $S(0, \infty), p > 1$. Then*

$$u(x) = M_p(x_n + \gamma)^{\frac{2}{1-p}} + v(x), \tag{4}$$

where

$$|v(x)| \le Ce^{-\beta x_n}, C, \beta, \gamma = \text{const} > 0, \tag{5}$$

M. Flato et al. (eds.), Physics on Manifolds, 235–251.

$$M_p = \left(\frac{2(1+p)}{(p-1)^2}\right)^{\frac{1}{p-1}}. \tag{6}$$

The paper [5] is not easy to read because of the technical misprints due to editors. We will repeat here the proof of Theorem 2, since this theorem is used in the proof of Theorem 3.

We consider classical solutions of problem (1), (2). This means that $u \in C^2(S(0,\infty)) \cap C^1(\bar{S}(0,\infty))$.

The following theorem is an essential improvement of Theorems 1 and 2 and it is a main result of this paper.

Theorem 3. *Let $u(x)$ be a solution of equation* (1) *in $S(0,\infty)$ with boundary condition* (2) *on $\sigma(0,\infty), p > 1$. Then*

$$u(x) = \pm K(x_n + h)^{\frac{2}{1-p}} + \sum_{i=1}^{m} A_i v_i(x') e^{-\sqrt{\lambda_2} x_n} + o\left(e^{-\sqrt{\lambda_2} x_n}\right), x_n \to \infty, \tag{7}$$

where the constant K is equal to zero or $K = M_p, h = \text{const}, A_i = \text{const}, i = 1, \dots, m; \lambda_2$ is the second eigenvalue of the eigenvalue problem:

$$\Delta v + \lambda v = 0 \quad \text{in} \quad \omega, \tag{8}$$

$$\frac{\partial v}{\partial \nu} = 0 \quad \text{on} \quad \partial\omega, \tag{9}$$

$v_i(x'), i = 1, \dots, m$, are the system of orthogonal eigenfunctions, corresponding to λ_2; m is the multiplicity of λ_2. The constants $h, A_i (i = 1, \dots, m)$ depend on $u(x)$.

Before proving of Theorem 3 we prove Theorem 2 and some auxiliary propositions.

Proof of Theorem 2. In paper [4] the following inequality is proved:

$$M_p(x_n + \gamma)^{\frac{2}{1-p}} \le u(x) \le M_p x_n^{\frac{2}{1-p}} \tag{10}$$

From the inequality (10) it follows that the set of nonnegative numbers δ such that

$$u(x) \le M_p(x_n + \delta)^{\frac{2}{1-p}}$$

for $x_n \ge N_\delta$ is not empty. We denote $\delta_0 = \sup \delta$.

Consider at first the case, when $\delta_0 < \infty$. From the definition of δ_0 it follows that $u(x) \le M_p(x_n + \delta_0 - \varepsilon)^{\frac{2}{1-p}}$ for any $\varepsilon = \text{const} > 0$ and $x_n > X_\varepsilon$ and in addition

$$u(x^\varepsilon) \ge M_p(x_n^\varepsilon + \delta_0 + \varepsilon)^{\frac{2}{1-p}}, x_n^\varepsilon \to \infty. \tag{11}$$

Consider the function $v(x) = u(x) - M_p(x_n+\delta_0)^{\frac{2}{1-p}}$. It is evident that

$$v(x) \le M_p(x_n+\delta_0-\varepsilon)^{\frac{2}{1-p}} - M_p(x_n+\delta_0)^{\frac{2}{1-p}} \tag{12}$$

for $x_n > X_\varepsilon$. It follows from (12) that

$$v(x) \le M_p(x_n+\delta_0)^{\frac{2}{1-p}}\left[\left(1-\frac{\varepsilon}{(x_n+\delta_0)}\right)^{\frac{2}{1-p}} - 1\right].$$

Therefore

$$v_+(x) = o(x_n^{\frac{1+p}{1-p}}) \quad \text{as} \quad x_n \to \infty, \tag{13}$$

where $v_+(x) = v(x)$, if $v(x) \ge 0$, and $v_+(x) = 0$, if $v(x) < 0$. It is easy to see that

$$v_-(x) = O(x_n^{\frac{2}{1-p}}) \quad \text{as} \quad x_n \to \infty, \tag{14}$$

where $v_-(x) = v(x)$, if $v(x) \le 0$, and $v_-(x) = 0$, if $v(x) > 0$.

We define the function $Z(x)$ in the following way:

$$Z(x)(x_n+\delta_0)^{\frac{2}{1-p}} = u(x) - M_p(x_n+\delta_0)^{\frac{2}{1-p}}. \tag{15}$$

The function Z satisfies the equation

$$\begin{aligned} &\Delta Z + 4[(1-p)(x_n+\delta_0)]^{-1}\frac{\partial Z}{\partial x_n} + \frac{2(p+1)}{(1-p)^2(x_n+\delta_0)^2}\, Z \\ &+ A_p(x,Z)[Z(x_n+\delta_0)^{\frac{2}{1-p}}]^{-1}Z = 0, \end{aligned} \tag{16}$$

where

$$A_p(x,Z) = 2M_p(1-p)^{-2}(1+p)(x_n+\delta_0)^{-2+\frac{2}{1-p}} - [M_p(x_n+\delta_0)^{\frac{2}{1-p}} + Z(x_n+\delta_0)^{\frac{2}{1-p}}]^p.$$

For the function Z we have an estimate from below

$$Z(x) = u(x)(x_n+\delta_0)^{-\frac{2}{1-p}} - M_p \ge M_p\left[\left(\frac{x_n+\gamma}{x_n+\delta_0}\right)^{\frac{2}{1-p}} - 1\right] = O(x_n^{-1})$$

and an estimate from above

$$Z(x) \le M_p\left[\left(1-\frac{\varepsilon}{x_n+\delta_0}\right)^{\frac{2}{1-p}} - 1\right] = o(x_n^{-1}). \tag{17}$$

Therefore we get

$$|Z(x)| \le O(x_n^{-1}), Z_+(x) \le o(x_n^{-1}), x_n \to \infty. \tag{18}$$

Let us prove that $Z_-(x^\varepsilon) = o\left(\frac{1}{x_n^\varepsilon}\right)$, where x^ε is defined by (11). According to (11) we have

$$Z(x^\varepsilon) \ge M_p\left[1 - \frac{\varepsilon}{x_n^\varepsilon + \delta_0}\right]^{\frac{2}{1-p}} - M_p \ge \frac{2\varepsilon M_p}{(1-p)(x_n^\varepsilon + \delta_0)} \quad . \tag{19}$$

It follows from (19) that $Z_-(x^\varepsilon) = o(\frac{1}{x_n^\varepsilon})$. Let us estimate the coefficients of the equation (16) for Z. We have

$$A_p(x, Z) = M_p\frac{2(1+p)}{(1-p)^2}(x_n + \delta_0)^{\frac{2p}{1-p}}(1 - (1 - \frac{Z}{M_p})^p).$$

Thus the function $Z(x)$ satisfies the equation

$$\Delta Z + a_1(x_n)\frac{\partial Z}{\partial x_n} + a_2(x_n)Z = 0,$$

where

$$a_1(x_n) \to 0, a_2(x_n) \to 0, a_1(x_n) = O(x_n^{-1}), a_2 = O(x_n^{-2}) \quad \text{as} \quad x_n \to \infty\,. \tag{20}$$

We shall prove that $|Z(x)| = o(x_n^{-1})$ as $x_n \to \infty$. We have $Z_+(x) = o(x_n^{-1})$ according to (18). Consider the function

$$w(x) = -Z(x) + \frac{\varepsilon}{x_n}.$$

It is easy to see that

$$w(x) \ge \frac{\varepsilon}{2x_n} \quad \text{for} \quad x_n \ge X_\varepsilon, \tag{21}$$

since $Z_+(x) = o(x_n^{-1})$. For the sequence x_n^ε we have

$$w(x_n^\varepsilon) \le 2\varepsilon(x_n^\varepsilon)^{-1}. \tag{22}$$

The function w satisfies the equation

$$\Delta w + a_1 w_{x_n} + a_2 w - 2\varepsilon x_n^{-3} + a_1\varepsilon x_n^{-2} - a_2\varepsilon x_n^{-1} = 0 \tag{23}$$

in $S(0,\infty)$ with the boundary condition $\frac{\partial w}{\partial \nu} = 0$ on $\sigma(0,\infty)$. Let w_0 be a solution of the equation (23) in $S(T-2, T+2)$ with the boundary conditions

$$w_0(x', T-2) = 0, w_0(x', T+2) = 0, \frac{\partial w_0}{\partial \nu} = 0 \text{ on } \sigma(T-2, T+2)\,. \tag{24}$$

The solution w_0 exists for T sufficiently large, since $a_2(x_n) \to 0$ as $x_n \to \infty$. Consider the function

$$W = w - w_0$$

in $S(T-2, T+2)$. We have

$$\Delta W + a_1 W_{x_n} + a_2 W = 0 \text{ in } S(T-2, T+2), \tag{25}$$

$$\frac{\partial W}{\partial \nu} = 0 \text{ on } \sigma(T-2, T+2), W = w \text{ for } x_n = T+2 \text{ and } x_n = T-2 \,. \tag{26}$$

In order to prove that $W > 0$ in $S(T-2, T+2)$ we use the following well-known proposition:

Let $u(x)$ be a solution of the equation

$$\Delta u + \sum_{j=1}^{n} a^j u_{x_j} + au = f$$

in the domain $S(T-2, T+2)$ with the boundary condition $\frac{\partial u}{\partial \nu} = 0$ on $\sigma(T-2, T+2)$ and with $a(x)$ sufficiently small in $S(T-2, T+2)$. Then

$$|u(x)| \le C[\sup_{S(T-2,T+2)} |f| + \max_{\substack{x_n = T-2 \\ x_n = T+2}} |u|], \tag{27}$$

C is a constant which does not depend on u.

From (21) and the estimate (27), applied to the solution w_0 of the problem (23), (24), we get

$$W(x) \ge \frac{\varepsilon}{2x_n} - O(x_n^{-3}) \ge \frac{\varepsilon}{4x_n} \text{ in } S(T-2, T+2).$$

For function W which is a solution of the problem (25), (26) we apply the Harnack inequalitiy (see [7]): for any positive solution W of the problem (25), (26) in $S(T-2, T+2)$ the inequality

$$W(x) \le CW(x^0) \tag{28}$$

is valid, where $x \in S(T-1, T+1), x^0 \in S(T-1, T+1)$. The constant C does not depend on W, x^0, x.

This proposition can be proved in the same way as the Harnack theorem is proved in [7].

It follows from (28) that

$$W(x) \le CW(x^\varepsilon) \le C_1 \varepsilon x_n^{-1}, \tag{29}$$

since $W > 0$ in $S(T-2, T+2)$ and $W(x^\varepsilon) = w(x^\varepsilon) - w_0(x^\varepsilon) \le 2\varepsilon(x_n^\varepsilon)^{-1} + O((x_n^\varepsilon)^{-3}) \le 3\varepsilon(x_n^\varepsilon)^{-1}$.

Therefore we get $w = W + w_0$,

$$|w(x)| \leq C_2\varepsilon x_n^{-1} + C_3 x_n^{-3} \leq o(x_n^{-1}) \text{ for } x_n > X_\varepsilon, \tag{30}$$

according to (29) and the estimate (27) for w_0. Since $Z = \varepsilon x_n^{-1} - w(x)$, we obtain from (30) that

$$Z(x) \geq \varepsilon x_n^{-1} - o(x_n^{-1}) = o(x_n^{-1}), Z_-(x) = o(x_n^{-1}). \tag{31}$$

From (31) and (18) we get

$$|Z(x)| = o(x_n^{-1}).$$

According to (31) we have

$$v(x) = u(x) - M_p(x_n + \delta_0)^{\frac{2}{1-p}} = Z(x)(x_n + \delta_0)^{\frac{2}{1-p}} = o(x_n^{\frac{1+p}{1-p}}). \tag{32}$$

Set

$$\theta(x_n) = M_p(x_n + \delta_0)^{\frac{2}{1-p}}.$$

For function $v(x)$ we have the equation

$$\Delta v = g(x)v,$$

where

$$g(x) = (|u|^{p-1}u - \theta^p)(u - \theta)^{-1}, \tag{33}$$

It is evident that $g(x) = p|\tilde{u}|^{p-1}$, where

$$\tilde{u} = \theta + o(x_n^\lambda), \lambda = (p+1)(1-p)^{-1}, \text{ as } x_n \to \infty,$$

according to (32). Therefore we have

$$g(x) = p(M_p)^{p-1}x_n^{-2}(1 + \delta_0 x_n^{-1})^{-2}(1 + o(x_n^{-1})) = p(M_p)^{p-1}x_n^{-2} + O(x_n^{-3}).$$

Let us consider the function $V(x) = S(x_n)v(x)$, where $S(t) = 0$ for $t < \tau, \tau = \text{const} > 0, S(t) = 1$ for $t > \tau + 1, S \in C^\infty(\mathbf{R}^1)$. The function V satisfies the equation

$$\Delta V - g_1 V = f \text{ in } S(-\infty, +\infty), \tag{34}$$

$\frac{\partial V}{\partial \nu} = 0$ on $\sigma(-\infty, +\infty)$, f has a compact support, $g_1 = g$ for $x_n > \tau + 1$. In the paper [4] it is proved that the equation (34) has in $S(-\infty, +\infty)$ a solution V_0 with such properties:

$$V_0(x) = O(e^{-\alpha x_n}), \alpha = \text{const} > 0, \text{ as } x_n \to \infty, \tag{35}$$

$$V_0(x) = ax_n + b + O(e^{\alpha x_n}), a, b, \alpha = \text{const}, \alpha > 0, x_n \to -\infty, \tag{36}$$

and for $V_0(x)$ the following estimate is valid

$$\int_{S(-\infty,+\infty)} \sum_{|\beta|\le 2} |\mathcal{D}^\beta V_0|^2 e^{2\alpha x_n} dx \le C_1 \int_{S(-\infty,+\infty)} |f|^2 e^{2\alpha x_n} dx \le C_2, C_1, C_2 = \text{const}. \quad (37)$$

It is well-known that for the solution of the equation (34) with the boundary condition $\frac{\partial V_0}{\partial \nu} = 0$ on $\sigma(-\infty,+\infty)$ for $t > \tau + 1$ we have

$$\max_{x\in S(t-1,t+1)} |V_0(x)|^2 \le C_3 \int_{S(t-2,t+2)} |V_0|^2 dx,$$

where C_3 is a constant and C_3 does not depend on t, V_0. (This is the DeGiorgi type estimate (see [8], [9]).)

For the solution $V_0(x)$ we get

$$\max_{x\in S(t-1,t+1)} |V_0(x)|^2 \le C \int_{S(t-2,t+2)} |V_0(x)|^2 dx \le$$
$$Ce^{-2\alpha(t-2)} \int_{S(t-2,t+2)} |V_0(x)|^2 e^{2\alpha x_n} dx \le C_1 e^{-2\alpha t}.$$

Therefore

$$|V_0(x)|^2 \le C_2 e^{-2\alpha x_n}, C, C_1, C_2 = \text{ const} > 0. \quad (38)$$

Let us consider $V_1 = V - V_0$. It follows from (32) and (38) that $V_1 = o(x_n^\lambda)$ as $x_n \to \infty$ and according to (36) $V_1(x) = ax_n + b + O(e^{\alpha x_n})$ as $x_n \to -\infty, a, b, \alpha = \text{const}, \alpha > 0$. We shall prove that $a = 0, b = 0$. Suppose that $a < 0$. For V_1 we have the equation

$$\Delta V_1 - g_1 V_1 = 0 \text{ in } S(-\infty,+\infty).$$

Let us consider the equation

$$Y''(x_n) - h(x_n)Y(x_n) = 0, -\infty < x_n < +\infty, h(x_n) = p(M_p)^{p-1} x_n^{-2}. \quad (39)$$

It is easy to verify that $Y(x_n) = x_n^\lambda$, where $\lambda = (p+1)(1-p)^{-1}$, is a solution of equation (39) for $x_n > 0$. Since $a < 0$ and $V_1(x) \to 0$ as $x_n \to \infty$, we have $V_1 > 0$ in $S(-\infty,+\infty)$. We introduce the function

$$Y_1(x_n) = Y(x_n) + kx_n^{\lambda-\frac{1}{2}}, k = \text{const} > 0.$$

It is easy to see that

$$\Delta Y_1 - g_1 Y_1 \ge 0 \text{ and } Y_1(x_n) > 0 \text{ in } S(0,\infty),$$

since $\Delta Y_1 - g_1 Y_1 = kC_0 x_n^{\lambda - \frac{5}{2}} + O(x_n^{-3+\lambda}) \geq 0$, if $x_n > 0$ and k is sufficiently large; $C_0 =$ const > 0. Let us consider the function

$$E = mV_1 - Y_1.$$

We have $\Delta - g_1 E \leq 0$ in $S(1,\infty), E(x) \to 0$ as $x_n \to \infty$, $E(x',1) > 0$, if m is sufficiently large, since $V_1 > 0$ in $S(-\infty,+\infty)$. The function $E(x)$ cannot have negative values in $S(1,\infty)$, since $E(x)$ cannot attain a negative minimum in $S(1,\infty)$, on $\sigma(1,\infty)$, and for $x_n = 1$. Therefore $E(x) \geq 0$ in $S(1,\infty)$,

$$mV_1 \geq x_n^{\lambda} + kx_n^{\lambda-\frac{1}{2}} \text{ in } S(1,\infty). \tag{40}$$

The inequality (40) contradicts $V_1(x) = o(x_n^{\lambda})$ as $x_n \to \infty$. In the same way we can get a contradiction, if we suppose $a > 0$ or $b \neq 0$. Thus, $V_1(x) \to 0$ as $x_n \to -\infty$. According to the maximum principle $V_1 \equiv 0$ in $S(-\infty,+\infty)$. We have $V = v$ for $x_n > \tau + 1, V_0(x) = O(e^{-\alpha x_n})$ as $x_n \to \infty$, $u(x) = v(x) + M_p(x_n + \delta_0)^{\frac{2}{1-p}}$. The theorem is proved in the case $\delta_0 < \infty$. Consider the case $\delta_0 = \infty$. Since $u > 0$, according to (10) $u(x) \geq M_p(x_n + \gamma)^{\frac{2}{1-p}}, \gamma =$ const > 0. Therefore, for any $\delta > 0$, which satisfies the inequality $u(x) \leq M_p(x_n + \delta)^{\frac{2}{1-p}}$, we have $\delta < \gamma$. It means that $\delta_0 < \infty$. Theorem 2 is proved.

We use Theorem 2 for the proof of the main Theorem 3. For this we use also the following lemmas.

Lemma 1. *Assume that $u(t)$ is a solution of the equation*

$$\frac{d^2u}{dt^2} - \frac{k^2}{t^2}u = f(t),\ 0 < t_1 < t < \infty, k = const \geq 0, \tag{41}$$

$|f(t)| \leq C_1 e^{-\alpha_1 t}, \alpha_1 = const > 0, |u(t)| \leq C_2 e^{-\alpha_2 t}, \alpha_2 = const > 0$. *Then*

$$|u(t)| \leq C_3 e^{-\alpha_1 t}, C_1, C_2, C_3 = \text{const} > 0.$$

Proof. It is easy to verify that the function

$$u_1(t) = t^{r_3} \int^{\infty} s^{r_2} \left(\int_s^{\infty} \tau^{r_1} f(\tau) d\tau \right) ds$$

is a solution of the equation (41), if $r_3 + r_2 + r_1 = 0$, $r_3(r_3 - 1) - k^2 = 0, 2r_3 + r_2 = 0$. If $k = 0$, we set $r_1 = r_2 = r_3 = 0$. It can be easily proved that $|u_1(t)| \leq Ce^{-\alpha_1 t}$, since $r_1 + r_2 + r_3 = 0$. The function $y(t) = u(t) - u_1(t)$ is a solution of the equation

$$y'' - \frac{k}{t^2}y = 0$$

and

$$|y(t)| \leq C_4 e^{-\alpha t}, \alpha = \min\{\alpha_1, \alpha_2\}. \tag{42}$$

This equation has two solutions $y_1(t) = t^{q_1}, y_2(t) = t^{q_2}$, where q_1, q_2 are roots of the equation $g^2 - q = k^2$. Therefore, the only solution, which satisfies the inequality $|y(t)| \leq C_4 e^{-\alpha t}$ is $y(t) \equiv 0$. We have $u(t) = u_1(t), |u(t)| \leq Ce^{-\alpha_1 t}$. Lemma is proved.

Lemma 2. *Let $u(x)$ be a solution of the equation*

$$\Delta u - \frac{k^2}{x_n^2} u = f(x) \text{ in } S(1, \infty), \tag{43}$$

$$\frac{\partial u}{\partial \nu} = 0 \text{ on } \sigma(1, \infty), \tag{44}$$

$$k \geq 0, |f(x)| \leq C_1 e^{-\gamma_1 x_n}, |u(x)| \leq C_2 e^{-\gamma_2 x_n},$$

where

$$0 < \gamma_2 < \gamma_1 < \sqrt{\lambda_2}. \tag{45}$$

Then

$$|u(x)| \leq C_3 e^{-\gamma_1 x_n} \quad , \quad C_j = \text{const} \, . \tag{46}$$

Proof. We can write

$$u(x) = v(x) + v_1(x_n), \tag{47}$$

$$v_1(x_n) = \frac{1}{\text{meas}\omega} \int_\omega u(x) dx', \int_\omega v(x) dx' = 0,$$

$$f(x) = F(x) + F_1(x_n), \tag{48}$$

$$F_1(x_n) = \frac{1}{\text{meas}\omega} \int_\omega f(x) dx', \int_\omega F(x) dx' = 0.$$

Integrating the equation (43) over ω, we get

$$v_1'' - \frac{k^2}{x_n^2} v_1 = F_1(x_n), \tag{49}$$

$$\Delta v - \frac{k^2}{x_n^2} v = F(x), \tag{50}$$

From the well-known inequality

$$|\nabla u(x)|^2 \leq C_4 [\max_{S(T-2,T+2)} |f| + \max_{S(T-2,T+2)} |u|], x \in S(T-1, T+1), \tag{51}$$

where the constant C_4 does not depend on u and $T > 2$, it follows that

$$|\nabla u(x)| \leq C_5 e^{-\gamma_2 x_n}, |\nabla v(x)| + |v(x)| \leq C_6 e^{-\gamma_2 x_n}. \tag{52}$$

Multiplying the equation (50) by $v(x)$ and integrating it over $S(t,\infty)$, after the integration by parts we get

$$\mathcal{T}(t) \equiv \int_{S(t,\infty)} |\nabla v|^2 dx = -\int_{\omega(t)} v\frac{\partial v}{\partial x_n} dx' - \int_{S(t,\infty)} vF dx - \int_{S(t,\infty)} \frac{k^2}{x_n^2} v^2 dx, \tag{53}$$

where $\omega(t) = S(t,\infty) \cap \{x : x_n = t\}$. Since $\int_{\omega(t)} v dx' = 0$, we can apply the Poincaré inequality. Using the Poincaré inequality and the elementary inequality $\sqrt{a}\sqrt{b} \le \frac{1}{2}(a+b)$, we obtain from (53) that

$$\begin{aligned} \mathcal{T}(t) \le \frac{1}{\sqrt{\lambda_2}} &\left(\int_{\omega(t)} \sum_{j=1}^{n-1} \left(\frac{\partial v}{\partial x_j}\right) dx'\right)^{1/2} \left(\int_{\omega(t)} \left(\frac{\partial v}{\partial x_n}\right) dx'\right)^{1/2} \\ &+ C_\varepsilon \int_{S(t,\infty)} |F|^2 dx + \varepsilon \int_{S(t,\infty)} v^2 dx \\ &\le \frac{1}{2}\lambda_2^{-1/2} \int_{\omega(t)} |\nabla v|^2 dx' + C_\varepsilon \int_{S(t,\infty)} |F|^2 dx + \varepsilon \int_{S(t,\infty)} |v|^2 dx. \end{aligned} \tag{54}$$

It follows from (54) that

$$\mathcal{T}(t) \le -\frac{1}{2\sqrt{\lambda_2}} \mathcal{T}'(t) + C_\varepsilon \int_{S(t,\infty)} |F|^2 dx + \varepsilon \frac{1}{\lambda_2} \int_{S(t,\infty)} |\nabla v|^2 dx,$$

$$\left(2\sqrt{\lambda_2} - 2\varepsilon\lambda_2^{-1/2}\right) \mathcal{T}(t) + \mathcal{T}'(t) \le 2\sqrt{\lambda_2} C_\varepsilon \int_{S(t,\infty)} |F|^2 dx.$$

Let us take ε in such a way that $2\lambda_2^{1/2} - 2\varepsilon\lambda_2^{-1/2} = 2\gamma_1 + \delta$, where $\delta = \text{const} > 0$. Then we have

$$2\gamma_1 \mathcal{T} + \delta \mathcal{T} + \mathcal{T}' \le C_7 e^{-2\gamma_1 t}, C_7 = \text{const} > 0. \tag{55}$$

From (55) we get, if the constant A is sufficiently large, that

$$\begin{aligned} (Ae^{-2\gamma_1 t} - \mathcal{T})' + (2\gamma_1 + \delta)(Ae^{-2\gamma_1 t} - \mathcal{T}) = \delta A e^{-2\gamma_1 t} \\ - (\mathcal{T}' + (2\gamma_1 + \delta)\mathcal{T}) \ge \delta A e^{-2\gamma_1 t} - C_7 e^{-2\gamma_1 t} > 0, \end{aligned} \tag{56}$$

$$(Ae^{-2\gamma_1 t} - \mathcal{T})|_{t=t_1} > 0. \tag{57}$$

From (56) and (57) we obtain that

$$Ae^{-2\gamma_1 t} - \mathcal{T} \ge 0, \mathcal{T} \le Ae^{-2\gamma_1 t} \text{ for } t > t_1. \tag{58}$$

According to the Poincaré inequality and (58) we have

$$\int_{S(t,\infty)} |v|^2 dx \le A_1 e^{-2\gamma_1 t}, A_1 = \text{const}. \tag{59}$$

From the E DeGiorgi estimate [8] [9] and (59) it follows that

$$|v(x)| \leq C_8 e^{-\gamma_1 x_n}. \tag{60}$$

To function $v_1(x)$ we can apply Lemma 1 and get

$$|v_1(x_n)| \leq C_9 e^{-\gamma_1 x_n}. \tag{61}$$

From (60) and (61) we get (46). Lemma is proved.

Lemma 3. *Let $u(x)$ be a solution of the equation (43) with the boundary condition* (44) *and*

$$|f(x)| \leq C_1 e^{-\gamma_1 x_n}, |u(x)| \leq C_2 e^{-\gamma_2 x_n}, \gamma_1 > \sqrt{\lambda_2}. \tag{62}$$

Then

$$|u(x)| \leq C_3 e^{-\sqrt{\lambda_2} x_n}, C_j = \text{const}, j = 1, 2, 3. \tag{63}$$

Proof. From (62) and Lemma 2 it follows that

$$|u(x)| \leq C_\varepsilon e^{(-\sqrt{\lambda_2}+\varepsilon)x_n}, C_\varepsilon = \text{const},$$

for any $\varepsilon = \text{const} > 0$. As in the proof of Lemma 2 we consider the solution $u(x)$ in the form (47). From (53) we get for $v(x)$

$$\mathcal{T}(t) \leq \frac{1}{2\sqrt{\lambda_2}} \int_{\omega(t)} |\nabla v|^2 dx' + \int_{S(t,\infty)} |\mathcal{F}||v| dx$$
$$\leq -\frac{1}{2\sqrt{\lambda_2}} \mathcal{T}'(t) + C_4 e^{(-\gamma_1 - \sqrt{\lambda_2}+\varepsilon)x_n} \leq -\frac{1}{2\sqrt{\lambda_2}} \mathcal{T}'(t) + C_5 e^{(-2\sqrt{\lambda_2}-\delta_1)x_n}, \tag{64}$$

where $\delta_1 = \text{const} > 0$. It follows from (64) that

$$(e^{2\sqrt{\lambda_2} x_n} \mathcal{T}(x_n))' \leq C_6 e^{-\delta_1 x_n}, e^{2\sqrt{\lambda_2} x_n} \mathcal{T}(x_n) < C_4.$$

Then applying the Poincaré inequality, the E. DiGiorgi theorem and Lemma 1 as in the proof of Lemma 2 we obtain (63).

Lemma 4. *Suppose that $u(x)$ is a solution of problem (1), (2) and conditions of Theorem 1 are satisfied. Then*

$$u(x) = O(e^{-\sqrt{\lambda_2} x_n}), x_n \to \infty. \tag{65}$$

If the conditions of Theorem 2 are satisifed, then

$$u(x) = M_p(x_n+\gamma)^{\frac{2}{1-p}} + O(e^{-\sqrt{\lambda_2}x_n}), x_n \to \infty, \tag{66}$$

γ is a constant which depends on $u(x)$.

Proof. Let α be a constant in (3). Consider the sequence

$$\sigma_1 = \alpha, \sigma_2 = p\alpha, \dots, \sigma_n = p^{n-1}\alpha.$$

If $\alpha > \sqrt{\lambda_2}$, then (65) is proved. Suppose that $\sigma_m \le \sqrt{\lambda_2}, \sigma_{m+1} > \sqrt{\lambda_2}$. From the equation (1) we have

$$\Delta u = |u|^{p-1}u = f(x), |f(x)| \le C_1 e^{-\sigma_2 x_n}. \tag{67}$$

Applying successively Lemma 2 we obtain that $|u(x)| \le C_m e^{(-\sigma_m+\varepsilon)x_n}$ for any $\varepsilon = \text{const} > 0$. Therefore we get

$$\Delta u = |u|^{p-1}u = f(x), |f(x)| \le C_{m+1}e^{(-p\sigma_m+\varepsilon p)x_n} = C_{m+1}e^{(-\sigma_{m+1}+\varepsilon p)x_n},$$

$\sigma_{m+1} - p\varepsilon > \sqrt{\lambda_2}$, if ε is sufficiently small. Then (65) follows from Lemma 3.

Assume that (4) is valid for $u(x)$. We set

$$u(x) = M_p(x_n+\gamma)^{\frac{2}{1-p}} + w(x).$$

From (1) we get

$$\Delta w = [w + M_p(x_n+\gamma)^{\frac{2}{1-p}}]^p - \Delta(M_p(x_n+\gamma)^{\frac{2}{1-p}}). \tag{68}$$

We can assume that $\gamma = 0, x_n > 1$. From (68) it is easy to obtain that

$$\Delta w - \frac{2p(1+p)}{(p-1)^2}\,\frac{w}{x_n^2} = f(x), \tag{69}$$

where $|f(x)| \le Cw^2, C = \text{const}$. Consider the sequence

$$\sigma_1 = \beta, \sigma_2 = 2\beta, \dots, \sigma_m = 2^{m-1}\beta, \sigma_m \le \sqrt{\lambda_2}, \sigma_{m+1} > \sqrt{\lambda_2},$$

β is the constant in (5). From (5) it follows that $|w(x)| \le C_1 e^{-\sigma_1 x_n}$ and therefore $|f(x)| \le C_2 e^{-\sigma_2 x_n}$. Applying successively Lemma 2 we obtain $|u(x)| \le C_\varepsilon e^{-(\sigma_m-\varepsilon)x_n}$ and $|f(x)| \le Ce^{-2(\sigma_m-\varepsilon)x_n}$, where $2\sigma_m - 2\varepsilon > \sqrt{\lambda_2}$. Then the estimate (66) follows from Lemma 3.

Lemma 5. *The equation*

$$y'' - \alpha^2 y = f(t)\,, \tag{70}$$

where $\alpha = \text{const} > 0, |f(t)| \le C_1 e^{-\beta t}, \beta > \alpha, t > 0,$ *has a unique solution such that*

$$|y(t)| \le C_2 e^{-\beta t}, \sup |y(t)e^{\beta t}| \le C_3 \sup |f(t)e^{\beta t}|. \tag{71}$$

Proof. It is easy to see that the function

$$z(t) = e^{\alpha t} \int_t^\infty e^{-2\alpha s} \left(\int_s^\infty f(\tau) e^{\alpha\tau} d\tau \right) ds$$

is a solution of the equation (70), which satisfies conditions (71). If $Z_1(t)$ and $Z_2(t)$ are two solutions with properties (71), then $Z_0 = Z_1 - Z_2$ satisfies the equation $Z_0'' - \alpha^2 Z_0 = 0$ and $|Z_0(t)| \le C_4 e^{-\beta t}$. Since $\beta > \alpha, Z_0 \equiv 0$ for $t > 0$.

Lemma 6. *The equation*

$$y'' - \alpha^2 y = g(t)y + f(t), \tag{72}$$

where $|f(t)| \le C_1 e^{-\beta t}, \alpha > 0, \beta > \alpha$, *has a unique solution, which satisfies conditions (71), if* $|g(t)| < \varepsilon$ *and* ε *is sufficiently small.*

Proof. The equation (72) is equivalent to the integral equation

$$y(t) = e^{\alpha t} \int_t^\infty e^{-2\alpha s} \left(\int_s^\infty e^{\alpha\tau} f(\tau) d\tau \right) ds + e^{\alpha t} \int_t^\infty e^{-2\alpha s} \left(\int_s^\infty g(\tau) y(\tau) d\tau \right) ds. \tag{73}$$

The existence of a solution of (73) with properties (71) can be proved by the method of successive approximations.

The uniqueness of such a solution follows from the fact that any nontrivial solution $Z(t)$ of the equation $y'' - \alpha^2 y - g(t)y = 0$ satisfies the inequality $Z(t) \ge C_2 e^{-(\alpha+\delta)t}$, if ε is sufficiently small, $\delta = \text{const} > 0, \alpha + \delta < \beta$. Indeed, let us take C_2 such that $Z(0) \ge C_2$ and such δ that $(\alpha + \delta)^2 - \alpha^2 + g < 0$. Then $v = Z(t) - C_2 e^{-(\alpha+\delta)t} \ge 0$ for $t \ge 0$, since v can not attain a negative minimum because $v'' - (\alpha^2 - g)v < 0$ and $v \to 0$ as $t \to \infty, v(0) \ge 0$.

Lemma 7. *Let* $y(t)$ *be a solution of the equation*

$$y'' - \alpha^2 y - \frac{k^2}{t^2} y = f(t) \tag{74}$$

where $\alpha \ne 0, k \ge 0, |f(t)| \le C_1 e^{-\beta t}, \beta > \alpha, |y(t)| \le C_2 e^{-\alpha_1 t}, \alpha_1 < \beta.$

Then

$$y(t) = C_3 e^{-\alpha t} + o(e^{-\alpha t}), t \to \infty. \tag{75}$$

Proof. According to Lemma 6 the equation (74) has a solution $y_1(t)$ such that $|y_1(t)| \le C_4 e^{-\beta t}$. The function $Y_1(t) = y(t) - y_1(t)$ satisfies the equation

$$Y'' - \alpha^2 Y - \frac{k^2}{t^2} Y = 0. \tag{76}$$

It is known [10] that any solution of the equation (76) has the form

$$Y(t) = C_4(1 + o(1))e^{\alpha t} + C_5(1 + o(1))e^{-\alpha t} \text{ as } t \to \infty, \tag{77}$$

where C_4, C_5 are constants. Since $Y_1(t) \to 0$ as $t \to \infty$, we have $C_4 = 0$ for $Y_1(t)$. Therefore, for $y(t)$ (75) is valid.

Lemma 8. *Assume that $u(x)$ is a solution of the equation*

$$\Delta u - \frac{k^2}{x_n^2} u = f(x) \text{ in } S(1, \infty) \tag{78}$$

with the boundary condition

$$\frac{\partial u}{\partial \nu} \text{ on } \sigma(1, \infty), \tag{79}$$

$|u(x)| \le C_1 e^{-\sqrt{\lambda_2} x_n}, |f(x)| \le C_2 e^{-\beta x_n}, \beta > \sqrt{\lambda_2}$. *Then*

$$u(x) = \sum_{i=1}^{m} A_i v_i(x') e^{-\sqrt{\lambda_2} x_n} + o(e^{-\sqrt{\lambda_2} x_n}), x_n \to \infty, \tag{80}$$

where $v_1, \ldots, v_m$ is a system of orthonormal eigenfunctions, corresponding to the eigenvalue λ_2 and m is the multiplicity of $\lambda_2, A_i = \text{const}, i = 1, \ldots, m$.

Proof. We set

$$u(x) = \sum_{i=0}^{m} B_i(x_n) v_i(x') + u^*,$$

$$f(x) = \sum_{i=0}^{m} F_i(x_n) v_i(x') + F^*,$$

where

$$v_0 = \text{const}, \int_\omega u^* v_i dx' = 0, \int_\omega F^* v_i dx' = 0, i = 0, 1, \ldots, m. \tag{81}$$

From the equation (78) we have

$$-\lambda_2 \sum_{i=1}^{m} B_i(x_n)v_i + \sum_{i=0}^{m} B_i''(x_n)v_i - \frac{k^2}{x_n^2}\sum_{i=0}^{m} B_i(x_n)v_i$$
$$+\Delta u^* - \frac{k^2}{x_n^2}u^* = \sum_{i=0}^{m} F_i(x_n)v_i + F^*, \tag{82}$$

Multiplying (82) by v_i and integrating over ω, after the integration by parts, we get

$$B_0''(x_n) - \frac{k^2}{x_n^2}B_0(x_n) = F_0(x_n), \tag{83}$$

$$B_i''(x_n) - \lambda_2 B_i(x_n) - \frac{k^2}{x_n^2}B_i(x_n) = F_i(x_n), i = 1, \dots, m, \tag{84}$$

$$\Delta u^* - \frac{k^2}{x_n^2}u^* = F^* \text{ in } S(1,\infty) , \tag{85}$$

$$\frac{\partial u^*}{\partial \nu} = 0 \qquad \text{on} \qquad \sigma(1,\infty). \tag{86}$$

In order to estimate $B_0(x_n)$ we use Lemma 1. Since $|f(x)| \le C_2 e^{-\beta x_n}, \beta > \sqrt{\lambda_2}$, and therefore $|F_0(x_n)| \le C_3 e^{-\beta x_n}, |u(x)| \le C_1 e^{-\sqrt{\lambda_2}x_n}, |B_0(x_n)| \le C_4 e^{-\sqrt{\lambda_2}x_n}$, according to Lemma 1 we have $|B_0(x_n)| \le C_5 e^{-\beta x_n}$. Using Lemma 7 we have

$$B_i(x_n) = A_i e^{-\sqrt{\lambda_2}x_n} + o(e^{-\sqrt{\lambda_2}x_n}), i = 1, \dots, m. \tag{87}$$

Let us estimate $u^*(x)$. Multiplying (85) by u^*, integrating it over $S(t,\infty)$, transforming the first integral by the integration by parts and taking into account (86), we obtain

$$T(t) \equiv \int_{S(t,\infty)} |\nabla u^*|^2 dx = -\int_{\omega(t)} u^* \frac{\partial u^*}{\partial x_n} dx' - \int_{S(t,\infty)} \frac{k^2}{x_n^2}(u^*)^2 dx - \int_{S(t,\infty)} F^* u^* dx. \tag{88}$$

It follows from (88) that

$$T(t) \le \left(\int_{\omega(t)} (u^*)^2 dx'\right)^{\frac{1}{2}} \left(\int_{\omega(t)} \left(\frac{\partial u^*}{\partial x_n}\right) dx'\right)^{\frac{1}{2}} + C_6 \left(\int_{S(t,\infty)} (u^*)^2 dx\right)^{1/2} e^{-\beta t}. \tag{89}$$

Since u^* satisfies the condition (81), according to the variational theory of eigenvalues we have

$$\int_{\omega(t)} (u^*)^2 dx' \le \lambda_3^{-1} \int_{\omega(t)} \sum_{i=1}^{n-1} \left(\frac{\partial u^*}{\partial x_i}\right)^2 dx', \tag{90}$$

where λ_3 is the third eigenvalue of the problem (8), (9). From (89), (90) we obtain

$$\mathcal{T}(t) \leq -\frac{1}{2}\lambda_3^{-\frac{1}{2}}\mathcal{T}'(t) + C_7 e^{(-\sqrt{\lambda_2}-\beta)t}, C_7 = \text{const}.$$

Integrating this inequality from t_0 to t, we get

$$\mathcal{T}(t) \leq C_8 e^{-2\gamma t},$$

where $\gamma = \min(\sqrt{\lambda_3}, \frac{1}{2}(\sqrt{\lambda_2}+\beta))$. Using the Poincaré inequality and the DiGiorgi theorem, as in the proof of Lemma 2, we get

$$|u^*(x)| \leq C_9 e^{-\gamma t}, \gamma > \sqrt{\lambda_2} \quad . \tag{91}$$

From the estimate for $B_0(x_n)$ and (87), (91) we obtain (80). Lemma is proved.

Now we can prove Theorem 3.

Proof of Theorem 3. From Lemma 4 it follows that

$$u(x) = \pm K(x_n + h)^{\frac{2}{1-p}} + w,$$

where $|w(x)| \leq C_1 e^{-\sqrt{\lambda_2} x_n}$. From equation (1) we obtain the equation for w of the form

$$\Delta w = |w \pm K(x_n + h)^{\frac{2}{1-p}}|^{p-1}(w \pm K(x_n + h)^{\frac{2}{p-1}})$$
$$\mp \Delta(K(x_n + h)^{\frac{2}{1-p}}) .$$

Using the Tailor formula and the estimate for w we get

$$\Delta w - pK^{p-1}(x_n + h)^{-2} w = \mathcal{F}(x),$$

where $|\mathcal{F}(x)| \leq C_2 e^{-2\sqrt{\lambda_2} x_n}$ or $|\mathcal{F}(x)| \leq C_3 e^{-p\sqrt{\lambda_2} x_n}$, if $K = 0$. Then (7) follows from Lemma 8. Theorem is proved.

The approach, which is used in this paper, can be applied to a more general class of nonlinear elliptic equations, which will be considered in the next paper.

References

[1] A.N. Kolmogorov, I.G. Petrovsky, N.S. Piscunov, The study of the equation, joint with a growth of quantity of substance. Bull MGU, Math, Mech. 1937, v. 1, p. 1–26.

[2] A.I. Volpert, On propogation of waves, described by nonlinear parabolic equations. In: I.G. Petrovsky selected papers. Differential equations. Theory of probability. Moscow, Nauka, 1987, p. 333–358.

[3] H. Berestycki, L. Nirenberg, Some qualitative properties of solutions of semilinear elliptic equations in cylindrical domains. Analysis, ed. by P. Rabinovitz, Academic Press, 1990, p. 114–164.

[4] V.A. Kondratiev, O.A. Oleinik, On asymptotic behaviour of solutions of some nonlienar elliptic equations in unbounded domains. Partial differential equations and related subjects. Proceedings of the Conference dedicated to Louis Nirenberg. Longman, 1992, p. 163–195.

[5] V.A. Kondratiev, O.A. Oleinik, Some results for nonlinear elliptic equations in cylindrical domains. In: Operator calculus and spectral theory. Birkhäuser Verlag, 1992, p. 185–195.

[6] A. Brada, Comportement asymptotique de solutions d'equations elliptiques semi - lineaires dans un cylindré. Université Francois Rabelais, Tour. Publications laboratoire de mathématiques et applications, n 44/92.

[7] E.M. Landis, Second order elliptic and parabolic equations. Moscow, Nauka, 1971.

[8] E. DeGiorgi, Sulla differenziabilita e l'analicita delle estremali degli integrali, Mem. Acc. Sci. Torino, 1957, p. 1–19.

[9] J. Moser, A new proof of DeGiorgi theorem concerning the regularity problem for elliptic differential equations, Comm. Pure and Appl. Math. v. 13, no. 3, 1960, p. 457–468.

[10] R. Bellman, Stability theory of differential equations, McGraw, New York, 1953.

FUNDAMENTAL PHYSICS IN UNIVERSAL SPACE-TIME

By Irving Segal
M.I.T., Cambridge, MA 02139, USA

Conference on Analysis, Geometry, and Physics,
College de France, Paris, June 3-5, 1992

ABSTRACT

Universal space-time is a natural candidate for the 'bare' arena of the fundamental forces, being the maximal 4-dimensional manifold having physically indicated properties of causality and symmetry. It is locally conformal to Minkowski space, and globally conformal to the Einstein Universe $\mathbf{E} \sim R^1 \times S^3$. The Einstein energy exceeds that in the canonically imbedded Minkowski space, and the difference has been proposed by the chronometric theory to represent the redshift. Although this eliminates adjustable cosmological parameters, the directly observable implications of this proposal have been statistically quite consistent with direct observations in objective samples of redshifted sources. These developments represent a mathematical specification of proposals by Mach, Einstein, Minkowski, and Hubble and Tolman. They suggest that the fundamental *forces* of Nature are conformally invariant, but that the *state* of the Universe breaks the symmetry down to the Einstein isometry group. This provides an alternative to the Higgs mechanism, and otherwise has implications for particle physics, including the elimination of ultraviolet divergences in representative nonlinear quantum fields, the formulation of a unified invariant interaction Lagrangian, assignments of observed elementary particles to irreducible unitary positive-energy representations of the conformal group, and the correlation of the S-matrix with the action in $\mathbf{E}$ of the generator of the infinite cyclic center of the simply-connected form of the conformal group.

INTRODUCTION

Universal space-time $\mathbf{M}^{\sim}$ is the unique space-time other than that of Minkowski that enjoys comparable physically indicated properties,- of causality, isotropy, homogeneity, and separability in appropriate frames into time and space. At the same time it represents an abstract form of the model proposed by Einstein in his paper founding modern theoretical cosmology (1917), in which the large-scale gravitational structure of the universe is represented as $R^1 \times S^3$ (with the obvious metric). It is remarkably effective for mathematical purposes connected with

M. Flato et al. (eds.), Physics on Manifolds, 253–264.

nonlinear wave equations and quantized fields, and suggestive of new departures in the physics of extreme distances (both large and small). For related reasons, $\mathbf{M}^{\sim}$ is also known as the 'Universal', or 'Einstein-Maxwell', cosmos.

It may be intrinsically characterized in a variety of mathematical ways, among which are as:

i) the unique maximal causal isotropic space-time in 4 dimensions (maximal in the sense of Leray, spatially and temporally isotropic in the sense of Tits);

ii) the universal cover of the conformal compactification $\mathbf{M}$ of Minkowski space $\mathbf{M_0}$;

iii) the maximal spacetime to which the Maxwell (or Yang-Mills, etc.) equations, and their solutions, canonically extend;

iv) the homogeneous space $\mathbf{G}^{\sim}/\mathbf{P}^{\sim}$, where tilde denotes the universal cover, $\mathbf{G}$ = SU(2,2), and $\mathbf{P}$ is the maximal parabolic subgroup of $\mathbf{G}$. (In physical terms, $\mathbf{G}$ is the usual conformal group, and $\mathbf{P}$ is the group generated by the usual Poincare group together with the one-parameter group of 4-dimensional scaling transformations.);

v) the simply connected form of the extension to a full group of the local group of causal transformations in the vicinity of a point in $\mathbf{M_0}$, modulo the subgroup fixing the point;

vi) the Einstein Universe $\mathbf{E}$ as a causal (or conformal) manifold. Note however that the isometry group $\mathbf{K}^{\sim}$ of $\mathbf{E}$ is only 7-dimensional, whereas its conformal group is 15-dimensional. $\mathbf{K}^{\sim}$ is the maximal essentially compact (i.e., having compact image in the adjoint group) subgroup of the simple Lie group $\mathbf{G}^{\sim}$, within conjugacy. Any such subgroup is the $\mathbf{K}^{\sim}$ for a unique metric of the Einstein form, dt^2-ds^2, where ds is the element of arc length on S^3 in radians. It follows that $\mathbf{M}^{\sim}$ may be represented in the form $R^1 x S^3$, or time x space, in an 8-parameter family of distinct but $\mathbf{G}^{\sim}$-conjugate ways.

Possible physical interpretation: $\mathbf{M}^{\sim}$ = empty (or reference, or free space-time); $\mathbf{E}$= space-time as 'clothed' by the energetic contents of the physical universe and their interaction (hence a model for the large-scale gravitational structure of the universe, as proposed by Einstein). On the small scale, $\mathbf{E}$ will be modified in accordance with local fluctuations in the mass and energy distribution, in accordance with Einstein's equation, adjusted on the right-hand side for the curvature difference between R^3 and S^3 (an unobservably small effect on the scale of local observation).

General theme: $\mathbf{M}^{\sim}$ together with its effective causal symmetry group $\mathbf{G}^{\sim}$ (where 'effective' here means the universal cover of the connected group of causality-preserving transformations, which is $\mathbf{G}^{\sim}/Z_2$), is a natural and terminal 'deformation' (in the sense of Lichnerowicz or Faddeev, or in intuitive terms, of Minkowski in his seminal adddress in 1908) of relativistic quantum mechanics. The relevant parameter is the ($\mathbf{G}^{\sim}$-invariant !) radius R of $\mathbf{E}$, a fundamental length, and as $R \to \infty$ the latter theory is recovered. Finite values of R lead to the 'chronometric' theory of the physics of extreme distances, whose energy differs from that of relativistic theory just as the generator of transformation to moving axes in relativistic theory differs from the corresponding Galilean generator, for finite values of c. The three fundamental constants required for fundamental physics are provided by R, c, and h, all of which may be regarded as deformation parameters in generally similar ways. Thus:

as $c \to \infty$, Special Relativity $\Rightarrow$ Galilean-Newtonian theory;

as $h \to 0$, Quantum Mechanics $\Rightarrow$ Classical Mechanics;

as $R \to \infty$, Chronometric Theory $\Rightarrow$ Relativistic Quantum Mechanics.

Chronometric theory may be summarized as to the effect that $\mathbf{M}^{\sim}$ is the true fundamental reference space-time of physics, rather than $\mathbf{M}_0$ or some other submanifold. $\mathbf{E}$ arises from $\mathbf{M}^{\sim}$ by the selection of the (quasi-inertial) frame in which the contents of the universe have minimal energy. Energy here naturally corresponds to the generator of global temporal evolution in $\mathbf{M}^{\sim}$, which is represented by infinitesimal temporal evolution $\partial/\partial t$ in $\mathbf{E}$, where t is the Einstein time. We call this energy the 'Einstein' energy; it differs from that corresponding to infinitesimal temporal evolution $\partial/\partial x_0$ in $\mathbf{M}_0$, i.e. the usual relativistic energy. Present direct laboratory measurements are of Minkowski space quantities such as the latter energy and other Poincare group quantum numbers; unlike corresponding quantum numbers associated with the Einstein isometry group, which differ by terms of order 1/R, the former are in practice determined by the structure of the wave function in the most immediate vicinity of the point of observation.

<u>Possible chronometric effects:</u>

1) the cosmic redshift (of galaxies, etc.);
2) 'gravity', as observed;
3) a certain class of the so-called weak interactions, including e.g. mixing of neutral kaons.

<u>Theoretical applications of the chronometric formalism:</u>

i) Strong indication in work of Pedersen, Segal, and Zhou (1992) for the nontriviality of massless nonlinear scalar quantum field theories in Minkowski spaces of even dimension ≥ 4, whose Lagrangians are proportional to even powers of the quantized field. This applies in particular to the conformally invariant case of the 4th power in 4-dimensional space-time, which underlies the application of the Higgs mechanism in electroweak theory. More specifically, if $\phi(x)$ denotes the massless quantized free field at the point x, then the integral over m-dimensional Minkowski space,

$\int :\phi(x)^q: dx$,

exists as a nonvanishing hermitian operator on the quantized field Hilbert space, assuming q and m are even and ≥ 4. This operator has a self-adjoint extension, and in the conformally invariant cases is already essentially self-adjoint, on the domain E(H) of entire vectors for the Einstein energy operator H of the quantized field.

Earlier, Aizenman and Frohlich gave arguments showing triviality in more than 4 dimension under various somewhat technical, rather than physical, assumptions regarding the applicability of the imaginary-time ('euclidean') formalism to the putative quantized theory. Later work by Segal and Zhou (1992) provides reason to doubt that this formalism is applicable in Minkowski space, but may well apply in terms of the Riemannian symmetric space corresponding to M, or G/K, in place of the complexification of Minkowski space.

N. B. 1. It is noteworthy that although the theorem quoted refers to the Minkowski space action integral, it appears provable in a natural and efficient way only by going outside Minkowski space into universal space-time.

With appropriate treatment of mass, by an adaptation of Mach's principle, the theorem quoted extends also to massive fields. For example,

ii) If $\phi(x)$ denotes the conformally invariant positive-energy massive field in Minkowski space, in the sense e.g. of Greenberg's generalized free fields, and q is an even integer ≥ 4, the the integral over Minkowski space $\int :\phi(x)^q: dx$ exists as an hermitian operator, is non-vanishing, and has a self-adjoint extension. In the case q = 1, this operator (essentially the quantized action) is essentially self-adjoint on E(H).

For fixed mass, we consider fields in E that satisfy the Klein-Gordon equation in E, which however 'live' on $\mathbf{M_0}$ as imbedded in E, while satisfying the Klein-Gordon equation in E. A theorem similar to i) is then true, but moreover, the interaction Hamiltonian, and also the total hamiltonian, are essentially self-adjoint. This eliminates the classic ultraviolet divergences at the slight cost of introducing a dispersion in mass (or 'mass width') of the order of 1/R, where R is the radius of the universe in laboratory units. (Note: R is of order ≥ 1040 fermi, on the basis of redshift measurements, so 1/R is unobservably small in the particle context.)

But some explanation for expressions such as $\int :\phi(x)^q: dx$ is in order here; the definitions of $:\phi(x)^q:$ in the physics literature appear opportunistic and informal in mathematical terms. Our rigorous work on this matter has been in collaboration with J. C. Baez, J. Pedersen, S. M. Paneitz, and Z. Zhou. $\phi(x)$ is itself and operator-valued distribution, so to treat renormalized powers such as $:\phi(x)^q:$, and their integrals, a regularity theory for quantized fields is needed. The stronger results are dependent on the following property of the Einstein energy operator B in the single particle space (e.g., solution manifold of the wave or Maxwell equations): $\exp(-tB)$ is a trace-class operator for all $t > 0$.

More specifically, if **H** is a given complex Hilbert space, in which B is a given selfadjoint operator with this property, a very convenient space of 'test functions' in the corresponding quantized field Hilbert space **K** (the context will prevent confusion with the maximal compact subgroups earlier referred to). In the familiar particle (Fock) representation, **K** is the direct sum of the symmetrized tensor powers of **H**, in the case of boson fields, to which we here limit consideration. For any unitary operator U on **H**, there is a corresponding unitary operator on **K**, i.e. the direct sum of its symmetrized tensor powers, which we denote as $\Gamma(U)$.

The field energy operator H is then $\partial\Gamma(B)$, i.e. the self-adjoint generator of the one-parameter unitary group $\Gamma(\exp(isB))$. In the complex wave representation of the free boson field over **H**, **K** is represented as the space of antiholomorphic entire functions on **H** that have uniformly bounded L_2 norm with respect to Gaussian measure on restriction to arbitrary finite-dimensional subspaces. There is then a precise characterization of the continuous sesqulinear forms on the space E(H) of entire vectors for H, in its natural topology. Any such form F can be represented by a kernel $K(z,z')$ that is antiholomorphic in z, holomorphic in z', and respectively anti-entire or entire, such that

$$|K(u,u')| \leq c \exp[||e^{tB}u||^2 + ||e^{tB}u'||^2]$$

for suitable $c, t > 0$:

$$F(f,g) = \iint K(z,z')f(z')g(z)^c dz dz',$$

the integral being over **H** + **H** with respect to the Gaussian measures denoted as dz and dz' (as earlier, such infinite-dimensional integrals are defined as the limits of the integrals over finite-dimensional subspaces, as they increase indefinitely to **H**). In these spaces, the 1-cohomology of the infinite Heisenberg group is trivial, and from this derives a unique covariant quantization map, which may be characterized as follows.

<u>Definition</u>. A quasi-n-cocycle is a continuous symmetric multilinear map from $E(B)^n$ to the space **F**(E(H)) of all continuous sesquilinear forms on E(H) (where H = $\partial\Gamma(B)$) such that

$$[F(z_1,z_2,\ldots,z_n),\phi(z')] = [F(z',z_2,\ldots,z_n),\phi(x)]$$

for arbitrary z' and $z_1,\ldots,z_n$, where $\phi(z)$ denotes the 'self-adjoint boson field over **H**', z in E(H), definable by the equation: $\exp(is\phi(z)) = W(sz)$ for arbitrary real s, where W denotes the Weyl system associated with the quantized field.

THEOREM. Every quasi-n-cocycle is exact: there exists F in **F**(E(H)) such that

$$F(z_1,\ldots,z_n) = \partial(z_1)\ldots\partial(z_n)F,$$

where $\partial(z)$ denotes the map $F \to [F,\phi(z)]$.

Moreover, there is a canonical F, characterized by vanishing expectation value properties, and the corresponding canonical map **Q** from quasi-n-cocycles to forms is continuous in natural topologies on the respective spaces, and covariant with respect to transformations represented by unitary operators on **H** that leave $F(z_1,\ldots,z_n)$ invariant.

The proof involves analysis of functions of z, z' that are anti-entire in z and entire in z, and includes a version of the Poincare lemma, for operators in place of forms, in infinitely many variables.

EXAMPLE: powers of the quantized wave equation field. For stronger results, we work in **E**, which has the stronger property than $\mathbf{M}_0$ that if A denotes the single-particle hamiltonian, then $\mathrm{tr}(\exp(-tA)) < \infty$ for all $t > 0$.

(<u>Remark</u>: it suffices to work with analytic rather than entire vectors for A, but there is a problem with the case of differentiable vectors, and in practice, the results based on entire vectors are often the most convenient.)

So consider the conformal wave equation on $R^1 \times S^3$:

$$(\partial_t^2 - \Delta + 1)\phi = 0;$$

H is the space of all solutions of finite conformally invariant norm, which solutions are automatically periodic of period 2π in t. Let B denote the self-adjoint generator of time evolution in **H.** For arbitrary $\phi_1, \ldots, \phi_n$ in E(B), let

$$F(\phi_1,\ldots,\phi_n) = \int \phi_1(u)\ldots\phi_n(u)du,$$

where the integration is over $S^1 \times S^3$ as the 2-fold cover $\mathbf{M}^{(2)}$ of **M.** Then F is a quasi-n-cocycle. Hence it has a quantization **Q**(F), whose canonical form defines $\int :\phi(u)^n: du$, integration being over $\mathbf{M}^{(2)}$. Alternatively, the n-fold commutator of the last expression with $\phi(f_1),\ldots,\phi(f_n)$, for suitable smooth test functions $f_1,\ldots f_n$ has the value expected from the canonical quantized field commutation relations, which together with the vanishing of expectation values, such as

$$\langle(\int :\phi(u)^n: du)v,v\rangle\ 0 \quad (v = \text{vaccum vector in } \mathbf{K}),$$

uniquely determines $\int :\phi(u)^n: du$.

For the Minkowski space hamiltonian H_0, exp $(-tH_0)$ is far from being trace class, for any value of t, and there are problems in the precise treatment of Wick powers in the massless case, and of their global integrals in the massive case. Note that for the 'number of particles operator', exp$(-tN)$ has discrete spectrum but is not trace class, and defines topologies that appear too weak for present purposes. Also: S^3 could be replaced by any compact manifold, in conjunction with any standard type of wave equation.

A decade ago, Professor Choquet-Bruhat, Stephen Paneitz, and I showed that the Minkowskian action integral for the Yang-Mills equation was absolutely convergent, by moving out of $\mathbf{M_0}$ into **E**. There is no visible practical way to prove this result by analysis totally in $\mathbf{M_0}$, and even the less singular elliptic ('euclidean)') case treated by Uhlenbeck and others is quite nontrivial. This indicates the power of the underlying technique, which is moreover physically suggestive, and I outline its use to prove the

THEOREM. For arbitrary even $q \geq 4$, if $\mathbf{\Phi_0}$ is the free quantized wave equation field on $\mathbf{M_0}$, then the action integral over $\mathbf{M_0}$, $\int :\mathbf{\Phi_0}(x)^q: d_4x$, exists as a nonzero operator that maps $C^\infty(H)$ continuously into itself, where H denotes the Einstein energy. The interaction energy $\int_{R^3} :\mathbf{\Phi_0}(0,x)^q: d_3x$ exists as a continuous sesquilinear form on $C^\infty(H)$.

The 'conformal connection' involved here is a more sophisticated version of that used early on by Prof. Choquet and her then student, Prof. Christodoulou in an essential way. More recently the Minkowski scattering operator S has been found to represent, in a natural way, the generator of the infinite cyclic central subgroup of $\mathbf{G^\sim}$, notwithstanding that the action of this generator on $\mathbf{M_0}$ is trivial. We return to this later for the treatment of the S-matrix in a massive Klein-Gordon field. A, perhaps the, key technical point may be summarized as the

LEMMA. The free wave equation field $\mathbf{\Phi_0}(x)$, $x \in \mathbf{M_0}$, is unitarily equivalent to the free conformal wave equation field on **E**, $\mathbf{\Phi}(u)$, $u \in \mathbf{E}$, via the following mapping. First, imbed $\mathbf{M_0} \to \mathbf{E}$ by

$$T: (x_0, x_1, x_2, x_3) \to (t, u_1, u_2, u_3, u_4)$$

where $e^{it}(u_1 i\sigma_1 + u_2 i\sigma_2 + u_3 i\sigma_3 + u_4) = (1 + ½\iota X)(1 - ½\iota X)^{-1}$, $X = x_0 + x_1\sigma_1 + x_2\sigma_2 + x_3\sigma_3$, the σ_j being the usual 'Pauli' matrices. Let $\mathbf{H_0}$ denote the (conformally invariant) Hilbert space of solutions of the wave equation on $\mathbf{M_0}$, and let **H** denote the same for **E**, and let T' denote the mapping on **H**: $\phi(u) \to p(Tx)\phi(Tx) = \phi_0(x)$. Then T' is unitary from **H** onto $\mathbf{H_0}$, and its induced action from $\mathbf{K}(\mathbf{H})$ onto $\mathbf{K}(\mathbf{H_0})$ implements a unitary equivalence of the quantized fields, $\boldsymbol{\Phi}(u)$ with $\boldsymbol{\Phi}_0(x)$ $(u = Tx)$.

Convergence. In 1958, after the high point of QED renormalizability, Schwinger wrote : 'We conclude that a convergent theory can not be consistently formulated within the framework of present space-time concepts.' Earlier, Heisenberg had proposed the introduction of a small fundamental length as a possible means of attaining a convergent theory. (Relatedly, physics needs three fundamental units, while h and c provide only two; alternatively, the Poincare group has a one-parameter group of outer automorphisms, corresponding to the absence of any natural length scale in $\mathbf{M_0}$.) This proposal was not successful, but a combination of a variant involving rather a very large fundamental length, with a variant of the ansatz proposed by Minkowski (1908), consisting in modern terms of the introduction of a deformation whose parameter is this length, appears to attain a convergent theory. This has so far has been detailed only in the case of massive scalar fields, and may be summarized as follows.

We define the Einstein mass M of a Klein-Gordon field on **E** by setting M^2 for the integral eigenvalue of the conformal analog to the usual relativistic mass operator, namely $\Delta - \partial_t^2 + 1$; the integrality results from the physical constraint that the wave function should represent a free particle in the usual physical sense, i.e. one that is carried into itself by the S-matrix. In these terms we may state the

THEOREM. Let $\boldsymbol{\Phi}(u)$ denote the quantized Klein-Gordon field on **E** of Einstein mass M. Then the putative interaction energy operator

$$V = g\int_{S^3} :\boldsymbol{\Phi}(0,x)^q: d_3x \qquad (g > 0)$$

is essentially self-adjoint on the domain A(H) of analytic vectors for the free hamiltonian H. The total hamiltonian H+V is also such, and its closure H'(g) has a unique lowest eigenvector, within a constant factor.

Moreover, the 'Einstein' S-matrix, which work of Baez, Segal and Zhou (1990) shows is definable coherently with relativistic theory as that representing the generator ζ of the infinite cyclic center of $\mathbf{G}^\sim$ (in the unitary action of the Einstein isometry group $\mathbf{K}^\sim$ on the interacting quantum field corresponding to the total hamiltonian), is

$$S = \exp(i\pi H'(g))\exp(-i\pi H).$$

In fact, the space of wave functions of given Einstein mass is finite-dimensional. The dimension is however of order $\geq 10^{200}$ in the case of a particle of mass m_e, on the basis of present estimates of the cosmic distance scale R. It is interesting that this comes about because gravity in **E** appears quite reasonably modelled as the difference between the Einstein and Minkowski energies. For this represents an attractive force that transforms under scaling and euclidean transformations in the same way as the Newtonian gravitational potential, is extremely weak in the small, and is coherent with the very limited empirical constraints on gravity, including a relation between Newton's G and the average mass density of the universe. It produces an effective cutoff on the energy of a massive particle, of the order of $>\approx 10^{37}$, corresponding to an estimate of $R \geq \approx 160$ Mpc.

PARTICLE ASSIGNMENTS AND FUNDAMENTAL FORCES

The universal cosmos has a naturally associated Lagrangian candidate for the mathematical description of the fundamental forces and particles of Nature. Its causal group $\mathbf{G}^\sim$ has two basic finite-dimensional representations. These are the spin representation Σ:

$$L_{ij} \to \tfrac{1}{2}\omega_i\omega_j$$

where the ω's are Clifford numbers for the form $x_{-1}^2+x_0^2-x_1^2-x_2^2-x_3^2-x_4^2$; and L_{ij} = generator of $\mathbf{G}^\sim$ representing infinitesimal transformations in the x_i-x_j plane;

and the adjoint representation, which can be regarded as acting on the space $\mathbf{A}$ of linear transformations A on the spinors. There is a corresponding $\mathbf{G}^\sim$-invariant (necessarily indefinite) inner product $\langle\langle\psi,\psi'\rangle\rangle$ for $\mathbf{G}$-spinors ψ and ψ', and Lagrangian density $\langle\langle A\psi,\psi\rangle\rangle$, $A \in \mathbf{A}$, leading to the $\mathbf{G}^\sim$-invariant action integral

$$\int_{\mathbf{M}} \langle\langle A(u)\psi(u),\psi(u)\rangle\rangle du,$$

if the conformal weights are appropriate.

This leads to the proposal that the fundamental fermions are the sections of the direct product bundle of the scalar fields of weight 2 with the spin representation Σ, and that the fundamental bosons are the sections of the direct product bundle of the scalars of weight 0 with the adjoint representation $\mathbf{A}$. (The so-called 'kinetic' terms do not require separate designation since they are implicit in the $\mathbf{G}^\sim$ transformation properties of these bundles.)

In the limit $R \to \infty$, these fermions become the direct sum of Dirac spinors of weights 3/2 and (the dual weight) 5/2. The bosons become the direct sum of vectors of weight 1 (or 1-forms) with vectors of weight -1, together with a space of weight 0 consisting of a scalar and a 2-tensor. In the exact form before the limit is taken, the section spaces are indecomposable. The $\mathbf{G}^\sim$-invariant irreducible composition factors are the same both before and after the $R \to \infty$ limit, and include an exceptional number of positive-energy unitary factors. These have assignments to 'bare' versions of observed particles, which are in part tentative, as follows (w denotes the weight, which is a feature of the bundle and the interaction, rather than the abstract representation, which e.g. is the same for both the electron and the cygnet):

Fermions	Bosons
electron (w=3/2)	neutral kaon (w=-1)
2 neutrinos (w= 3/2 and 5/2)	neutral pion; rho? (w=0)
cygnet (w=5/2)	photon, W and Z (w=1)

The section spaces have composition series in which the factors occur in the order shown, with the photon subspace at the bottom of the 1-forms. Joint work with D. A. Vogan Jr. shows that the bare W and Z that occur here are explicitly identifiable with factors that occur in the bundle (i.e., local) duals of the electron x neutrino spaces and the neutrino x neutrino spaces. This is in accord with the general ideas of electroweak theory, but no Higgs boson is needed or occurs in the present formulation. The relevant 'broken symmetry' is rather the restriction from $\mathbf{G}^{\sim}$-invariance to $\mathbf{K}^{\sim}$-invariance, enforced by the state of the background universe, from which stable particle masses derive in accordance with Mach's principle. The masses of unstable particles are constrained by those of the stable particles and theoretically determined modulo these by local considerations based on the indicated $\mathbf{G}^{\sim}$-invariant interaction Lagrangian.

The physical ('clothed') W and Z are more complicated, and the possible mixing of the neutrinos and decays of the various electrons (including muon and taon) further modify the naive format of current relativistic particle theory, in which the ensemble is merely a direct sum of irreducible representations corresponding to the 'elementary' types. Thus when indecomposability is taken into account, the observed particles are represented by linear combinations of sections, more specifically those which are approximate eigensections of the Minkowski mass operator. This serves to explain universality of the massive leptons, but the computation of their wave functions on the basis of this and related constraints of relative stability and localization becomes several orders of magnitude more difficult than in the less fundamental approaches current in conventional relativistic theory. The theoretical cygnet, for which there are substantial indications for physical existence in cosmic rays emanating from certain distant X-ray stars, but which as a particle devoid of electromagnetic interactions is not readily observed, eliminates the need for quarks or related particles such as gluons.

Each of the indicated irreducible irreducible unitary representations of $\mathbf{G}^{\sim}$ is known. Most of these remain irreducible on restriction to $\mathbf{P}^{\sim}$, but some appear to split, in particular that for K^0, into two distinct $\mathbf{P}^{\sim}$-irreducible components, of different spins. These are necessarily mixed by the excess of the Einstein over the Minkowski energy, i.e. the quasi-gravitational energy, which is definable Lorentz-covariantly by its transformation properties under scaling, relative to any given observer. This may explain apparent CP violation, which is deduced from the assumption that the two components are both of spin 0, whereas the spin can not be measured directly and is inferred from statistical considerations for the decay products of the long-lived form K_L, and basically merely assumed to be the same for the short-lived form K_S, which may however have spin 1. Measurements that may suffice to establish more conclusively the spin of the K_S appear to be at least an order of magnitude more refined than those available.

COSMOLOGY AND THE REDSHIFT

The interpretation of the redshift as the difference between the Einstein and the Minkowski energies has been supported by detailed statistical analyses of direct observations on discrete redshifted sources. Believers in the Big Bang have however been content to validate their fundamental assumption of the Hubble Law by measurements of galaxies selected in large part (if not totally and explicitly, in some major cases) under the assumption of this law. Deviations in other samples have been ascribed to assumed effects incapable of substantiation by direct model-independent observation, e.g. 'evolution' in the case of quasars, or a plethora of perturbations at lower redshifts.

But in the past few years, large samples have been observed or compiled based on the automated IRAS observations that indicate the Hubble Law to be irreconcilable with direct observation. E.g., the sample of Strauss et al. (1990-92) has 2551 galaxies brighter than its flux limit in the redshift range from 500-30,000 km s^{-1}, with an average redshift hardly more than 0.01, which is too small for 'evolution' to be more than a purely exculpatory assumption (quasar redshifts are typically 100 times greater). The Hubble Law predicts that the dispersion in log flux will be much larger than is observed; and unobserved perturbations would only tend to to make the prediction even more deviant, by contributing to the dispersion. Two other large samples, one in the infrared ('QDOT'), and the other in the X-ray band (the 'EMSS' AGN sample) show similar effects. In all of these samples, the comparable directly observable implications of the chronometric (or Lundmark) law (that at low redshifts the redshift varies basically as the square of the distance) are rather precisely on the mark.

Since an expert such as S. van den Bergh considers the strongest case for the Hubble Law to be based on a sample of 116 galaxies (of Hoessel et al., 1980) selected from a catalog (that of Abell) that explicitly assumes the Hubble Law, the positive case for the law is at best quite weak. By normal statistical standards, the negative case against it is extremely strong. In any event, the chronometric theory is consistent with all clearly objective redshift samples, and is interpretable as a curvature effect, which even Hubble (and his collaborator, Tolman), regarded as scientifically more satisfactory than Expansion of the Universe.

REFERENCES

1. The redshift-distance relation, I. E. Segal, Invited address, Colloquium on Physical Cosmology sponsored by National Academy of Sciences, USA, March, 1992. In press. (Cosmological survey.)
2. Convergence of nonlinear massive quantum field theory in the Einstein Universe. I. E. Segal & Z. Zhou, Ann. Phys. 218(1992) 279-292.
3. Massless $\phi^q{}_d$ quantum field theoryies and the nontriviality of $\phi^4{}_4$, J. Pedersen, I. E. Segal, and Z. Zhou. Nucl. Phys. B376 (1992), 129-142.
4. Nonlinear quantum fields in ≥4 dimensions and 1-cohomology of the infinite Heisenberg group. J. Pedersen, I. E. Segal, and Z. Zhou, preprint, 1992.
5. Apparent superluminal sources, comparative cosmology, and the cosmic distance scale, I. E. Segal, M. N. R. A. S. 242(1990), 423-427. (Estimate of R.)
6. Is the cygnet the quintessential baryon?, I. E. Segal, Proc. Natl. Acad. Sci. USA 88(1991), 994-998. (Possible particle assignments.)
7. Singular operators on boson fields, S. M. Paneitz, J. Pedersen, I. E. Segal, and Z. Zhou, J. Funct. Anal. 93 (1990), 239-269.
8. The mathematical implications of fundamental physical principles, I. E. Segal, in Proc. Symp. Pure Math. 50 (1990), Amer. Math. Soc., 151-178.
9. The global Goursat problem and scattering for nonlinear wave equations, J. C. Baez, I. E. Segal, Z. Zhou, J. Funct. Anal. 93 (1990), 239-269.
10. Statistically efficient testing of the Hubble and Lundmark Laws on IRAS galaxy samples, I. E. Segal, J. F. Nicoll, P. Wu, and Z. Zhou, preprint, 1992.
11. The Yang-Mills equations on the universal cosmos, Y. Choquet-Bruhat, S. M. Paneitz, and I. E. Segal, J. Funct. Anal. 53 (1983), 112-150.
12. Analysis in space-time bundles, I-IV, S. M. Paneitz, I. E. Segal, and D. A. Vogan, Jr., J. Funct. Anal. 47(1982) 78-142, 49(1982),335-414, 54(1983), 18-112, 75(1987), 930-948. (Conformal fields and particles.)
13. *Introduction to algebraic and constructive quantum field theory*. J. C. Baez, I. E. Segal, and Z. Zhou, Princeton University Press, 1992. (Textbook.)
14. Causal symmetries and the physics of extreme distances, I. E. Segal, in *Infinite-dimensional Lie algebras and quantum field theory*, ed. H. Doebner et al., World Scientific, 1988. (The physical interpretation of the Einstein equation, and group-theoretic interpretation of Mach's principle.)

Interaction of Gravitational and Electromagnetic Waves in General Relativity

A. H. Taub
Mathematics Department
University of California
Berkeley, CA 94720
U.S.A.

May 26, 1992

Résumé: Les équations d'Einstein-Maxwell, dans un espace-temps V admettant deux vecteurs de Killing de type espace qui commutent, font intervenir les composantes du tenseur métrique $g_{\mu\nu}$ et celles du potentiel vecteur A_μ. Les solutions dont on discute sont telles que $g_{\mu\nu}$ et A_μ ont des dérivées premières discontinues à travers chacune des deux ondes planes en collision, mais pas de fonction delta en le tenseur de Ricci de V, ni en le tenseur de Maxwell $F_{\mu\nu}$.

Choquet-Bruhat, dans son travail fondamental sur les solutions radiatives approchées des équations d'Einstein-Maxwell, a fait remarquer qu'un champ électromagnétique à haute fréquence peut "créer" un champ gravitationnel à haute fréquence et inversement. Lichnérowicz a montré qu'un phénomène similaire a lieu lorsqu'un espace-temps admet des ondes de choc gravitationnelles et électromagnétiques.

M. Flato et al. (eds.), Physics on Manifolds, 265–287.

Dans et article, on détermine des solutions des équations d'Einstein-Maxwell qui ont la propriété que les composantes du tenseur métrique $g_{\mu\nu}$ sont des fonctions explicites des composantes du potentiel vecteur A_μ. Ainsi, on voit que, dans cette solution, l'observation de Choquet-Bruhat a lieu indépendamment des fréquences qui interviennent et la relation de Lichnéro-wicz entre les fronts de choc électrique et gravitationnel s'étend à la région d'interaction toute entière.

Abstract

The Einstein-Maxwell equations in a space-time V that admits two commmuting space-like Killing vectors involve the components of the metric tensor, $g_{\mu\nu}$ and those of the four vector potential A_μ. The solutions discussed are such that $g_{\mu\nu}$ and A_μ have discontinuous first derivatives across each of two colliding plane waves but no delta functions in the Ricci tensor of V nor in the Maxwell field tensor $F_{\mu\nu}$.

Choquet-Bruhat, in her basic work on approximate radiative solutions of the Einstein-Maxwell equations, pointed out that a high-frequency electromagnetic field can "create" a high frequency gravitational field and conversely. Lichnerowicz showed that a similar phenomenon takes place when a space-time admits gravitational and electromagnetic shock waves.

In this paper solutions of the Einstein-Maxwell equations are determined that have the property that the components of the metric tensor $g_{\mu\nu}$ are explicit functions of the components of the vector potential A_μ. Thus in this solution Choquet-Bruhat's observation is seen to hold irrespective of the frequencies that may be involved and Lichnerowicz's relation between the electric and gravitational shock fronts extends to the entire interaction region.

1 Introduction

In her study of approximate radiative solutions of the Einstein-Maxwell equations [1] by means of the W.K.B. method of expansion discussed in [2] Professor Choquet-Bruhat showed that a high-frequency electromagnetic field can "create" a high frequency gravitational field and conversely. This result was derived from the study of the propogation of the high frequency perturbations of the four vector potential of the electromatic fields $\dot{A}_\mu^{(1)}$ and those of the gravitational field $\dot{g}_{\mu\nu}^{(1)}$. It was shown that

$$W = W_A + W_g \tag{1}$$

formed from

$$W_A = W(\dot{A}_\mu^{(1)}) \tag{2}$$

$$W_g = W(\dot{g}_{\mu\nu}^{(1)}) \tag{3}$$

is conserved along a ray of the perturbing fields. That is

$$(Wl^\mu)_{:\mu} = 0$$

even when

$$((l^\mu W_A)_{:\mu})((l^\mu W_g)_{:\mu}) \neq 0 \tag{4}$$

where the colon denotes the covariant derivative with respect to the unperturbed metric, $g_{\mu\nu}^{(0)}$,

$$l^\mu = g^{(0)\mu\nu} l_\nu = g^{(0)\mu\nu} \partial\varphi / \partial x^\nu \tag{5}$$

the null hypersurfaces

$$\varphi(x) = \text{ constant} \tag{6}$$

are the wavefronts of the perturbations.

In reference [3] Choquet-Bruhat and I discussed high-frequency, self-gravitational charged massive scalar fields by the WKB method and showed that a high-frequency scalar field can create, through an electromagnetic field, a high-frequency gravitational field and conversely.

In his study of space times V admitting electro-magnetic and gravitational shock waves Lichnerowicz [4] defined a strength of each type of shock wave and showed that the sum of these is conserved along rays defining the full hypersurface in V that describes the history of the shock-wave's motion.

Lichnerowicz's discussion treats weak solutions of the Einstein-Maxwell equations for the metric tensor $g_{\mu\nu}$ of V and an electromagnetic field determined by a vector potential vector A_μ. He assumes that there exists a hypersurface Σ in V across which $g_{\mu\nu}$ and A_μ are continuous but have discontinuous partial derivatives. The Einstein-Maxwell equations are generalized so as to involve distribution valued curvature tensors, Maxwell field tensors and energy-momentum tensors. However since the latter tensor for an electromagnetic field is a quadratic function of the Maxwell field tensor both tensors are required to be free from delta functions. Then the generalized Einstein tensor must be free of delta functions.

The discontinuities in the first derivation of $g_{\mu\nu}$ and A_μ are described by

$$[g_{\mu\nu,\lambda}] = l_\lambda b_{\mu\nu} \tag{7}$$

and

$$[A_{\mu,\lambda}] = l_\lambda b_\mu \tag{8}$$

where $[T^{\alpha\cdots}_{\beta\ldots}] = T^{\alpha\ldots+}_{\beta\ldots} - T^{\alpha\ldots-}_{\beta\ldots}$ and $T^{\alpha\ldots+}_{\beta}$ and $T^{\alpha\ldots-}_{\beta\ldots}$ are the values of the tensor $T^{\alpha\cdots}_{\beta\ldots}$ on the two sides of the hypersurfaces where $b_{\mu\nu}$ and b_μ

are a symmetric tensor and a vector defined on Σ, where l_μ satisfies (5) and φ satisfies equation (6), the equation defining the hypersurface Σ in V.

It is a consequence of the requirement that the Ricci tensor and the Maxwell tensor

$$F_{\mu\nu} = A_{\nu,\mu} - A_{\mu,\nu} \tag{9}$$

be free of delta functions that Σ is null hypersurface and hence

$$g^{\mu\nu} l_\mu l_\nu = 0 = l^\mu l_\mu \tag{10}$$

In addition, one finds that

$$2[\Gamma^\alpha_{\beta\gamma}] = l_\beta b^\alpha_\gamma + l_\gamma b^\alpha_\beta - l^\alpha b_{\beta\gamma} \tag{11}$$

$$l^\beta b'^\alpha_\beta = l^\beta (b^\alpha_\beta - \frac{1}{2}\delta^\alpha_\beta b) = 0 \tag{12}$$

where $b = g^{\alpha\beta} b_{\alpha\beta}$. Hence

$$l^\gamma [\Gamma^\alpha_{\beta\gamma}] = 0, \tag{13}$$

and $l^\gamma \Gamma^\alpha_{\beta\gamma}$ is defined on Σ, e.g.

$$l^\gamma \Gamma^{\alpha+}_{\beta\gamma} = l^\gamma \Gamma^{\alpha-}_{\beta\gamma} = l^\gamma \bar{\Gamma}^\alpha_{\beta\gamma} = \frac{1}{2} l^\gamma (\Gamma^{\alpha-}_{\beta\gamma} + \Gamma^{\alpha+}_{\beta\gamma}). \tag{14}$$

Thus it is a consequence of (10) that

$$l^\beta l_{\beta;\alpha} = l^\beta l_{\alpha;\beta} = 0 \tag{15}$$

where the semi-colon denotes the covariant derivative with respect to either of the three connections listed above.

It follows from equations (8) and (9) that

$$[F_{\mu\nu}] = l_\mu b_\nu - l_\nu b_\mu \tag{16}$$

It is a consequence of the requirement that the distribution valued Ricci tensor be free of delta functions, that equations (10) and (11) hold and that

$$\tau_1 = \frac{1}{4} b^{\alpha\beta} b'_{\alpha\beta} = \frac{1}{4}(b^{\alpha\beta} b_{\alpha\beta} - \frac{b^2}{2}) \tag{17}$$

satisfies the equation

$$(l^\alpha \tau_1)_{;\alpha} = 0 \tag{18}$$

in case there is no electromagnetic field present.

When one imposes the condition that the distribution valued Maxwell field tensor defined as

$$F^{\mu\nu D} = \theta F^{\mu\nu +} + (1-\theta) F^{\mu\nu -}$$

where θ is the Heaviside function is such that its divergence does not contain a delta function one shows that

$$l^\lambda [l_\lambda b_\mu - l_\mu b_\lambda] = 0$$

and that equation (10) holds. Hence it follows that

$$l^\alpha b_\alpha = 0.$$

The quantity

$$\tau_2 = -b^\alpha b_\alpha$$

may be considered as the strength of the electromagnetic shock. Lichnerowicz showed that

$$((\kappa \tau_1 + \tau_2) l^\alpha)_{;\alpha} = 0.$$

That is, the sum of the strengths of the gravitational and electromagnetic shocks is conserved along a ray that generates the hypersurface Σ.

It is the purpose of this paper to discuss the Einstein-Maxwell equations in a space-time V that contains two colliding impulsive gravitational waves that are also electromagnetic shock waves with wave fronts described by two intersecting plane null hypersurfaces. Sample solutions of these equations for the metric tensor $g_{\mu\nu}$ of V and the vector potential A_μ of the Maxwell field are obtained. In the sample solution the components of $g_{\mu\nu}$ and those of A_μ in various regions of V will be shown explicit functions of each other. Thus in this case the interaction between the gravitational waves and the electromagnetic ones is determined by the Einstein-Maxwell field equations irrespective of the frequencies involved.

2 The Metric of V

The evolution of V will be assumed to take place under the following assumptions.

(1) V admits two commuting space-like Killing vectors ξ^μ_A ($A = 1,2$; $\mu = 0,1,2,3$).

The coordinate system may be chosen so that

$$\xi^\mu_A = \delta^\mu_A \tag{19}$$

(2) It is further assumed that in addition the metric of V is invariant under the transformation

$$x^{*i} = x^i \quad (i = 0,3); \qquad x^{*A} = -x^A \quad (A = 1,2) \tag{20}$$

The metric of V may then be put in the form

$$ds^2 = e^{\omega}\eta_{ij}dx^i dx^i - e^{\mu}(e^{-\sigma}(dx^1 - qdx^2)^2 + e^{\sigma}(ds^2)^2) \tag{21}$$

where

$$\eta_{ij} = \delta_{ij} - 2\delta^3_i\delta^3_j \tag{22}$$

and ω, μ, σ and q are functions of x^0 and x^3 alone. Equation (21) may be written as

$$ds^2 = e^{\omega} dudv - e^{\mu}(e^{-\sigma}(dx^1 - qdx^2)^2 + e^{\sigma}(dx^2)^2) \tag{23}$$

where the null coordinates u and v are given by

$$u = x^0 - x^3 \qquad v = x^0 + x^3 \tag{24}$$

and the metric coefficients are functions of u and v alone.

The null hypersurfaces $u = 0$ and $v = 0$ will be used to divide V into four subregions:

Region I where ($u > 0$ and $v > 0$), the interaction region,

Region II where $u > 0$ and $v < 0$

Region III where $u < 0$ and $v > 0$

Region IV where $u < 0$ and $v < 0$.

The hypersurface $u = 0(v = 0)$ will be interpreted as the wave front of a gravitational wave and also an electromagnetic shock wave travelling in the $x^3(-x^3)$ direction. The variables

$$2x^0 = (v + u) \tag{25}$$

and

$$2x^3 = (v - u) \tag{26}$$

are measures of the time from ($u = 0$, $v = 0$) the instant of collisino of these waves and the distance between the wave fronts respectively.

(3) It is assumed that if $g_{\mu\nu}$ is an exact solution of the Einstein field equations holding in Region I, the metric tensor in Region II (III) is given by $g_{\mu\nu}(u, 0)(g_{\mu\nu}(0, v))$. That in Region IV is $g_{\mu\nu}(0, 0)$ so that Region IV is flat. The vector potential of the Maxwell field $A_\mu(u, v)$ holding in Region I is extended similarly, i.e., in Region II (III) $A_\mu = A_\mu(u, 0)(A_\mu(0; v))$ and in Region IV $A_\mu = 0$.

This method of extending the solution of Region I produces metrics which are continuous across the hypersurfaces $u = 0$ and $v = 0$ but may have discontinuous first derivitatives across these hypersurfaces. If so the curvature tensor derived from $g_{\alpha\beta}$ and the Maxwell tensor $F_{\mu\nu}$ will be distribution valued, i.e., may contain delta functions with support on these null hypersurfaces.

In addition space-times with metric obtained as above are said to contain impulsive gravitational waves only if the components of the Einstein tensor (equivalently to the Ricci tensor) do not contain such delta functions.

It is a consequence of assumption (3) that in Regions II (III) the only non-vanishing component of the Ricci tensor can be $R_{uu}(R_{vv})$. Thus Regions II and III are either vacuum regions or contain null dust, i.e., a medium with stress-energy tensor of the form

$$T_{\mu\nu} = E u_{,\mu} u_{,\nu} \qquad (E v_{,\mu} v_{,v})$$

in Region II (III), $f_{,\mu} = \partial f / \partial x^{\mu}$. In addition we have

$$R_{AB} = R_{\mu\nu} \delta^{\mu}_{A} \delta^{\nu}_{B} = 0 \tag{27}$$

in Regions II, III and IV.

(4) It is assumed that $R^1_1 + R^2_2 = 0$ in Region I.

(5) It is assumed that the components of $R_{\mu\nu}$ and $F_{\mu\nu}$ do not involve delta functions.

It has been shown in an earlier paper [5] that the latter assumption implies that $\mu_{,u} = 0 (\mu_{,v} = 0)$ on the hypersurface $u = 0 (v = 0)$.

3 The Ricci Tensor

It follows from equation (23) that

$$R^{i}_{A} = 0 \tag{28}$$

$$e^{\mu+\omega} R^A_A = e^{\mu+\omega}(R^1_1 + R^2_2) = 4(e^{de})_{,uv} \tag{29}$$

$$2e^{\mu+\omega} R^1_1 = e^{\mu+\omega}(R^A_A - \{e^{\mu}(\sigma_{,i} + e^{-2\sigma} q_{,i} q)]_{ij}\} \eta^{ij} \tag{30}$$

$$2e^{\mu+\omega} R^2_1 = (e^{\mu-2\sigma} q_{,i})_{,j} \eta^{ij} \tag{31}$$

$$2e^{\mu+\omega} R^2_2 = e^{\mu+\omega} \{R^A_A + [e^{\mu}(\sigma_{,i} + e^{-2\sigma} q_{,i} q)_{,j}] \eta^{ij}\} \tag{32}$$

$$R_{uu} = \mu_{,uu} + \frac{1}{2}\mu^2_{,u} - \mu_{,u}\,\omega_{,u} - S_{uu}, \tag{33}$$

$$R_{uv} = \mu_{,uv} + \frac{1}{2}\mu_{,u}\,\mu_{,v} + \omega_{,uv} - S_{uv}, \tag{34}$$

$$R_{vv} = \mu_{,vv} + \frac{1}{2}\mu^2{}_{,v} - \mu_{,v}\,\omega_{,v} - S_{uv}, \tag{35}$$

where

$$S_{ij} = -\frac{1}{2}(\sigma_{,i}\,\sigma_{,j} + q_{,i}\,q_{,j}\,e^{-2\sigma}) \tag{36}$$

4 Maxwell Energy-Momentum Tensor

This tensor is given in terms of the Maxwell field tensor

$$F_{\mu\nu} = A_{\nu,\mu} - A_{\mu,\nu} \tag{37}$$

by the equation

$$4\pi T_{\mu\nu} = F_{\mu\alpha}F^{\alpha}_{\nu} + \frac{1}{4}g_{\mu\nu}F_{\alpha\beta}F^{\alpha\beta} \tag{38}$$

It follows that

$$T = g^{\mu\nu}T_{\mu\nu} = 0. \tag{39}$$

The vector potential A_μ and $F_{\mu\nu}$ have components that are functions of u and v alone when the Lie derivatives of these quantities with respect to the Killing vectors δ^{μ}_{A} vanish. Hence

$$F_{AB} = 0. \tag{40}$$

It is a consequence of equations (38) and (40) that

$$4\pi T^{A}_{A} = 4\pi(T^{1}_{1} + T^{2}_{2}) = F_{uv}F^{uv} = (F_{uv})^2(g^{uv})^2. \tag{41}$$

Hence the only non-vanishing component of F_{ij} is

$$F_{uv} = 0 = F_{ij}, \tag{42}$$

when

$$T^{A}_{A} = 0 \tag{43}$$

and conversely. Since

$$T = T^{\mu}_{\mu} = T^{i}_{i} + T^{A}_{A} = 0 \tag{44}$$

as follows from equation (38), equation (43) also implies that

$$T^{i}_{i} = g^{ij}T_{ij} = 2g^{uv}T_{uv} = 0 \tag{45}$$

An additional consequence of equation (38) is

$$4\pi T^i_A = F^i_j F^j_A = 0 \tag{46}$$

Thus the only non-vanishing components of $F_{\mu\nu}$ are F_{iA}. In the coordinate system in which (23) obtains we have

$$F_{u1} = \phi_{,u} = -F_{1u} \tag{47}$$

$$F_{v1} = \phi_{,v} = -F_{1v} \tag{48}$$

$$F_{u2} = \psi_{,u} = -F_{2u} \tag{49}$$

$$F_{v2} = \psi_{,\sigma} = -F_{2v} \tag{50}$$

where

$$\xi^\mu_1 A_\mu = \delta^\mu_1 A_\mu = A_1 = \phi(u,v), \qquad \xi^\mu_2 A_\mu = \delta^\mu_2 A_\mu = A_2 = \psi_{u,v} \tag{51}$$

It follows that

$$F^{\alpha\beta} F_{\alpha\beta} = -8e^{-(\mu+\omega)}[e^\sigma \phi_{,u}\phi_{,v} + e^{-\sigma}\theta_u\theta_v] \tag{52}$$

where

$$\theta_i = \psi_{,i} + q\phi_{,i}\ , \qquad (i = u, v) \tag{53}$$

hence

$$4\pi T^u_u = 4\pi T^v_v = 0, \tag{54}$$

$$4\pi T^u_v = 2e^{-(\mu+\omega)}[e^{-\sigma}\theta^2_v + e^\sigma \phi^2{}_{,v}] \tag{55}$$

$$4\pi T^v_u = 2e^{-(\mu+\omega)}[e^{-\sigma}\theta^2_u + e^\sigma \phi^2{}_{,u}] \tag{56}$$

$$4\pi T_1^1 = 2e^{-(\mu+\omega)}[(e^\sigma + q^2 e^{-\sigma})\phi_{,u}\,\phi_{,u} - e^{-\sigma}\psi_{,u}\,\psi_{,v}] = -4\pi T_2^2 \quad (57)$$

$$4\pi T_1^2 = 2e^{-(\mu+\omega)}[\phi_{,u}\,\theta_v + \phi_{,v}\,\theta_u] \quad (58)$$

where θ_i, $(i = u, v)$ is defined by equation (53). Hence

$$4\pi(T_1^1 - qT_1^2) = 2e^{-(\mu+\omega-\sigma)}[\phi_{,u}\,\phi_{,v} - e^{-2\sigma}\theta_{,u}\,\theta_v] \quad (59)$$

Thus

$$4\pi T_{uu} = 2\pi e^\omega T_u^v = e^{-\mu}[e^{-\sigma}\theta_u^2 + e^\sigma \phi_{,u}^{\,2} \quad (60)$$

$$4\pi T_{vv} = 2\pi e^\omega T_v^u = e^{-\sigma}[e^{-\sigma}\theta_v^2 + e^\sigma \phi_{,v}^{\,2} \quad (61)$$

$$4\pi T_{uv} = 2\pi e^\omega T_v^v = 4\pi T_{va} = 0 \quad (62)$$

5 The Maxwell Equations

These equations are

$$F^{\mu\nu}{}_{;\nu} = J^\mu \quad (63)$$

where J^μ is the electric current vector and the semi-colon denotes the covariant derivative with respect to the metric $g_{\mu\nu}$ of V. Since the only non-vanishing components of $F^{\mu\nu}$ are F^{Ai} $(A = 1, 2,\ i = 0, 3$ or

$i = u, v$) and these components are functions of u and v (x^0 and x^3) alone we must have $J^i = 0$ and

$$(\sqrt{-g}F^{Ai}),_{;} = \sqrt{-g}J^A \tag{64}$$

In view of equations (47–50) and (23) we must have

$$-\sqrt{-g}J^1 = (e^{\sigma}\phi_{,k} + qe^{-\sigma}\theta_k)_{,i}\,\eta^{ki} \tag{65}$$

$$-\sqrt{-g}J^2 = (e^{-\sigma}\theta_k)_{,j}\,\eta^{kj} \tag{66}$$

Thus when the current vanishes the Maxwell equations become

$$(e^{\sigma}\phi_{,u})_{,v} + (e^{\sigma}\phi_v)_{,u} + e^{-\sigma}(q_{,u}\theta_v + q_{,v}\theta_u) = 0 \tag{67}$$

and

$$(e^{-\sigma}\theta_u)_{,v} + (e^{-\sigma}\theta_v)_{,u} = 0 \tag{68}$$

The last equation is the integrability condition for the equations

$$H_{,u} = \phi_{,u} - ie^{-\sigma}\theta_u \tag{69}$$

$$H_{,v} = \phi_{,v} + ie^{-\sigma}\theta_v \tag{70}$$

where $H(u, v)$ is a complex function of u and v. These equations and equation (67) are equivalent to the equation

$$2H_{uv} + (\sigma_{,v} + iq_v e^{-\sigma})H_{,u} + (\sigma_{,u} - iq_{,u}e^{-\sigma})H_{,v} = 0 \tag{71}$$

Equations (69–70) may be written as

$$\phi_{,u} = \frac{1}{2}(H_{,u} + \bar{H}_{,u}), \qquad \phi_v = \frac{1}{2}(H_{,v} + \bar{H}_{,v}) \tag{72}$$

$$e^{-\sigma}\theta_u = \frac{i}{2}(H_n - \bar{H}_u), \qquad e^{-\sigma}\theta_v = \frac{i}{2}(H_{,v} - \bar{H}_{,v}) \tag{73}$$

6 The Einstein Equations

These equations may be written as

$$R^{\mu}_{\nu} = -8\pi(T^{\mu}_{\nu} - \frac{1}{2}\delta^{\mu}_{\nu}T) = 8\pi T^{\mu}_{\nu} \tag{74}$$

as follows from equation (41) and the choice of units such that $G = c = 1$, where G is the Newtonian constant of gravitation and c is the special relativistic velocity of light. It is a consequence of assumption (4) and equation (28) that

$$(e^{\mu})_{,uv} = e^{\mu}(\mu_{,uv} + \mu_{,u}\,\mu_{,u}) = 0 \tag{75}$$

Hence we may write

$$e^{\mu} = 1 + U(u) + V(v) \tag{76}$$

where $U(V)$ is a function of $u(v)$ alone.

Equations (74), (75), (31) and (58) imply that

$$(e^{\mu-2\sigma}q_{,u})_{,v} + (e^{\mu-2\sigma}q_{v})_{,u} - 2ie^{\sigma}(H_u\bar{H}_v - \bar{H}_vH_u) = 0 \tag{77}$$

It follows from equations (74), (75), (30), (31) and (59) that

$$(e^{\mu}\sigma_{,u})_{,v} + (e^{\mu}\sigma_{,v})_{,u} + 2e^{\mu-2\sigma}q_{,u}\,q_{,v} - 2e^{\sigma}(H_n\bar{H}_v + \bar{H}_uH_v) = 0 \tag{78}$$

From equations (33) through (36) and (60) to (62) one obtains

$$\mu_{,uu} + \frac{1}{2}\mu^2{}_{,u} - \mu_{,u}\,\omega_{,u} - S_{uu} = -2e^{\sigma-\mu}H_{,u}\,\bar{H}_{,u} \tag{79}$$

$$\mu_{,uv} + \frac{1}{2}\mu_{,u}\,\mu_{,v} + \omega_{,uv} - S_{uv} = 0 \tag{80}$$

$$\mu_{,vv} + \frac{1}{2}\mu^2{}_{,v} - \mu_{,v}\,\omega_{,v} - S_{vv} = -2e^{\sigma-\mu}H_v\bar{H}_{,v} \tag{81}$$

where S_{ij} is given by equation (36).

7 The Colinear Case

When the approaching waves have aligned linear polarizations one may choose the coordinates x^A so that the line element given by equation (23) reduces to

$$ds^2 = e^{\omega} du dv - e^{\mu}(e^{-\sigma}(dx^1)^2 + e^{\sigma}(dx^2)^2) \tag{82}$$

that is, so that

$$q = 0. \tag{83}$$

Then equation (53) becomes

$$\theta_i = \psi_{,i} \tag{84}$$

and equations (69,70) become

$$H_{,u} = \phi_{,u} - ie^{-\sigma}\psi_{,u} \tag{85}$$

$$H_{,v} = \phi_{,v} + ie^{-\sigma}\psi_{,v} \tag{86}$$

The Maxwell equations for vanishing J^A, equations (67) and (68) become

$$(e^{\sigma}\phi_{,u})_{,v} + (e^{\sigma}\phi_{,v})_{,u} = 0 \tag{87}$$

and

$$(e^{-\sigma}\psi_{,u})_{,v} + (e^{-\sigma}\psi_{,v})_{,u} = 0. \tag{88}$$

The Einstein equations become the following ones: e^{μ} is still given by

$$(e^{\mu})_{,uv} = 0, \tag{89}$$

thus

$$e^{\mu} = 1 + U(u) + V(v), \tag{90}$$

equation (77) becomes

$$4e^{\sigma}(\psi_{,u}\phi_{,v} + \psi_{,v}\phi_{,u}) = 0 \tag{91}$$

and equation (78) becomes

$$(e^{\mu}\sigma_{,u})_{,v} + (e^{\mu}\sigma_{,v})_{,u} = 4e^{\sigma}(\varphi_{,u}\varphi_{,v} - e^{-2\sigma}\psi_u\psi_{,v}) \tag{92}$$

Equations (79) through (81) apply with

$$S_{ij} = -\frac{1}{2}(\sigma_{,i}\sigma_{,j}) \tag{93}$$

Hence

$$\mu_{,uu} + \frac{1}{2}\mu^2{}_{,u} - \mu_{,u}\omega_{,u} = -\frac{1}{2}\sigma^2{}_{,u} - 2e^{\sigma-\mu}(\phi^2{}_{,u} + e^{-2\sigma}\psi^2{}_{,u}) \tag{94}$$

$$\mu_{,u\varphi} + \frac{1}{2}\mu_{,u}\mu_{,v} + \omega_{,uv} = -\frac{1}{2}\sigma_{,u}\sigma_{,\sigma} \tag{95}$$

$$\mu_{,vv} + \frac{1}{2}\mu^2{}_{,\sigma} - \mu_{,v}\omega_{,v} = -\frac{1}{2}\sigma^2{}_{,v} - 2e^{\sigma-\mu}(\phi^2{}_{,v} + e^{-2\sigma}\psi^2{}_{,v}) \tag{96}$$

Equation (95) follows from equation (94) (or (96)) and equations (86), (87), (88) and (92). Thus the integrability conditions of equations (94) and (96) for the function $\omega(u, v)$ are satisfied as a result of the Maxwell equations and the remaining Einstein field equations.

8 The Case $q = \psi = 0$

In this case the Maxwell equations reduce to equation (87) and the Einstein equations become, equation (89), (92) and equations (94) and

(96) in which ψ is set equal to zero. Charach and Malin have pointed out [6] that the equations for the functions σ and ϕ have solutions given by

$$\sigma = \mu + 2 \ln \cosh \Theta \tag{97}$$

$$\phi = \tanh \Theta \tag{98}$$

where $\Theta(u, v)$ is a solution of the wave equation

$$(g^{\mu\nu} \Theta_{,\mu})_{;\nu} = 0 \tag{99}$$

or equivalently a solution of

$$(e^{\mu} \Theta_{,u})_{,v} + (e^{\mu} \Theta_{,v})_{,u} = 0 \tag{100}$$

in the coordinate system in which (23) obtains.

In this coordinate system $e^{\mu-\sigma}$ and $e^{\mu+\sigma}$ describe the transverse gravitational waves present and ϕ determines the transverse electromagnetic wave. Equations (97) and (98) together with equation (90) then determine the relation between the two tranverse waves via the solution of equation (100). This relationship is independent of the frequency of the waves and holds throughout region I in addition to holding on the boundaries of this region namely the hypersurfces $u = 0$ and $v = 0$.

The function $\omega(u, v)$ which is determined by equations (94) and (96) when σ and ϕ are given via equations (97), (98) and (99) describes the longitudinal gravitational field due to the functions σ and ϕ.

It should be noted that under the transformation

$$\omega = \omega^* - \frac{1}{2}\mu + ln(U'V') \tag{101}$$

where

$$A = U'(0) \tag{102}$$

and

$$B = V'(0) \tag{103}$$

and the prime denotes the derivative of the function U or V with respect to its argument the line element described by equation (23) becomes

$$ds^2 = \frac{e^{\omega *} U'V' du dv}{(1+U+V)^{\frac{1}{2}}} - (1+U+V)(e^{-\sigma}(dx^1)^2 + e^{\sigma}(dx^2)^2) \tag{104}$$

The transformation to coordinates U and V instead of u and v simplifies the metric tensor and equations (94) to (96). These become equations for the function ω^*.

It follows from equations (97) and (98) that

$$1 - \phi^2 = e^{\mu - \sigma} \tag{105}$$

the equation which relates the components of the electromagnetic potential, and the components of the transverse gravitational field when $q = \psi = 0$.

If $q = 0$ and ψ is assumed to be a function of $\phi(u, v)$ so that

$$\phi_{,u}\, \psi_{,v} - \phi_{,v}\, \psi_{,u} = 0, \tag{106}$$

then it follows from equation (91) that $\psi = constant$. After a gauge transformation of the electromagnetic potential ψ may be set equal to zero.

Thus the only solutions of the Einstein-Maxwelll equations for which σ and ϕ are such that $q = 0$ and

$$\sigma = \sigma(\phi) \tag{107}$$

are given by equations (97), (98) and (99) and the equation

$$\psi = 0 \tag{108}$$

9 The Ernst Equations

Equation (77) may be written as

$$(e^{\mu-2\sigma}q_{,u} - i(H_u\bar{H} - \bar{H}_u H))_{,v} + (e^{\mu-2\sigma}q_{,v} + i(H_v\bar{H} - \bar{H}_v H))_{,u} = \tag{109}$$

Hence there exists a function Φ such that

$$e^{\mu-2\sigma}q_{,u} - i(H_u\bar{H} - \bar{H}_u H), = \Phi_{,u} \tag{110}$$

and

$$e^{\mu-2\sigma}q_{,v} + i(H_v\bar{H} - \bar{H}_v H) = -\Phi_{,v} \tag{111}$$

Solving these equations for $q_{,u}$ and $q_{,v}$ we find

$$q_{,u} = e^{\mu}\Psi^{-2}(\Phi_{,u} + i(H_u\bar{H} - \bar{H}_u H)) \tag{112}$$

$$q_{,v} = -e^{\mu}\Psi^{-2}(\Phi_{,v} + i(H_v\bar{H} - H_v\bar{H})) \tag{113}$$

where

$$\Psi = e^{\mu-\sigma} \tag{114}$$

Let

$$\mathbb{Z} = \Psi + i\Phi + H\bar{H} \tag{115}$$

then the Einstein-Maxwell equations may be written as two complex equations for the functions $\mathbb{Z}(u, v)$ and $H(u, v)$. These equations are

$$(Re\mathbb{Z} - H\bar{H})\nabla^2\mathbb{Z} = (\nabla\mathbb{Z})^2 - 2\bar{H}(\nabla H \cdot \nabla\mathbb{Z}) \tag{116}$$

$$(Re\mathbb{Z} - H\bar{H})\nabla^2 H = \nabla\mathbb{Z} \cdot \nabla H - 2\bar{H}(\nabla H)^2 \tag{117}$$

where

$$F = F(u,v) \tag{118}$$

$$\sqrt{-g}\,\nabla^2 F = (\sqrt{-g}g^{ij}F_{,i})_{,j} \tag{119}$$

$$\nabla F = F_{,i}\,; \qquad (\nabla F)^2 = F_{,i}\,F_{,j}\,g^{ij} \tag{120}$$

$$\nabla F \cdot \nabla G = F_{,i}\,G_{,j}\,g^{ij} \tag{121}$$

Equations (116) and (117) are the complex Ernst equations for an Einstein-Maxwell field.

In the coordinate system in which equation (23) obtains we have

$$\sqrt{-g} = e^{\mu+\omega} \tag{122}$$

and

$$g^{ij} = e^{-\omega}\eta^{ij} = 2e^{-\omega}\delta^i_\mu\delta^i_v \tag{123}$$

$$\nabla^2 F = 2e^{-\omega}(2F_{,uv} + \mu_{,u}\,F_{,\sigma} + \mu_{,v}\,F_{,u}) \tag{124}$$

$$(\nabla F)^2 = 4e^{-\omega}F_{,u}\,F_{,v} \tag{125}$$

$$\nabla F \cdot \nabla G = 2e^{-\omega}(F_{,u}\,G_{,v} + F_{,v}\,G_{,u}). \tag{126}$$

In the case discussed in §8 we have $\Phi = 0$ and equations (97) and (98) implythat

$$\Psi = \cosh^{-2}\Theta. \tag{127}$$

and

$$H = \phi = \tanh\Theta \tag{128}$$

Hence

$$\mathbb{Z} = 1, \tag{129}$$

a solution of equation (116). Equation (117) reduces to equation (100). Chandrasekhar and Xanthopoulos [7] discussed a number of solutions of equations (116) and (117) generated from the solution for which equation (123) holds.

Another class of solutions of these equations have been generated by these authors [8] from a solution satisfying

$$H = Q(\mathbb{Z} + 1) \tag{130}$$

It should be noted that if one assumes that

$$H = F(\mathbb{Z}) \tag{131}$$

where F is a complex function of the complex variable $\mathbb{Z}$, then equations (116) and (117) imply that

$$\frac{d^2F}{d\mathbb{Z}^2} = 0 \tag{132}$$

that is

$$H = A\mathbb{Z} + B \tag{133}$$

where A and B are complex constants.

References

1. Choquet-Bruhat, Y., in *Gravitation*, papers in honor of M. Rosen, edited by V. C. Kuper and A. Peres, Gordon & Breach, New York; and in (1976) *Marcell Grossman Meeting*, edited by R. Ruffini, North-Holland, Amsterdam.
2. Choquet-Bruhat, Y., Commun. Math. Phys. **12** (1969), 16.
3. Choquet-Bruhat, Y. and Taub, A. H., Gen. Rel. Grav. **8** (1977), 561.
4. Lichernowicz, A., *Symposia Mathematics*, vol. XII, Instituto Nazionale di Alta Mathematica, 1973, p. 93.
5. Taub, A. H., J. Math. Phys. **29** (1988), 2622.
6. Charoch, Ch. and Malin, S., Phys. Rev. **21** (1988), 3284.
7. Chandrasekhar, S. and Xanthopolous, B. C., Proc. Roy. Soc. **410** (1987), 311–336.
8. Chandrasekhar, S. and Xanthopolous, B. C., Proc. Roy. Soc. **398** (1985), 223–259.

Anti-self dual conformal structures on 4-manifolds

Clifford TAUBES
Department of Mathematics, Harvard University
Cambridge, MA 02138, USA

Abstract :

The Weyl curvature tensor is the conformally invariant part of the Riemann curvature tensor. On an oriented 4-dimensional manifold, the Weyl tensor is a direct sum of two pieces, W_+ and W_-, its self and anti-self dual parts. Here is a theorem : for a compact, oriented 4-manifold, M, the connect sum of M with a large enough number of (orientation reversed) CP^2's has a metric with $W_+ = 0$. by the way, Penrose's twistor space for such a metric is a complex 3-fold. And so, there are a zoo of bizarre 3-folds.

M. Flato et al. (eds.), Physics on Manifolds, 289.

Chaotic Behavior in Relativistic Motion †

Esteban Calzetta

Universidad de Buenos Aires, Argentina

Our objective in this talk is, in technical terms, to employ the so - called "Melnikov Method" [1] to show that the geodesic motion of massive test particles near the unstable circular orbit surrounding a Schwarzschild Black Hole becomes chaotic under most gravitational perturbations of the metric. Our motivation is, beyond the intrinsic interest of this result, to provide a concrete example of the application of the Melnikov Method to problems issued from General Relativity. Indeed, while chaotic behavior is widespread in relativistic problems (e. g., the presence of chaos in the geodesic flow on compact spaces of negative curvature), the methods available for the detection of chaos (such as numerical computation of Liapunov exponents) have not allowed so far a systematic, wide range analysis of the implications of chaotic behavior for Einstein's Theory. The Melnikov Method is an important acquisition for our theoretical toolbox, for, it being purely analytical and relatively simple to use, it can provide a quick check on the presence of chaos in a class of problems, which can then be studied by more sophisticated methods.

The Melnikov Method addresses the question of whether an integrable Hamiltonian system, containing unstable fixed points, becomes chaotic under a given, time - periodic perturbation. In the unperturbed motion, the different unstable fixed points are connected to each other by "doubly asymptotic solutions" or separatrices. These separatrices can survive the perturbation, or be destroyed, "splitting" into trajectories asymptotic to the fixed points connected by the separatrix. If the split trajectories moreover cross each other transversely, then a stochastic layer will appear surrounding the old separatrix. Motion in the stochastic layer is extremely complex [2]; in particular, it is equivalent to a Bernoulli flow, and therefore impossible to tell from a purely random process, in a suitable invariant set.

The complexity of the motion resulting from "split separatrices" has been recognized by Poincaré [3]; the magic of the Melnikov Method is that it provides us with a simple, purely analytical test of the formation of a stochastic layer. Indeed, let us consider the

M. Flato et al. (eds.), Physics on Manifolds, 291–293.

"homoclinic" case, in which the separatrix connects an unstable fixed point to itself. Let us call H_0 the unperturbed, integrable Hamiltonian, ϵG the time - periodic perturbation, and introduce the integral, taken along the unperturbed separatrix,

$$I(t_0) = \int dt \, \{H_0(t), G(t + t_0)\} \tag{1}$$

Where {} denotes Poisson brackets. Then, if $I(t_0)$ shows isolated, first order zeroes, a stochastic layer will form under the perturbation.

Let us apply the Melnikov Method to the motion of a test particle of mass μ and angular momentum L orbiting a Schwarzschild Black Hole of mass M. For $L^2 \geq 12M^2\mu^2$ there will be an unstable circular orbit, which shall be our unstable "fixed" point (it is an unstable fixed point in the effective problem describing the evolution of the Schwarzchild radial coordinate r as a function of the static time t). If moreover $L^2 \leq 16M^2\mu^2$, there will also be a trajectory which slowly unwinds off the unstable circular orbit, bounces off the Newtonian tail of the potential and winds back again around the unstable orbit. This shall be our separatrix (if $L^2 \geq 16M^2\mu^2$, this orbit escapes to infinity). The unperturbed Hamiltonian is (minus) the conjugated momentum associated to t, which is connected to the other canonical variables by the mass shell constraint $g^{\rho\sigma}p_\rho p_\sigma = -\mu^2$.

The most natural perturbations to consider in connection to this problem are gravitational ones, since they are physically unavoidable, and involve no new parameters. The gravitational perturbations of a Schwarzschild background were classified by Regge and Wheeler and others [4]. In the high frequency limit, accurate asymptotic forms are known; in this limit, the oscillatory Melnikov Integral (1) can be simply computed by saddle point methods. The result is that indeed isolated zeroes appear, and therefore that, under the perturbation, the doubly asymptotic orbit shall be replaced by a stochastic layer. (Because of technical reasons certain gravitational perturbations have not been analyzed; details are given in [5]).

The Melnikov Method thus allows us to achieve in a relatively simple way a rather non trivial result; moreover, the set up in which the method applies is by no means exceptional. Indeed, Poincaré has shown that stable and unstable fixed points, and their attending separatrices, are a generic occurrence in the wake of disintegrating KAM tori [3].

It is to be expected, therefore, that the Melnikov Method will find manyfold applications in the study of chaos in relativistic systems. We plan to explore some of these applications in the near future.

Hopefully, an awareness of the incidence of chaos will increase our understanding of the mathematics of relativistic physics, a field where Mme. Choquet - Bruhat has left an undeletable mark. It has been a great honor for us to participate in this timely meeting, to celebrate Mme. Choquet - Bruhat's contributions to our discipline.

Acknowledgments: I wish to thank in the first place the organizers of the Colloquium on "Analysis, Manyfolds and Physics", for their kind invitation to deliver this talk, and the hospitality of the RGGR group at the Free University (Brussels), where this work was completed.

This work was supported by the Directorate General for Science, Research and Development of the Comission of the European Communities under contract CI1-0540-M(TT), and by CONICET, UBA and Fundación Antorchas.

References

† Work done in collaboration with Luca Bombelli, RGGR, ULB (Belgium).

[1] J. Guckenheimer and P. Holmes, *Non - linear Oscillations, Dynamical Systems, and Bifurcations of Vector Fields*, Springer, Berlin (1983); S. Wiggins, *Global Bifurcations and Chaos*, Springer, Berlin (1988).

[2] G. M. Zaslavsky, R. Z. Sagdeev, D. A. Usikov and A. A. Chernikov, *Weak Chaos and Quasi Regular Patterns*, Cambridge University Press, Cambridge (1991).

[3] V. I. Arnold and A. Avez, *Ergodic Problems of Classical Mechanics*, Benjamin, New York (1968).

[4] T. Regge and J. A. Wheeler, Phys. Rev. **108**, 1063 (1957).

[5] L. Bombelli and E. Calzetta, RGGR/GTCRG preprint (1991).

SOME RESULTS ON NON CONSTANT MEAN CURVATURE SOLUTIONS OF THE EINSTEIN CONSTRAINT EQUATIONS

James Isenberg
Department of Mathematics
and Institute for Theoretical Science
University of Oregon
Eugene, OR 97403

Vincent Moncrief
Department of Physics
Yale University
P.O. Box 6666
New Haven, CT 06511

ABSTRACT: While the set of constant mean curvature solutions of Einstein's constraint equations is fairly well-understood, those solutions which have nonconstant mean curvature have resisted study. Here, we discuss a result obtained by the authors, and independently Choquet-Bruhat, which shows how to obtain a large set of nonconstant mean curvature solutions. In [1], a Leray-Schauder version of the proof of our existence theorem is presented. Here, we sketch a more constructive proof, which is based on sub and super solution theory.

During the late 1950's, Yvonne Bruhat proved that Einstein's vacuum field equation, which lies at the heart of the theory of general relativity, has a well-posed Cauchy formulation [2]. It follows from this work that if one can find a 3-manifold Σ^3 with a Riemannian metric γ_{ab} and a symmetric tensor K^{cd} on Σ^3 which satisfy the constraint equations

M. Flato et al. (eds.), Physics on Manifolds, 295–302.

$$R - K^{cd}K_{cd} + (trK)^2 = 0 \tag{1a}$$

and

$$\nabla_c K^c_d - \nabla_d(trK) = 0, \tag{1b}$$

then there exists a spacetime (M^4, g) which satisfies the Einstein vacuum field equation $G_{\mu\upsilon}= 0$, and has Σ^3 embedded in M^4 as a Cauchy hypersurface with intrinsic geometry γ and extrinsic geometry K. Hence the set of solutions (Σ^3, γ, K) of the constraints (1) is very important, both for understanding the space of solutions of the spacetime field equation, $G_{\mu\upsilon}= 0$, and for explicitly finding such solutions.

In their standard form, as presented in eq. (1), the constraint equations do not tell us very much about the nature of their set of solutions; nor is it clear how one might go about finding explicit solutions of the constraints. This situation is remedied by doing a conformal transverse-traceless decomposition of the data (γ, K), as discussed by Lichnerowicz [3], Choquet-Bruhat [4], and York [5]. That is, one writes

$$\gamma_{ab} = \phi^4 \lambda_{ab} \tag{2a}$$

and

$$K^{cd} = \phi^{-10}\left(\sigma^{cd} + LW^{cd}\right) + \tfrac{1}{3}\phi^{-4}\lambda^{cd}\tau, \tag{2b}$$

where λ_{ab} is a Riemannian metric, σ^{cd} is a symmetric tensor which is traceless $(\lambda_{cd}\sigma^{cd} = 0)$ and transverse $(\nabla_m\sigma^{md} = 0)$ with respect to λ_{cd} and its covariant derivative ∇, τ is a scalar field, W^a is a vector field, L is the conformal Killing operator (given by $LW_{ab} = \nabla_a W_b + \nabla_b W_a - \frac{2}{3}\lambda_{ab}\nabla_c W^c$) and ϕ is a positive definite scalar field. In terms of these functions, the constraints (1) may be written as

$$\nabla^2\phi = \tfrac{1}{8}R\phi - \tfrac{1}{8}\left(\sigma^{ab} + LW^{ab}\right)\left(\sigma_{ab} + LW_{ab}\right)\phi^{-7} + \tfrac{1}{12}\tau^2\phi^5 \tag{3a}$$

and

$$\nabla_a(LW)^a_b = \tfrac{2}{3}\phi^6\nabla_b\tau. \tag{3b}$$

The idea now is to freely choose λ_{ab}, σ^{cd}, and τ, and then solve eqs. (3) for W^a and ϕ. Note that eqs. (3) are roughly a coupled nonlinear elliptic system for ϕ and W^a [5], so the scheme has a chance.

Does it work? Not always, as one can see from a simple example: Let $\Sigma^3 = S^3$, choose λ_{ab} to be any metric with positive definite scalar curvature and no conformal Killing fields, and then pick $\sigma^{ab} = 0$ and $\tau = \tau_0$ (a constant). Since τ is constant, the right hand side of eq. (3b) vanishes, and then since the operator ∇L is elliptic, eq. (3b) requires that LW_{ab} vanish. Equation (3a) is now left in the form

$$\nabla^2\phi = \tfrac{1}{8}R\phi + \tfrac{1}{12}\tau_0^2\phi^5 \tag{4}$$

Recall that we are only interested in solutions with $\phi(x) > 0$ everywhere on S^3. Hence the right hand side of eq. (4) is positive definite. But the maximum principle doesn't allow there to be any functions ϕ for which $\nabla^2\phi > 0$ on a closed manifold. So there are no solutions to eqs. (3) with this choice of $(\Sigma^3, \lambda, \sigma, \tau)$.

While this example shows that for some choices of $(\Sigma^3, \lambda, \sigma, \tau)$ there are no solutions to the constraints, for many other choices of the data, there are. What one would like is a set of criteria which can identify those sets of data $(\Sigma^3, \lambda, \sigma, \tau)$ which lead to solutions and those choices which do not.

For closed manifolds and constant mean curvature (CMC) data -- i.e., if one chooses τ, and hence trK, to be a constant -- such criteria were found some years ago [6], [7]. The criteria are fairly simple when described in terms of Yamabe classes. The Yamabe theorem [8], recall, says that on a fixed closed manifold Σ^3, every Riemannian metric γ can be conformally deformed to a new metric $\tilde{\gamma}$ which has constant scalar curvature. The sign of that constant scalar curvature is fixed within each conformal class of metrics, so each metric γ on Σ^3 may be labeled as belonging to one of three Yamabe classes $\mathcal{Y}^+$, $\mathcal{Y}^0$, or $\mathcal{Y}^-$, depending on the sign of $R[\tilde{\gamma}]$. To determine if a given set of CMC data $(\Sigma^3, \lambda, \sigma, \tau)$ leads to a solution, all one needs to know besides the Yamabe class of the metric λ is whether the constant τ is zero or not, and whether the tensor field σ is zero everywhere on Σ^3 or not. We summarize the results in Table I.

Table 1: Existence of Solutions for Constant Mean Curvature Data

$\tau = 0$	$\sigma \equiv 0$	$\sigma \not\equiv 0$
$\lambda \in \mathcal{Y}^+$	No	Yes
$\lambda \in \mathcal{Y}^0$	Yes	No
$\lambda \in \mathcal{Y}^-$	No	No

$\tau \neq 0$	$\sigma \equiv 0$	$\sigma \not\equiv 0$
$\lambda \in \mathcal{Y}^+$	No	Yes
$\lambda \in \mathcal{Y}^0$	No	Yes
$\lambda \in \mathcal{Y}^-$	Yes	Yes

While the story is fairly complete in the CMC case, much less is known about the existence of solutions to eqs. (3) if τ is chosen to be nonconstant (non CMC). The CMC case is much easier than the general non CMC case because if τ is picked constant, then the right hand side of eq. (3b) vanishes and hence the two equations (3a) and (3b) decouple. Indeed, eq. (3b) is solved by taking LW to vanish, and hence the CMC analysis reduces to the study of the single semi-linear elliptic equation

$$\nabla^2\phi = \tfrac{1}{8}R\phi - \tfrac{1}{8}\sigma^{ab}\sigma_{ab}\phi^{-7} + \tfrac{1}{12}\tau_0^2\phi^5 \tag{5}$$

(This is sometimes called the Lichnerowicz equation). In the non CMC case, one must deal with the fully coupled system (3).

This past year, the authors, and independently Yvonne Choquet-Bruhat, have finally made some progress in understanding when solutions exist for non CMC choices of $(\Sigma^3, \lambda, \sigma, \tau)$. Our result thus far (leaving out details concerning function spaces) may be stated as follows:

Theorem

Let Σ^3 be closed, and choose λ and σ on Σ^3 so that $\lambda \varepsilon \mathcal{Y}^-$ with no conformal Killing fields. There exists a constant M (depending on λ and *σ), such that if one chooses τ with $|\tau| > 0$ and $\frac{\nabla_a\tau\nabla^a\tau}{\tau^2} < M$, then for $(\Sigma^3, \lambda, \sigma, \tau)$ there exists a unique solution (ϕ, W) of eq. (4).*

Choquet-Bruhat proved this theorem using Leray-Schauder theory, as explained in [1]. Here, we wish to briefly sketch how we have proven the result using a method based on sub and super solution theory.

Let us first recall what the basic sub-super solution theorem says (again, leaving out function space details) [8]: Let Σ^n be a closed manifold, and let $f:\Sigma^n \times [0, \infty) \to \mathbb{R}$ be a sufficiently smooth function. Assume there exists a pair of functions $\phi_-:\Sigma^n \to \mathbb{R}$ (the sub solution) and $\phi_+:\Sigma^n \to \mathbb{R}$ (the super solution) such that (a) $0 < \phi_-(x) < \phi_+(x)$; (b) $\nabla^2\phi_-(x) \geq f(x,\phi_-)$; and (c) $\nabla^2\phi_+(x) \leq f(x,\phi_+)$. Then there exists a function $\phi:\Sigma^n \to \mathbb{R}$ such that $\nabla^2\phi = f(x,\phi)$, with $\phi_-(x) \leq \phi(x) \leq \phi_+(x)$.

This theorem may readily be used to prove that solutions to eq. (5) exist in the various "yes" cases indicated in Table I for CMC data. As a simple example, let us take $\lambda \varepsilon \mathcal{Y}^-$, $\tau = \tau_0$, and σ arbitrary on some closed Σ^3. We may use the Yamabe theorem to conformally transform λ to a metric $\tilde{\lambda} = \psi^4\lambda$ which has scalar curvature $R[\tilde{\lambda}]=-1$ on Σ^3. The Lichnerowicz eq. (5) then takes the form

$$\nabla^2\phi = -\tfrac{1}{8}\phi - \tfrac{1}{8}A_\sigma\phi^{-7} + \tfrac{1}{12}\tau_0^2\phi^5$$

$$=: f(x,\phi) \tag{6}$$

where $A_\sigma := \sigma^{ab}\sigma_{ab}$. One readily verifies that if one picks

$$\phi_- = |\tau_0|^{-1/2} - \varepsilon \tag{7a}$$

and

$$\phi_+ = Max\left\{1,\ \sqrt[4]{\tfrac{3}{2}}\,|\tau_0|^{-1/2}\left[1 + \underset{\Sigma^3}{Max}\,A_\sigma\right]^{1/4} + \varepsilon\right\} \tag{7b}$$

where ε is a positive number less than $|\tau_0|^{-1/2}$, then these constants ϕ_- and ϕ_+ serve as sub and super solutions for eq. (6). It follows that there is solution ϕ to (6), bounded between ϕ_+ and ϕ_-. Using $\tilde{\phi} = \psi^{-1}\phi$, one maps the data ($\Sigma^3$, λ, σ, τ) to a solution (Σ^3, γ, K) of the Einstein constraints.

Some of the cases indicated in Table I are not so easily proven as this example (e.g., the case $\lambda\varepsilon\mathcal{Y}^+$, $\tau = 0$, $\sigma \not\equiv 0$ seems to require nonconstant sub and super solutions).[1] But all of the "yes" cases in the table can be proven by applying the sub-super solution theorem. The "no" cases can all be verified using maximum principle arguments, as in our example earlier.

We cannot apply the sub-super solution theorem readily to the fully coupled (non CMC) system of equations (3). However, the theorem plays a crucial role in the proof of our non CMC theorem, as we now show.

Sketch of Proof

The key to the proof is the study of the sequence of fields $\{\underset{n}{\phi}, \underset{n}{W}\}$ which is defined iteratively by the semi-decoupled equations

$$\nabla_a(L\underset{n}{W})^a{}_b = \tfrac{2}{3}\underset{n-1}{\phi}{}^6\nabla_b\tau \tag{8a}$$

[1]This is the case which has never been verified by Leray-Schauder methods.

and $$\nabla^2 \underset{n}{\phi} = \tfrac{1}{8}R\underset{n}{\phi} - \tfrac{1}{8}(\sigma^{ab} + L\underset{n}{W}^{ab})(\sigma_{ab} + L\underset{n}{W}_{ab})\underset{n}{\phi}^{-7} + \tfrac{1}{12}\tau^2\underset{n}{\phi}^5 \tag{8b}$$

together with a choice of $\underset{0}{\phi}$. This sequence, we show, converges to a solution of eqs. (3). First, we show that the sequence is well-defined.

With λ and τ specified by the choice of data $(\Sigma^3, \lambda, \sigma, \tau)$, and with $\underset{n-1}{\phi}$ known from the previous iteration, eq. (8a) is a straightforward linear elliptic equation for $\underset{n}{W}$. Elliptic theory [9] tells us that a unique solution $\underset{n}{W}$ exists, and further it guarantees that for some constant C (depending on Σ^3 and λ), $L\underset{n}{W}$ satisfies the inequality

$$\| L\underset{n}{W} \| \leq C\|\phi\|^6\|\nabla\tau\| \tag{9}$$

where $\| \ \|$ are suitable norms.[2]

Once $\underset{n}{W}$ has been determined, we can substitute it into eq. (8b) and we have essentially the independent Lichnerowicz equation, to solve for $\underset{n}{\phi}$. Indeed, since by hypothesis, $\lambda \varepsilon \mathcal{Y}^-$, we may conformally transform to a metric $\tilde{\lambda}$ with $R[\tilde{\lambda}]=-1$, and so eq. (8b) is replaced by

$$\nabla^2 \underset{n}{\phi} = \tfrac{1}{8}\underset{n}{\phi} - \tfrac{1}{8}A_{\sigma,\underset{n}{W}}\underset{n}{\phi}^{-7} + \tfrac{1}{12}\tau^2\underset{n}{\phi}^5 \tag{10}$$

where $A_{\sigma,\underset{n}{W}} := (\sigma^{ab} + L\underset{n}{W}^{ab})(\sigma^{ab} + L\underset{n}{W}_{ab})$. This is just like eq. (6) -- with the notable difference that τ is not constant -- and as in the discussion following eq. (6), we readily find sub and super solutions which guarantee the existence of a solution to (10). Another argument shows that this solution $\underset{n}{\phi}$ is unique, and hence we see that the sequence $\{\underset{n}{\phi}, \underset{n}{W}\}$ is well-defined (up to choice of $\underset{0}{\phi}$).

We next need to show that the sequence converges . Ellipticity of the operator ∇L, and its consequent inequality (9), guarantees that if the sequence $\{\underset{n}{\phi}\}$ converges, then the sequence $\{\underset{n}{W}\}$ converges as well, so we focus on $\{\underset{n}{\phi}\}$.

A crucial step in the verification of the convergence of the sequence $\{\underset{n}{\phi}\}$ is the establishment of master sub and super solutions for eq. (10). That is, we show that there is a pair of constants ϕ_- and ϕ_+ (depending on λ, σ, τ) which are sub and super solutions for eq. (10) *independent of n*. It follows that, for all n,

$$\phi_- \leq \underset{n}{\phi}(x) \leq \phi_+. \tag{11}$$

---

[2] The norms may be chosen so that inequality (9) specifies a *pointwise* bound on $L\underset{n}{W}$.

The proof that the sequence $\{\phi_n\}$ converges follows now from a contraction map argument. Specifically, we use a fairly long calculation, involving inequalities (9) and (11), to establish the inequality

$$\left\| \phi_{n+1} - \phi_n \right\| \leq \mu \left\| \phi_n - \phi_{n-1} \right\| \tag{12}$$

where μ is a known (rather messy) function of λ, σ, and τ. Here is where the bound on $\frac{\nabla_a \tau \nabla^a \tau}{\tau^2}$ comes in. We find that if $\frac{\nabla_a \tau \nabla^a \tau}{\tau^2}$ is required to be less than a certain positive number M -- which can be expressed as a function of $\max_{\Sigma^3}|\tau|$, $\min_{\Sigma^3}|\tau|$, $\max_{\Sigma^3}(\sigma^{ab}\sigma_{ab})$, and the C in (9) -- then we have $\mu<1$. Thus the map $\phi_n \to \phi_{n+1}$ is a contraction mapping, and we conclude that the sequence $\{\phi_n\}$ converges (in C^0 norm) to some function ϕ_∞. As noted above, it follows that $\{W_n\}$ converges to some W_∞ as well.

Three steps remain. First, we use the iteration eqs. (8) together with some of the estimates arising in the contraction map argument to establish that (ϕ_∞, W_∞) is a weak solution of eqs. (3). Next, we use standard bootstrap arguments to show that ϕ_∞ and W_∞ are C^2 functions, and hence strong solutions of (3). Then finally, we use modified contraction mapping arguments to show that for every ϕ_0 which satisfies $\phi_- \leq \phi_0 \leq \phi_+$, we get the same (ϕ_∞, W_∞). This completes the proof sketch. A detailed treatment of this work is in preparation [10].

As noted above, the non CMC Theorem can be proven either in the manner we have just described, or by using Leray-Schauder methods. There are two possible advantages, however, to the sub and super solution scheme: First it produces a priori upper and lower bounds for ϕ_∞ and $\left\| LW_\infty \right\|$, hence providing estimates for the solutions (γ, K) of the constraints. Second, the proof is constructive in the sense that ϕ_∞ and W_∞ are limits of sequences of solutions of the decoupled eqs. (8), and further each ϕ_n may be obtained as the limit of solutions of certain linear elliptic eqs [11]. These considerations could be useful for obtaining solutions (ϕ, W) of eqs. (3) in practice.

Whichever way it is proven, the non CMC Theorem which we have discussed here is, we hope, just the first step in the analysis of non CMC initial data sets. Already, we have non CMC results for $\mathcal{Y}^+$ and $\mathcal{Y}^0$ data if electromagnetic fields with $E^2 + B^2 > 0$ on Σ^3 are included in the data. We hope to obtain results for metrics in $\mathcal{Y}^+$ and $\mathcal{Y}^0$ in the vacuum case as well. We also have some hope that we can obtain an improved version of the theorem which does not have the $|\tau| > 0$ condition, and which has a sharpened limit on $\frac{\nabla \tau}{\tau}$. Work continues toward these goals.

Acknowledgements

This work was supported in part by NSF grant PHY-9012301 to the University of Oregon, and by NSF grants PHY-8903939 and INT-9015153 to Yale University.

References

[1] Y. Choquet-Bruhat, J. Isenberg, and V. Moncrief (1992), "Solution of Constraints for Einstein Equations," C. R. Acad. Sci. Paris.

[2] Y. Bruhat (1962), "The Cauchy Problem" in Gravitation: An Introduction to Current Research [ed: L. Witten], Wiley, N.Y.

[3] A. Lichnerowicz (1944), "L'Integration des Equations de la Gravitation Relativiste et le Probleme de n Corps, J. Math. Pures et Ap. 23, 37-63.

[4] Y. Bruhat (1956), "Sur l'Integration des Equation d'Einstein," J. Rat. Mech. Anal. 5, 951-966.

[5] J. York (1972), "Role of Conformal Three-Geometry in the Dynamics of Gravitation," Phys. Rev. Lett. 28, 1082-1086.
J. York (1979), "Kinematics and Dynamics of General Relativity," in Sources of Gravitational Radiation [ed: L. Smarr], Cambridge U Press.

[6] Y. Choquet-Bruhat and J. York (1980), "The Cauchy Problem" in General Relativity and Gravitation [ed: A. Held], Plenum.

[7] J. Isenberg (1987), "Parametrization of the Space of Solutions of Einstein's Equations," Phys. Rev. Lett. 59, 2389-2392.

[8] J. Kazdan and F. Warner (1974), "Curvature Functions for Compact 2-Manifolds," Ann Math 99, 14-47.

[9] A. Besse (1987) Einstein Manifolds, Springer Verlag.

[10] J. Isenberg and V. Moncrief, unpublished.

[11] J. Isenberg, unpublished.

Levi condition for general systems

WAICHIRO MATSUMOTO

Résumé. *On donnera la condition nécessaire et suffisante sur un système d'opérateurs aux dérivées partielles pour que le problème de Cauchy pour ce système soit bien posé dans la classe C^∞ — la condition de Levi —.*

We consider the Cauchy problem in a real domain Ω for general system of partial differential equations;

$$(1) \qquad \begin{cases} P(t,x,D_t,D_x)u = f(t,x), \\ u(t_0,x) = \varphi(x), \end{cases}$$

where

$$(2) \qquad P(t,x,D_t,D_x) = I_K\, D_t - \sum_{|\alpha|\le\nu} A_\alpha(t,x)\, D_x^\alpha \qquad (\ \nu \in \mathbb{N}\ ).$$

In order to describe our theorems, we need introduce the following.

DEFINITION 1. *Let $\omega \times \Gamma$ be an open conic set in $\mathbb{C}_t^1 \times \mathbb{C}_x^\ell \times \mathbb{C}_\xi^\ell$. We say that a formal sum $a(t,x,\xi) = \sum_{i=0}^\infty a_i(t,x,\xi)$ is a meromorphic formal symbol, when $\{a_i(t,x,\xi)\}_i$ satisfies the following;*

There exist an analytic conic set Σ in $\omega\times\Gamma$ and $\kappa \in \mathbb{R}$, and it holds that

(3) *$a_i(t,x,\xi)$ is meromorphic in $\omega\times\Gamma$, holomorphic in $\omega\times\Gamma\setminus\Sigma$ and positively homogeneous of order $\kappa - i$ on ξ ($i \in \mathbb{Z}_+$).*

(4) *For an arbitrary conically compact subset $\widetilde{\omega}$ in $\omega\times\Gamma\setminus\Sigma$, there exist $C>0$ and $R>0$, and we have*

$$|a_i(t,x,\xi)| \le C\, R^i\, i!\, ||\xi||^{\kappa-i} \quad \text{on } \widetilde{\omega} \qquad (\ i \in \mathbb{Z}_+\ ).$$

We call κ the order of $a(t,x,\xi)$. Let us denote the set of the meromorphic formal symbols of order κ by $S_M^\kappa(\omega\times\Gamma)$ and set $S_M(\omega\times\Gamma) = \cup_\kappa S_M^\kappa(\omega\times\Gamma)$.

M. Flato et al. (eds.), Physics on Manifolds, 303–307.

When $a(t,x,\xi)=\sum_{i=0}^{\infty}a_i(t,x,\xi)$ has the order κ, $a_i(t,x,\xi)\equiv 0$ ($0\le i<i_0$) and $a_{i_0}(t,x,\xi)\not\equiv 0$, we say that $\kappa-i_0$ is the true order of $a(t,x,\xi)$ and denote it by t-ord a. Further, we call $a_{i_0}(t,x,\xi)$ the true principal part of $a(t,x,\xi)$.

For $a(t,x,\xi)=\sum_{i=0}^{\infty}a_i(t,x,\xi)$ and $b(t,x,\xi)=\sum_{i=0}^{\infty}b_i(t,x,\xi)$ in $S_M(\omega\times\Gamma)$, we set

$$(5)\quad a(t,x,\xi)\circ b(t,x,\xi)=\sum_{i=0}^{\infty}c_i(t,x,\xi),$$

$$c_i(t,x,\xi)=\sum_{j+k+|\gamma|=i,\gamma\in\mathbb{Z}_+^{\ell}}\frac{1}{\gamma!}a_j^{(\gamma)}(t,x,\xi)b_{k(\gamma)}(t,x,\xi).$$

By the usual sum and the above product, S_M becomes noncommutative field. By the same way, we can also define the meromorphic formal symbols on $\omega\times\widetilde{\Gamma}$, $\widetilde{\Gamma}$ in $\mathbb{C}_\tau\times\mathbb{C}_\xi^{\ell}$. $S_M(\omega\times\widetilde{\Gamma})$ is also a noncommutative field. Adopting the commutative multiplicative group of the true principal parts of the elements in $S_M(\omega\times\widetilde{\Gamma})\setminus\{0\}$ in stead of the factor commutator group, we can define the determinant of the matrix $P(t,x,\tau,\xi)$ on $S_M(\omega\times\widetilde{\Gamma})$: $\det_M P$. (See E.Artin[1].) This determinant coincides with that of M.Sato and M.Kashiwara[9] for the matrix of differential operator.

DEFINITION 2. *We say that* $\{v_{jk}(h)(t,x,\xi)\}_{1\le j\le d,1\le k\le\delta_j,1\le h\le n_{jk}}$ *is a basis of M-chains of order* μ *of* $P(t,x,D_t,\xi)$ *w.r.t.* $\{\lambda_j\}_j$ *when the followings hold.*

$$(6)\quad P(t,x,D_t,\xi)\circ v_{jk}(h)(t,x,\xi)|_{D_t=\lambda_j}=\xi_1^{\mu}\circ v_{jk}(h-1)(t,x,\xi)+b_{jk}(h)(t,x,\xi),$$

where $b_{jk}(h)$ *belongs to* $S_M^{\mu-1}$ *and* $v_{jk}(0)=\mathbf{0}$ ($1\le h\le n_{jk}, 1\le k\le\delta_j, 1\le j\le d$, $\sum_{k=1}^{\delta_j}n_{jk}=m_j$ *and* $\sum_{j=1}^{d}m_j=K$).

Theorem 3.1 in W.Matsumoto[5] is reformulated as follows.

THEOREM 0. *We consider* $P(t,x,D_t,\xi)=I_K D_t-A(t,x,\xi)$, A *in* $S_M^{\kappa}(\omega\times\Gamma)$. *We assume that* $A_0(t,x,\xi)$ *has holomorphic eigenvalues* $\{\lambda_j(t,x,\xi)\}_{1\le j\le d}$ *and* λ_j *has the constant multiplicity* m_j *on* $\omega\times\Gamma$ ($1\le j\le d$). *Then, there exists a basis of M-chains* $\{v_{jk}(h)\}_{j,k,h}$ *of order* κ *w.r.t.* $\{\lambda_j\}_j$.

By the same way as S.Mizohata[8], [7] and K.Kajitani[2], we have the following theorem.

THEOREM 1. *We assume that all coefficients of $P(t,x,D_t,D_x)$ are analytic in Ω. If the Cauchy problem (1) is C^∞ well-posed in Ω, the following three equivalent conditions hold.*

(α) *In the meromorphic formal symbol class, $P(t,x,D_t,\xi)$ is transformed to a hyperbolic system, that is, to a matrix $I_K D_t - \widetilde{A}$, $\widetilde{A}$ in $S^1_M(\widetilde{\Omega}\times\mathbf{C}^\ell_\xi)$ whose first order part has real eigenvalues on $\Omega\times\mathbf{R}^\ell_\xi$, where $\widetilde{\Omega}$ is an complex neighborhood of Ω.*

(β) *The determinant of $P(t,x,\tau,\xi)$: $\det_M P$ is a hyperbolic polynomial. (We say that $p(t,x,\tau,\xi)$ is hyperbolic if $\deg_\tau p$ is equal to $\deg_{\tau,\xi} p$ and it has only real roots on τ for all (t,x,ξ) in $\Omega\times\mathbf{R}^\ell_\xi$.)*

(γ) *$A_0(t,x,\xi) = \sum_{|\alpha|=\nu} A_\alpha(t,x)\xi^\alpha$ is nilpotent. There exists a basis of M-chains of order ν of $P(t,x,D_t,\xi)$ w.r.t. $\{0\}$ (by virtue of Theorem 0) and $b_{jk}(h)$ satisfies the following*

(7) $\text{t-ord}\, b_{jk}(h) \le 1-(\nu-1)(n_{jk}-h), \quad (1\le h\le n_{jk}\,, 1\le k\le \delta_j\,, 1\le j\le d).$

Further, $\lambda^{n_{jk}} + \sum_{h=1}^{n_{jk}} \hat{b}_{jk}(h)\xi_1^{\nu(n_{jk}-h)}\lambda^{h-1} = 0$ has real roots on $\Omega\times\mathbf{R}^\ell_\xi$ where $\hat{b}_{jk}(h)$ is the part of order $1-(\nu-1)(n_{jk}-h)$ of $b_{jk}(h)$.

Since P is a differential operator, $\det_M P$ is a polynomial on τ and ξ with holomorphic coefficients. (See M.Sato and M.Kashiwara[9].) Then, if a root on τ has a constant multiplicity, it is holomorphic. Assuming that every root of $\det_M P = 0$: λ_j has a constant multiplicity m_j ($1\le j\le d$), we can again apply Theorem 0 on the transformed operator in Condition (α) as $\kappa = 1$. We add these the tilde in order to distinguish these from those in Condition (γ).

THEOREM 2. *We assume that all coefficients of $P(t,x,D_t,D_x)$ are analytic and each connected component of the zero set of $\det_M P$ in $\mathbf{R}^1_t\times\mathbf{R}^\ell_x\times\mathbf{R}_\tau\times\mathbf{R}^\ell_\xi\setminus O$ has a constant multiplicity. Then the following three conditions are equivalent.*

(a) *The Cauchy problem (1) is C^∞ well-posed in Ω.*

(b) *In the meromorphic formal symbol class, $P(t,x,D_t,\xi)$ is transformed to a hyperbolic system with a diagonal principal part.*

(c) *Adding to Condition (γ), there exists a basis of M-chains of first order of $P(t,x,D_t,\xi)$ w.r.t. the roots of $\det_M P = 0$ and it holds that*

$$\text{(8)} \qquad \text{t-ord}\, \tilde{b}_{jk}(h) \le -(\tilde{n}_{jk} - h), (1 \le h \le \tilde{n}_{jk} - 1, 1 \le k \le \tilde{\delta}_j, 1 \le j \le d).$$

We prove the above theorem as (a) $\Rightarrow$ (c) $\Rightarrow$ (b) $\Rightarrow$ (a). We can show (a) $\Rightarrow$ (c) by the usual way and (c) $\Rightarrow$ (b) is trivial. Then, we need to show (b) $\Rightarrow$ (a). We have the formal fundamental solution $E(t;x,D_x) = \sum_{k=0}^{\infty}(k!)^{-1}t^k e(k)(x,D_x)$ for $f \equiv 0$, where $e(k)(x,D_x)$ is a differential operator of order at most νk with holomorphic coefficients in $\tilde{\Omega}$. However, under Condition (α), the order of $e(k)(x,D_x)$ is at most $k + \nu_0$. (See W.Matsumoto and H.Yamahara[6].)

By virtue of the assumption of the constant multiplicity of characteristics, $E(t;x,D_x)$ is expressed as a formal Fourier integral operator $\sum_{j=1}^{d} E_{\phi_j}^j$, $E^j(t;x,\xi) = \sum_{k=0}^{\infty}(k!)^{-1}t^k e^j(k)(x,\xi)$, where ϕ_j is a phase function with respect to a characteristic root λ_j and $e^j(k)(x,\xi)$ belongs to $S_M^{k+\nu_0}(\tilde{\Omega}\times\Gamma)$ with $\Sigma = \emptyset$, Γ is a conic complex neighnborhood of $\mathbf{R}_\xi^\ell$. (See, for example, H.Kumano-go[3].) Under Condition (b), we can show the following inequality;

$$\text{(9)} \qquad |e^j(k)(x,\xi)| \le C\, R^k\, k!\, ||\xi||^{\nu_0'},$$

on an arbitrary conically compact set in $\tilde{\Omega}\times\Gamma\setminus\Sigma$ by the same way as W.Matsumoto and H.Yamahara[6]. Since $e^j(k)(x,\xi)$ is a holomorphic function of x and ξ, (9) holds on $\tilde{\Omega}\times\Gamma$ by the maximum principle. This implies that $E(t;x,D_x)$ operates on the functions of x in $C^\infty(\Omega\cap\{t=t_0\})$ for small t .

Remark. In Theorem 2, if we remove the real analyticity of coefficients, (a) (= (b) = (c)) is necessary for C^∞ well-posedness but not sufficient.

Example.

$$L = I_3 D_t - \begin{pmatrix} 0 & 1 & 0 \\ & 0 & 0 \\ & & 0 \end{pmatrix} D_x + \begin{pmatrix} 0 & 0 & 0 \\ 0 & 0 & \mu(t) \\ \nu(t) & 0 & 0 \end{pmatrix}, \quad (x \in \mathbf{R}),$$

where $\mu(t)$ and $\nu(t)$ belong to $C^\infty(\mathbf{R})$, $\mu\nu \equiv 0$, the either boundary of $supp\,\mu$ and $supp\,\nu$ is included in $\{t \ge 0\}$ and has a unique accumulation point 0. It is shown that the Cauchy problem for L is C^∞ well-posed for the initial time $t_0 \neq 0$ but not C^∞ well-posed for $t_0 = 0$ by the same way as W.Matsumoto[4]. On the other hand, for the analytic μ and ν, $\mu\nu \equiv 0$ is equivalent to (a) (= (b) = (c)).

REFERENCES

[1] E.Artin, *Geometric algebra, Chap. 4*, Interscience Publishers (1957).

[2] K.Kajitani, *On the E-well posed evolution equations*, Comm. P.D.E. **4 (6)** (1979), 595-608.

[3] H.Kumano-go, *Pseudo-differential operators, Chap. 10*, The MIT Press (1981).

[4] W.Matsumoto, *On the conditions for the hyperbolicity of systems with double characteristics, I, 3.3°*, J. Math. Kyoto Univ. **21 (1)** (1981), 47-84.

[5] *Normal form of systems of partial and pseudo differential operators in formal symbol classes*, (to appear in J. Math. Kyoto Univ).

[6] W.Matsumoto and H.Yamahara, *On the Cauchy-Kowalevskaya theorem for systems*, Proc. Japan Acad. **67, Ser.A, (6)** (1991), 181-185.

[7] S.Mizohata, *Some remarks on the Cauchy problem*, J. Math. Kyoto Univ. **1 (1)** (1961), 109-127.

[8] *On evolution equations with finite propagation speed*, Israel J. Math. **13** (1972), 173-187.

[9] M.Sato and M.Kashiwara, *The determinant of matrices of pseudo-differential operators*, Proc. Japan Acad. **51 Ser.A** (1975), 17-19.

[10] J.Vaillant, *Conditions d'hyperbolicité des systèmes d'opérateurs aux dérivées partielles*, Bulletin Sci. Math. **114 (3)** (1990), 243-328.

[11] J.Vaillant, *Opérateurs de multiplicité constante*, (to appear).

Department of Applied Mathematics and Informatics
Ryukoku University
Seta, 520-21 Ohtsu, JAPAN

Conditions invariantes pour un système, du type conditions de Levi,

par Jean Vaillant

h is a square matrix of linear partial differential operators ; we define new invariant conditions on h ; we use these conditions to solve hyperbolic and holomorphic Cauchy's problems.

Pour un opérateur différentiel linéaire scalaire, on connaît la condition de bonne décomposition ; elle permet, en réel, de caractériser l'hyperbolicité ; les opérateurs holomorphes, pour lesquels elle est satisfaite, ont des solutions ramifiées du problème de Cauchy, à singularités polaires au logarithmiques, lorsque les données sont de ce type ; des conditions de bonne décomposition partielle permettent de caractériser les problèmes de Cauchy bien posés dans les classes de Gevrey. Il s'agit ici du problème beaucoup plus difficile de la généralisation de ces notions aux systèmes différentiels. Nous définissons des conditions L invariantes qui permettront ces généralisations, dont nous donnons quelques exemples.

I. Conditions L

(a) $x = (x_0,x') = (x_0,;x_1,\dots,x_n) \in \Omega$, Ω est un voisinage ouvert de 0 dans $\mathbb{R}^{n+1}$ (resp. $\mathbb{C}^{n+1}$).

On considère un opérateur différentiel h d'ordre 1, matriciel $m \times m$, à coefficients analytiques dans Ω. On note $\mathfrak{a}$ sa partie principale et b sa partie d'ordre 0, de sorte que :

$$h = \mathfrak{a} + b.$$

On note $\xi = (\xi_0,\xi') = (\zeta_0,\xi_1,\dots,\xi_n)$ la variable duale de x et on considère le déterminant caractéristique :

$$\det\mathfrak{a}(x,\xi)$$

on suppose qu'il n'est pas identiquement nul pour x=0. On peut décomposer $\det\mathfrak{a}(x,\xi)$ en polynômes irréductibles dans $\mathcal{O}[\xi]$, anneau des polynômes en ξ à coefficients les germes analytiques en x à l'origine ; pour simplifier les notations, nous supposerons qu'il n'y a qu'un facteur multiple H ; nous noterons m_1 sa multiplicité, de sorte que :

$$\det\mathfrak{a}(x,\xi) = H^{m_1}K\ ;$$

on suppose en fait la <u>multiplicité constante</u>, c'est-à-dire que, pour chaque x, les points ξ, $(\xi \neq 0)$ du cône :

$$HK(x,\xi) = 0,$$

M. Flato et al. (eds.), Physics on Manifolds, 309–314.

sont simples.

On note $\mathbb{A}$ la matrice des cofacteurs de a, telle que :

$$\mathfrak{a}\mathbb{A} = \mathbb{A}\mathfrak{a} = \text{dét}\,\mathfrak{a}\,I.$$

On considère l'anneau localisé de l'anneau $\mathcal{O}[\xi]$ par rapport à l'idéal premier défini par H ; cet anneau est principal et dans cet anneau $\mathfrak{a}$ est équivalente à la matrice diagonale [2]

$$\text{diag}\left[H^p, H^{q_1},\ldots,H^{q_\ell},1,\ldots,1\right],$$

où les entiers $p = q_0, q_1, \ldots, q_\ell$ sont tels que :

$$p \geq q_1 \geq \ldots \geq q_\ell > 0 \; ; \; p + q = m_1, \text{ où l'on a posé } q = q_1 + \ldots + q_\ell .$$

$\mathbb{A}$ est divisible par H^q ; on pose : $\mathcal{A} = \mathbb{A}/H^q$, de sorte que :

$$\mathfrak{a}\mathcal{A} = \mathcal{A}\mathfrak{a} = H^p\, K\, I.$$

Pour un opérateur matriciel ou scalaire $\Lambda'(x,D)$ d'ordre $\leq \mu$, on notera $\Lambda(x,\xi) = \sigma_\mu(\Lambda')$ le symbole homogène d'ordre μ égal à la partie principale de Λ' si celle-ci est d'ordre μ, à O sinon. On posera s = degré de H, χ = degré de K, $\gamma = s+\chi-1$. Inversement à une matrice $\Lambda(x,\xi)$ de polynômes ou de symboles homogènes d'ordre μ, on associera des opérateurs matriciels notés $\Lambda'(x,D)$, de sorte que : $\sigma_\mu(\Lambda') = \Lambda$. Ainsi on aura : $H'(x,D)$ tel que : $\sigma_s(H') = H\,I$; de même ;
$\sigma_\chi(K') = K$; $\sigma_{\mu_o}(\mathcal{A}') = \mathcal{A}$, où $\mu_o = ps+\chi-1$. On pose :

$$\mu_j = \mu_o + j\gamma + \left(\sum_{1\leq k\leq j} q_k - j\right)s, \text{ pour : } 0 \leq j \leq \ell$$

et

$$\mu_j = \mu_o + j\gamma + (q-\ell)s \text{ pour : } \ell+1 \leq j \leq m_1-1.$$

(b) <u>Conditions</u> L,

On définit par récurrence des conditions sur l'opérateur h.

<u>Condition</u> L_1 ; Lorsque $\ell \geq 1$, (resp. $\ell = 0$), il existe des opérateurs différentiels $\mathcal{A}'$, H', K' et une polynôme $\Lambda_1(x,\xi)$ homogène en ξ, à coefficients matriciels, de degré μ_1 ou nul tels que :

$$S_o \equiv \mathcal{A}\;\sigma_{\mu_o}(h\mathcal{A}' - H'^p K') = H^{p-q_1}\,\Lambda_1 \quad (\text{resp. } H^{p-1}\,\Lambda_1)$$

On a alors :

$$\mathfrak{a}\,\Lambda_1 = H^{q_1} K\,\sigma_{\mu_o}(h\mathcal{A}' - H'^p K') \qquad (\text{resp. } H\,K\,\sigma_{\mu_o}(h\mathcal{A}' - H'^p K').$$

On suppose L_1réalisée, $\mathcal{A}'$, H', K' et Λ_1 choisis.

Condition L_2 : Lorsque $\ell \geq 2$, resp. ($\ell = 1$), il existe un opérateur différentiel Λ_1' de symbole Λ_1 et un polynôme Λ_2 homogène en ξ de degré μ_2 ou nul tel que :

$$S_1 \equiv \mathcal{A}\ \sigma_{\mu_1} (h\ \Lambda_1' - h\mathcal{A}'\ H'^{q_1}\ K' + H'^p\ K'\ H'\ ^{q_1}\ K')$$
$$= H^{p-q_2}\ \Lambda_2\ (\text{resp.} = H^{p-1}\ \Lambda_2)$$

...

Condition L_ℓ : Il existe un opérateur différentiel $\Lambda'_{\ell-1}$ de symbole $\Lambda_{\ell-1}$ et un polynôme, Λ_ℓ homogène en ξ de degré μ_ℓ ou nul tel que :

$$S_{\ell-1} \equiv \mathcal{A}\ \sigma_{\mu_{\ell-1}} (h\Lambda'_{\ell-1} - h\Lambda'_{\ell-2}\, H'^{q_{\ell-1}}\ K' + \ldots + (-1)^{\ell-1}\, h\mathcal{A}'\ H'^{q_1}\ K'.\ H'^{q_2}\ K' \ldots H'^{q_{\ell-1}}\ K' + (-1)^{\ell}\ H'^p\ K'\ H'^{q_1}\ K' \ldots H'^{q_{\ell-1}}\ K') = H^{p-q_\ell}\ \Lambda_\ell\ .$$

Condition $L_{\ell+1}$: Il existe un opérateur différentiel Λ'_ℓ de symbole Λ_ℓ et un polynôme $\Lambda_{\ell+1}$ homogène en ξ de degré $\mu_{\ell+1}$ ou nul tel que :

$$S_\ell \equiv \mathcal{A}\ \sigma_{\mu_\ell} (h\Lambda'_\ell - h\Lambda'_{\ell-1} H'^{q_\ell}\ K' + \ldots + (-1)^\ell\ h\mathcal{A}'\ H'^{q_1}\ K' \ldots H'^{q_\ell}\ K'$$
$$+ (-1)^{\ell+1}\ H'^p\ K'\ H'^{q_1}\ K' \ldots H'^{q_\ell}\ K')$$
$$= H^{p-1}\ \Lambda_{\ell+1}$$

...

Condition L_{m_1} : Il existe un opérateur différentiel Λ'_{m_1-1} de symbole Λ_{m_1-1} et un polynôme Λ_{m_1} homogène en ξ de degré μ_{m_1} ou nul tel que :

$$S_{m_1-1} \equiv \mathcal{A}\ \sigma_{\mu_{m_1-1}} \big[h\Lambda'_{m_1-1} - h\Lambda'_{m_1-2}\ H'\ K' + \ldots$$
$$+ (-1)^{m_1-1}\ h\mathcal{A}'\ H'^{q_1}\ K' \ldots H'^{q_\ell}\ K'(H'K')^{m_1-\ell-1}$$
$$+ (-1)^{m_1}\ H'^p\ H'\ H'^{q_1}\ K' \ldots H'^{q_\ell}\ K'(H'\ K')^{m_1-\ell-1}\big]$$
$$= H^{p-1}\ \Lambda_{m_1}.$$

Remarques.

1°) Si $p = q_1 = \ldots = q_{r-1} \neq q_r$, les conditions $L_1,\ldots,L_{r-1}$ sont réalisées quelque soit b et les conditions L se réduisent aux conditions $L_r \ldots L_{m_1}$.

2°) Il peut arriver que L_{m_1}, par exemple soit une conséquence des conditions L_j qui la précèdent.

3°) On peut définir des conditions analogues qui auraient des propriétés semblables pour un opérateur h d'ordre t.

4°) Ces conditions dans le cas d'un opérateur scalaire se réduisent à la condition de bonne décomposition de l'opérateur.

5°) Les conditions L sont évidemment invariantes, puisqu'elles portent sur des symboles d'opérateurs différentiels.

Proposition 1 : Les conditions L ne dépendent pas du choix des opérateurs H', K', $\mathcal{A}'$, $\Lambda'_1,\ldots,\Lambda'_{m_1-1}$.

Remarque : On peut les expliciter en fonction des coefficients de l'opérateur h.

II. Applications.

On considère le problème de Cauchy :

$$h(x,D)u = f$$
$$u\big|_{x_0=t} = g_t(x').$$

f est donnée, g_t est la donnée de Cauchy sur $x_0 = t$ et u est l'inconnue; Les hyperplans $x_0 = t$ sont supposés non caractéristiques en chaque point, de sorte que l'on peut écrire :

$$h(x,D) = \mathrm{I}D_0 + \sum_{1\leq j\leq n} a_j D_j + b(x),$$

où les a_j, b_j sont des matrices m × m fonctions analytiques de x.

(a) $x \in \mathbb{R}^{n+1}$.

Définition. Le problème de Cauchy est bien posé pour h en $(\underline{t};\underline{x}') \in \Omega$ si, et seulement si, il existe un voisinage ouvert ω de $(\underline{t},\underline{x}')$, $\omega \subset \Omega$, tel que : $\forall f \in \mathcal{C}^\infty(\omega)$, $\forall g_t \in \mathcal{C}^\infty(\omega_t)$, où :
$\omega_{\underline{t}} = \omega \cap \{x_0 = \underline{t}\}$, il existe une solution $\mathcal{C}^\infty$, unique, du problème de Cauchy dans ω ; le problème de Cauchy est bien posé dans Ω, si il est bien posé en tout point de Ω. Il est localement bien posé, si il existe un voisinage ouvert de O, $\Omega_1 \subset \Omega$ où il est bien posé.

On considère le déterminant caractéristique :

$$\text{dét}\, a(x,\xi) = \text{dét}\left(\xi_0 I + \sum_{1\leq i\leq n} a_i(x)\xi_i\right);$$

on fait l'hypothèse qu'il est hyperbolique par rapport à la direction (1,0), c'est-à-dire que les m racines caractéristiques en ξ_0 sont réelles et, de plus, que ces racines sont de multiplicité constante.

Proposition 2 : Les conditions L sont invariantes lorsqu'on transforme l'opérateur h par un opérateur pseudo différentiel elliptique d'ordre 0, c'est-à-dire que les conditions L sont équivalentes aux conditions $\tilde{L}$ relatives à $\tilde{h} = \Delta_0^{-1} h \Delta_0$ cf. [2].

Théorème 1 : Si $m_1 \leq 7$, une condition nécessaire et suffisante pour que le problème de Cauchy soit localement bien posé est pour h est que h vérifie les conditions L.

Remarques

1°) Si $p = 1$, alors $\ell = m_1 - 1$, les conditions L sont toujours réalisées, quelque soit *b* et l'opérateur est fortement hyperbolique.

2°) Si $m_1 = p = 2$, les conditions L sont les conditions classiques sur la matrice sous caractéristique [5].

3°) Si $p = m_1$, on a le cas étudié dans [2].

(b) $x \in \mathbb{C}^{n+1}$.

On considère le problème de Cauchy. Les données de Cauchy sur $x_0 = 0$, sont ramifiées autour de $x_1 = 0$; les ramifications sont polaires ou logarithmiques [6].

Théorème 2 : Si $m_1 \leq 7$, les ramifications de la solution sont polaires au logarithmiques.

[1] W. MATSUMOTO. On the Levi conditions for first order systems with characteristics of constant multiplicity (à paraître).

[2] J. VAILLANT. Conditions d'hyperbolicité des systèmes d'opérateurs aux dérivées partielles. Bulletin des Sciences Mathématiques, 2ème sem. 114, 1990, p.243-328.

[3] J. VAILLANT. Condition d'hyperbolicité des systèmes. C.R. Acad. Sciences PARIS t.313, série I, p.227-30, 1991.

[4] V.M. PETKOV. Microlocal forms for hyperbolic systems. Math. Nachrichten t.93, 1979, p.117-131.

[5] J. VAILLANT. Données de Cauchy portées par une caractéristique double... Journal de Mathématiques Pures et Appliquées 47, 1968, p.1 à 40.

[6] J.C. DE PARIS. Problème de Cauchy à données singulières... Journal de Mathématiques Pures et Appliquées 51, 1972, p.465-488.

J. VAILLANT
Mathématiques, URA 213, SDI 6183 du CNRS
Tour 46-0, 5ème étage,
Université Pierre et Marie Curie (Paris VI)
4, place Jussieu
75252 PARIS Cedex 05

Black Holes in Supergravity

P.C. Aichelburg
Institut für Theoretische Physik
Universität Wien
A-1090 Vienna, Boltzmanngasse 5, Austria

Abstract

A review is given on what is known about stationary black hole configurations in $N = 1$ and $N = 2$ supergravity.

Soon after supergravity was discovered, the question arose how the causal structure of spacetime might be influenced by the gravitino (spin-3/2) field. It was Mme. Yvonne Choquet-Bruhat who helped to clarify this question by studying the Cauchy problem in supergravity [1], [2], [3]. It is for me a pleasure to summarize what we have learned since then about the existence of black hole solution in supergravity.

In classical relativity stationary black holes are characterized by a restricted number of "charges" that can be expressed as surface intergrals at spacial infinity (no-hair conjecture). In supergravity the invariance under local supersymmetry transformations give rise to a new conserved spinor charge (supercharge). It is therefore of interest to study whether black hole solution with supercharge exist.

Within the Einstein-Maxwell theory a generic stationary black hole is completely characterized by its mass (M), angular momemtum (a), electric (e) and/or magnetic (q) charge. This theory can be embedded in $N = 2$ supergravity whose field content is: The gravitational vierbein field $e^a = e^a{}_\mu dx^\mu$, the electromagnetic potential one-form $A = A_\mu dx^\mu$ and two Majorana spinor-valued one-forms ψ^j, combined to a complex (Dirac) field $\psi = \psi^1 + i\psi^2 = \psi_\mu dx^\mu$ (Rarita-Schwinger field). All fields are Grassmann-valued, the bosonic fields (e, A) being even elements while ψ is odd.

A possible way to study the existence of black hole solutions in supergravity is to consider pertubation of the spin- 3/2 field on a black hole background.If one retains from the full $N = 2$ field equations term linear in ψ, the system is reduced to the

M. Flato et al. (eds.), Physics on Manifolds, 315–320.

Einstein-Maxwell equations plus the (linearized) Rarita-Schwinger equation on a given background:

$$\gamma \wedge \widehat{D}\psi = 0 \tag{1}$$

where

$$\widehat{D} = d + \frac{1}{2}\omega^{ab}\sigma_{ab} - \frac{k}{2}F^{ab}\sigma_{ab}\gamma, \qquad \gamma = \gamma_a e^a. \tag{2}$$

(The explicit form of the field equations , notations and conventions may be found in Ref. 4, especially we have $k^2 = 4\pi G$ and signature $(+---)$.) The local supersymmetry transformation are generated by a complex spinor field $\varepsilon = \varepsilon^1 + i\varepsilon^2$ ($\varepsilon^1, \varepsilon^2$ Majorana). At the linearized level these transformations change the Rarita- Schwinger field by

$$\delta\psi = k^{-1}\widehat{D}\varepsilon \tag{3}$$

while the vierbein and the electromagnetic potential are invariant.

The global conserved quantity associated with this symmetry is the spinorial supercharge

$$\mathcal{S} = -\frac{i}{k}\oint_{S^2_\infty} \gamma_5\gamma \wedge \psi \tag{4}$$

which acts as a generator for the (asymptotic) global supersymmetry transformations. We consider only asymptotically flat configurations and ψ has to fall off like $O(r^{-2})$ in order to render $\mathcal{S}$ finite.

For the background one takes the Kerr-Newman black hole and tries to solve eq.(1) imposing the following conditions:

i) stationarity

ii) fall off at spatial infinity

iii) regularity at and outside the horizon.

Because of the local gauge freedom (3) any field of the form

$$\psi = k^{-1}\widehat{D}\varepsilon \tag{5}$$

is automatically a solution to eq. (1). Moreover, the supercharge (4) is only invariant under (proper) gauge transformations for which the gauge spinor ε tends to zero at spatial infinity. We shall therefore distinguish between gauge-generated and non-gauge configurations (see Fig. 1).

Non-Gauge Fields

Let us first consider non-gauge configurations. A detailed analysis of the spin-3/2 modes on a Kerr-Newman black hole [5] has shown that there are no solutions of Eq. (1) that satisfy the conditions i) – iii) unless the background parameters are such that

$$k^2M^2 = e^2 + q^2 \tag{6}$$

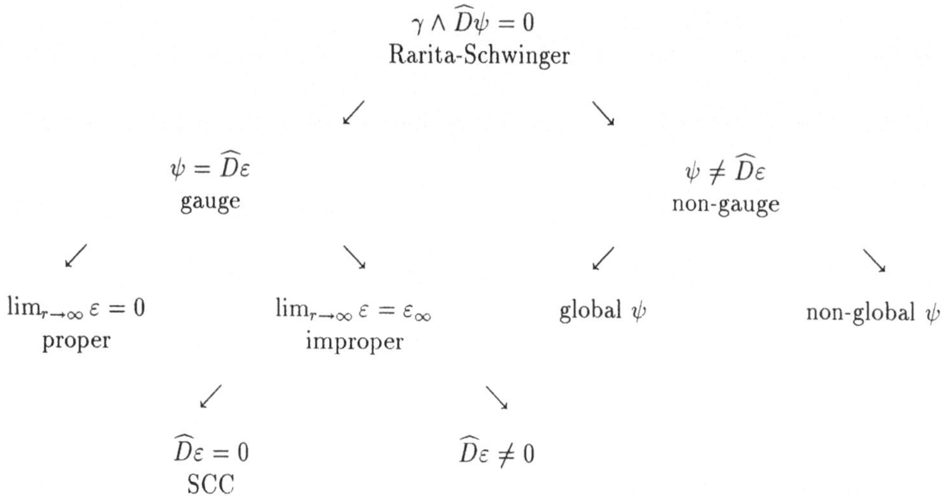

Figure 1: The classification of solutions to the Rarita-Schwinger equation are shown. Of interest are the non-gauge global and non-global (monopole-like) as well as the improper gauge generated (superpartners) configurations.

i.e. the charge of the hole equals its mass (in units where $4\pi G = 1$). This singles out the extreme Reissner-Nordstrøm metric, since a has to vanish in order that a horizon exists. It was shown [6,7] that the extreme Reissner-Nordstrøm field indeed admits spin-3/2 perturbations satisfying the conditions i) – iii). These perturbations are characterized by two complex constants and give rise to a non-zero supercharge. The linear solution was later generalized (by rather tedious calculations) to an exact one, taking into account the full back-reaction of the Rarita-Schwinger field onto the Einstein-Maxwell fields [8]. Embacher [9] has shown that this solution is the most genral one.

The uniqueness of the extreme Reissner-Nordstrøm metric in supporting a spin-3/2 supercharge has a geometrical explanation: Among the Kerr-Newman black holes the extreme Reissner-Nordstrøm metric is singled out by the existence of supercovariantly constant (SCC) spinors χ_{SCC} satisfying

$$\widehat{D}\chi_{SCC} = 0. \tag{7}$$

These spinors give rise to additional gauge invariant Rarita-Schwinger modes. Actually it is possible to construct the linear superhair solutions directly from χ_{SCC} [10].

Further insight into the nature of static configurations with supercharge was gained by considering Rarita-Schwinger fields on multi-black-hole backgrounds. Spacetimes of more than one black hole are expected to be non-static. There exists however a class of spacetimes, the Majumdar-Papapetrou fields, which represent static configurations of several extreme black holes in equilibrium under their mutual gravitational and electrical forces. Moreover, the Majumdar-Papapetrou black holes are the only solutions within the Einstein-Maxwell fields that admit SCC spinors.

Detailled calculations have shown that regular static spin-3/2 perturbations exist also on Majumdar-Papapetrou spacetimes. On a background with n black holes these solutions depend on $2n$ complex parameters which are however, strongly constraint.The constraints reduce the number of free parameter to two, indepent of the underlying spacetime. A direct consequence is that for a two-black hole system the supercharge is proportional to the difference in the mass parameters [11]. Consequently, for two equal black holes the supercharge is zero. One may interpret the constraints as equilibrium conditions for static configurations.This would imply that there is a new kind of interaction between black holes due to the existence of supercharge. A step towards such an interpretation was made in Ref. [12], by relating the supercharge to changes in the mass and charge parameters of the background.

Monopole-Like Solutions

There is another possibility to look at the constraints: Since they follow from global integrability conditions, violating them implies the non-existence of global solutions for ψ on the given background. However, ψ being a gauge field, it is conceivable to consider only its supercovariant field strength $\widehat{D}\psi$ as the fundamental object and to give up global existence of ψ. If one requires regularity for $\widehat{D}\psi$, then ψ may have line-singularities which may be shifted by gauge transformations of the form (3). This is analogous to the Dirac monopole in electrodynamics.

The existence of "strings" is associated with non-zero global quantities:

$$\mathcal{P} = \frac{1}{4\pi i} \oint_{S^2_\infty} \widehat{D}\psi$$

which we call "spinorial magnetic-type charges".

In a paper "Monopole-like solutions on Majumdar Papapetrou backgrounds" [13], we have studied these kind of solutions.At the linearized level it is possible to associate with each black hole a spinorial magnetic charge.Because of the Grassmann character of the fields it is not clear whether these magnetic type charges are of topological origin.

Gauge-Generated Solutions

Consider now ψ-configurations of the type (5) for which the gauge spinor ε tends to a constant ε_∞ at spatial infinity. The gauge-induced supercharge may then be written as:

$$\mathcal{S} = -\frac{i}{k^2} \oint_{S^2_\infty} \gamma_5 \gamma \wedge \widehat{D}\varepsilon = (-i\gamma^a P_a^{ADM} + \frac{i}{k} e + \frac{i}{k}\gamma_5 q)\varepsilon_\infty \tag{8}$$

whre P_a^{ADM} is the usual ADM four-momentum of the configuration, while e and q is the total electric and magnetic charge, respectively.

We wish to identify all configurations which can be obtained by "proper" gauge transformations ($\varepsilon \to 0$) from one another thereby dividing the solutions into equivalence classes, which may be parametrized by a constant spinor at infinity. If the background is extreme, satisfying condition (6), the matrix on the r.h.s. of eq.(8) is singular. As a consequence half of the asymptotic spinor space leaves the supercharge $\mathcal{S}$ invariant. This part is associated with the super covariantly constant spinor χ_{SCC}, eq. (7), at spatial infinity. The remaining transformations induce a supercharge via eq. (8).

It was pointed out by Gibbons [14], [15], that in $N = 2$ supergravity the extreme Reissner-Nordstrøm or, more generally, the Majumdar-Papapetrou spacetimes are partially supersymmetric: gauge transformations with χ_{SCC} leave the fields invariant. On the other hand, those transformations that induce a supercharge give rise to "superpartners" to the purely bosonic configurations. Hajicek has emphasized [16] that the extreme Reissner-Nordstrøm spacetimes should be considered also quantum mechanically as stable particle-like (solitons) configurations because they do not show Hawking radiation.

The linear gauge-generated solutions may be itereted to obtain an exact superpartner satisfying the full $N = 2$ supergravity equations [17]. The gauge induced supercharge gives rise to an "intrinsic" angular momentum.Similar to the Kerr-Newmann spacetimes the configurations is stationary with a magnetic moment.

In a series of papers [18], [19], [20], [21], we have formulated a dynamics for these superpartners. Generalizing the work of Hajicek [22] for the Einstein-Maxwell solitons to supergravity one describes the superpartner by collective coordinates associated with its asymptotic degrees of freedom: translational and supertranslational (spinorial) parameters.By freezing all degrees of freedom but these asymptotic ones, one is lead to a particle dynamics in rigid (flat) $N = 2$ superspace. Upon quantizing his theory, one arrives at the basic $N = 2$ hypermultiplet. The next step was to study the dynamics of these particle in an external field. This naturally gave rise to a formulation of the theory in curved superspace. The classical equation of motion is the supersymmetric generalization of the Mathison-Papapetrou equation for a spinning particle in a gravitational field.

The last step in this development was to formulate an effective soliton-soliton interaction in supergravity. This was achieved by choosing a superpartner configuration for the background in which the other particle moves and applying a slow-motion and long-distance approximation. The interaction potential was obtained to lowest order that incorporates the effect of the supercharge. A Dirac-like proceedure was then applied to quantize the supersymmetric two- particle system.

Summary

The study of classical particle-like solutions in supergravity has shown that stationary black holes can carry a new conserved quantity called supercharge. Within $N = 2$ supergravity only the extreme Reissner-Norstrøm can support a supercharge. Exact non-gauge and gauge- induced (superpartner) configurations are known. On the classical level the geometrical interpretation of these black holes are obscured by the fact that all the fields are Grassmann- valued. In a quantum theory especially the superpartners should play an essential role when considering non-perturbative effects.

References

[1] Choquet-Bruhat Y 1983 *Lett. Math. Phys.* **7** 459

[2] Choquet-Bruhat Y 1985 *J. Math. Phys.* **26** 329

[3] Choquet-Bruhat Y 1985 *Commun. Math. Phys.* **97** 541

[4] Embacher F 1985 *Gen. Rel. Grav.* **16** 909

[5] Aichelburg P C and Güven R 1981 *Phys. Rev.* **D24** 2066

[6] Güven R 1982 *Phys. Rev.* **D25** 3117

[7] Aichelburg P C and Güven R 1983 *Phys. Rev.* **D27** 456

[8] Aichelburg P C and Güven R 1983 *Phys. Rev. Lett.* **51** 1613

[9] Embacher F 1985 *Class. Quantum Grav.* **2** 323

[10] Aichelburg P C and Güven R 1984 *Phys. Lett.* **135B** 291

[11] Embacher F and Aichelburg P C 1985 *Phys. Rev.* **D30** 2457

[12] Aichelburg P C and Embacher F 1985 *Class. Quantum Grav.* **2** 65

[13] Aichelburg P C and Embacher F 1986 *Gen. Rel. Grav.* **18** 705

[14] Gibbons G 1982 in *Unified Theories of Elementary Particles*, edited by P. Breitenlohner and H.P. Dürr (Lecture Notes in Physics, Vol. 160) (Springer New York)

[15] Gibbons G 1985 *Black Holes, Solitons in Ungauged Extended Supergravity* in Proceedings of the XVth GIFT Seminar, eds. F. Aguila et al. (World Scientific)

[16] Hajicek P 1982 *Phys. Rev.* **D26** 3384

[17] Aichelburg P C and Embacher F 1986 *Phys. Rev.* **D34** 3006

[18] Aichelburg P C and Embacher F 1988 *Phys. Rev.* **D37** 338

[19] Aichelburg P C and Embacher F 1988 *Phys. Rev.* **D37** 911

[20] Aichelburg P C and Embacher F 1988 *Phys. Rev.* **D37** 1436

[21] Aichelburg P C and Embacher F 1988 *Phys. Rev.* **D37** 2132

[22] Hajicek P 1981 *Nucl. Phys.* **B185** 254

Low-Dimensional Behaviour in the Rotating Driven Cavity Problem

E.A.CHRISTENSEN[1], J.N.SORENSEN[2], M.BRONS[3] AND P.L.CHRISTIANSEN[1]

[1]Laboratory of Applied Mathematical Physics, Building 303,
[2]Department of Fluid Mechanics, Building 404,
[3]Mathematical Institute, Building 303,
The Technical University of Denmark, DK-2800 Lyngby, Denmark.

Abstract:

In searching for low-dimensional structures in the driven cavity problem, in this case a fluid flow in a closed cylinder created by a rotating lid, we use a Galerkin approximation to project the infinite Navier-Stokes equations into a finite dimensional subspace in amplitudes. Then, utilizing bifurcation analysis on the system of ODE's, we succeeded in establishing the early transition to an oscillatory motion as a supercritical Hopf-bifurcation, and in particular we also estimated the critical Reynolds number within 0.2% of the Reynolds number due to the full numerical system in 40000 degrees of freedom. An excellent result with a low-dimensional model consisting of only 25 degrees of freedom. Finally, we present the spectrum of the full numerical system in the range from stationary to chaotic fluid flow. This spectrum diagram will serve as the basic reference system through out all investigations.

1. Introduction

The onset of turbulence has always and will always be a challenge to the curiosity of the experimental working scientist in the laboratory and to the most abstract thinker in mathematics. The phenomena fascinate us every day with their beauty and regularity. We know the determining equations for these phenomena, but we can't control them. They are indeed reproducible, so some underlying structure in some sense must exist.

Stokes stated in 1843 that under certain circumstances steady motion looses its stability and becomes sinuous-like. Later, *Reynolds (1883)* found, by observing laboratory experiments (streaming in pipes), that breakdown occurs at the same value of a certain system parameter, called the Reynolds number (Re). Later again *Landau (1944)* and *Hopf (1948)* predicted, encouraged by their work about a first stable periodic solution, that the transition to turbulence is connected with an infinite number of quasi-periodic branching of stable solutions, compared with the Feigenbaum scenario, this sounds reasonable.

M. Flato et al. (eds.), Physics on Manifolds, 321–330.

However, experiments have proved that turbulence occurs after a finite number of bifurcations, if any, and also the prediction contradicts measurements in the laboratory which demonstrate turbulent fluid flow to be phase-mixed. Phase-mixing means that the autocorrelation function,g(τ), equation (1) tends to zero as τ tends to infinity,

$$g(\tau) = \lim_{T\to\infty} \frac{1}{T}\int_0^T f(t+\tau)\,\bar{f}(t)\,dt \to 0 \quad \text{as} \quad \tau \to \infty. \tag{1}$$

As *Bass (1962)* showed, quasi-periodic solutions do not comply with this property. Moreover, *Lorenz (1963)* obtained, by introducing a trigonometric expansion into a heating of a surface problem, a system of ODE's in 3 dimensions that, after only a finite number of transitions, turns to an almost turbulentlike motion. *Ruelle & Takens (1971)* derived a model in 4 dimensions, where a phase-mixing solution (turbulence) appears after two quasi-periodic bifurcations. Today, *Sirovich & Sirovich (1989)* have used a Proper Orthogonal Decomposition technique (POD) to reduce complex phenomena by several orders to treatable, low-dimensional models and even to reproduce the dynamic perfectly well, as observed in the full numerical system.

In the given problem, the first comprehensive, experimental results were elaborated by *Escudier (1984)*. By changing the Reynolds number, Re, and the radius height ratio, λ, he found three vortex breakdowns; two stationary and one instationary in the domain up to oscillatory motion. *Daube & Sorensen (1989)* confirmed some of these results numerically, and increasing the Reynolds number even further, they discovered that transition to turbulence in the axisymmetric, numerical system is associated with various bifurcations, among others a Hopf-bifurcation, a hysteresis and a period doubling. (See also *Lopez & Perry (1992)*). Recently, *Christensen et al. (1992)* used the mentioned POD method to identify the early transition as a supercritical Hopf-bifurcation and *Sorensen & Christensen (1992)* presented a frequency diagram for axisymmetric, numerical solutions from stationary to turbulentlike motion.

One of the main difficulties is the prohibitively expensive computer costs. We give some examples from the work in only 2D. In 3D you have to multiply by at least a factor 50 because:

- close to critical points it takes hours for a supercomputer (we used an Amdahl VP1100) to integrate to a stable attractor (no transients),
- taking 60 snapshots are entailed with storing of 3E06 data,
- 73.2E06 multiplications have to be done for calculating the covariance matrix,
- a model of 30 dimensions consists of 28860 coefficients.

Hence, the necessity to improve a technique to model a rotating fluid flow in few dimensions.

2. The Model

The rotating fluid flow is created by letting the cover at one end of a closed cylinder with height, H, and radius, R, rotate with a constant angular velocity, Ω. With scaling time Ω^{-1} and length R, the following driven cavity problem in dimensionless form has to be considered, see figure 1.

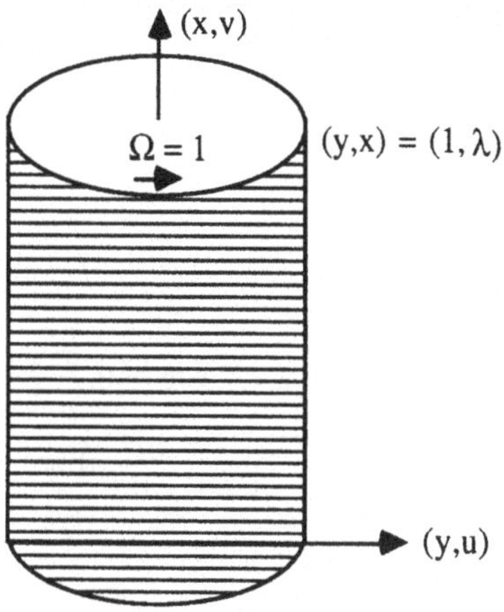

Figure 1. The driven cavity problem in dimensionless form with a rotating top.

The coordinate set (y,u) is the variable, the velocity component in the radial direction, and the coordinate set (x,v) the same but in the axial direction.

3. The Governing Equations

In the dimensionless form the constant angular velocity, Ω, and the radius, R, are contained in the Navier-Stokes equations by the Reynolds number, Re,

$$Re = \frac{R^2\Omega}{\nu}, \tag{2}$$

with ν as the kinematic viscosity. The height, H, and the radius, R, are contained in the boundary conditions by the aspect ratio, λ,

$$\lambda = H/R. \tag{3}$$

Only the Reynolds number, Re, is to be varied, the aspect ratio, λ, is fixed to 2.

Hence, in the circulation, Γ, vorticity, ω, and streamfunction, ψ, formulation the axisymmetric Navier-Stokes equations are given as, see *Daube & Sorensen (1989)*,

$$\frac{\partial\Gamma}{\partial t} = \frac{1}{Re}\left(\frac{\partial^2\Gamma}{\partial x^2} + \frac{\partial^2\Gamma}{\partial y^2} - \frac{1}{y}\frac{\partial\Gamma}{\partial y}\right) - \left(\frac{\partial(u\Gamma)}{\partial x} + \frac{\partial(v\Gamma)}{\partial y} - \frac{v\Gamma}{y}\right), \quad (4)$$

$$\frac{\partial\omega}{\partial t} = \frac{1}{Re}\left(\frac{\partial^2\omega}{\partial x^2} + \frac{\partial^2\omega}{\partial y^2} + \frac{1}{y}\frac{\partial\omega}{\partial y} - \frac{\omega}{y^2}\right) - \left(\frac{\partial(u\omega)}{\partial x} + \frac{\partial(v\omega)}{\partial y} - \frac{\partial}{\partial x}\left(\frac{\Gamma^2}{y^2}\right)\right), \quad (5)$$

$$\frac{\partial\psi}{\partial y} = -\ uy, \quad \frac{\partial\psi}{\partial x} = vy, \quad (6)$$

$$\frac{\partial^2\psi}{\partial x^2} + \frac{\partial^2\psi}{\partial y^2} - \frac{1}{y}\frac{\partial\psi}{\partial y} = \omega y. \quad (7)$$

Boundary conditions:

Symmetry axis; $y = 0, 0 \le x \le \lambda$: $\psi = v = \Gamma = \omega = 0, \dfrac{\partial^2\psi}{\partial y^2} = -\ u.$ (8a)

Cylinder wall; $y = 1, 0 \le x \le \lambda$: $\psi = u = v = \Gamma = 0, \quad \dfrac{\partial^2\psi}{\partial y^2} = \omega y.$ (8b)

Rotating endwall; $x = 0, 0 \le y \le 1$: $\psi = u = v = 0, \Gamma = y^2, \quad \dfrac{\partial^2\psi}{\partial x^2} = \omega y\ .$ (8c)

Fixed endwall; $x = \lambda, 0 \le y \le 1$: $\psi = u = v = \Gamma = 0, \quad \dfrac{\partial^2\psi}{\partial x^2} = \omega y\ .$ (8d)

4. The Dynamical System

Analysing time series of the stable, no transients, numerical solutions, we found the following dependences between basic frequencies and the Reynolds number, Re. Note that all of the harmonics have been removed (see figure 2).

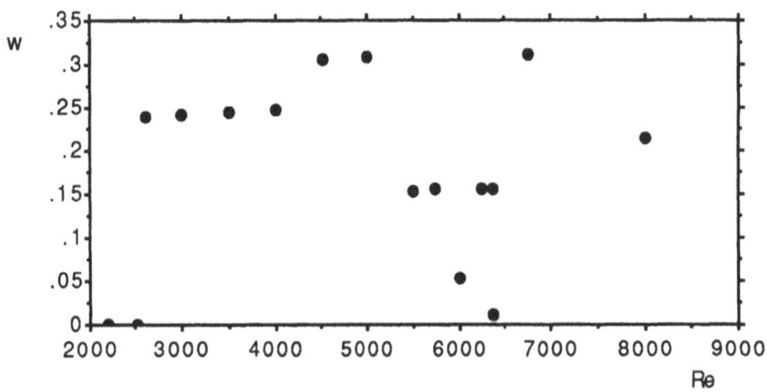

Figure 2. Spectrum diagram with basic frequencies.

The frequencies are unlocked, i.e. no driving force. Thus a frequency changes continuously between transitions (see the interval $2600<Re<4000$, figure 2). A transition is seen as a frequency jump, Re = 4250, or a frequency production, Re = 6300, (figure 2). The frequency just before and just after a transition is denoted by ω^- and ω^+, respectively. A solution to the system (4-8) will simply be written in the form,

$$x = x(Re, \omega_1 t, \dots, \omega_n t, x_0), \tag{9}$$

where the frequencies ω_i, $i = 1,\dots,n$, are nonrational related, numbered increasingly, and where x_0 is added, if more stable solutions have been discovered or are to be expected for the same Reynolds number, Re, as for example in hysteresis.
Next we present a first description of the transition from stationary to chaotic fluid flow, restricted to be axisymmetric because of the numerical code.

Stationary solutions: $0<Re<2544$
Stationary solutions, time independent,

$$x = x(Re). \tag{9a}$$

Due to experimental investigations done by *Escudier (1984)* and due to investigations of full numerical solutions by *Daube & Sorensen (1989)* the system undergoes two stationary bifurcations or vortex breakdowns at Re = 1500 and Re = 1750. Axisymmetric fluid flow is expected between Re = 0 and Re = 2544.

Stationary to periodic transition: Re = 2544
A Hopf-bifurcation, supercritical or subcritical with hysteresis, is generally expected, *Joseph (1981)*. *Christensen et al. (1992)* identified the transition to be a supercritical Hopf-bifurcation at the critical Reynolds number Re = 2544.

Periodic solutions: $2544<Re<6300$
Periodic solutions, one frequency-motion,

$$x = x(Re, \omega t, x_0). \tag{9b}$$

- Re: 4000 - 4500 Hysteresis, ω^-, ω^+ nondependent.
- Re: 5000 - 5500 $\omega^+ = 1/2\ \omega^-$, 2T bifurcation, periode doubling.
- Re: 5750 - 6000 $\omega^+ = 1/3\ \omega^-$, 3T bifurcation, periode tripling.
- Re: 6000 - 6250 $\omega^+ = 3\ \omega^-$, 1/3T bifurcation, periode back-tripling.

Periodic to quasi-periodic transition: Re = 6300

A Hopf-bifurcation of the periodic solution, seen as spiralization on the Poincare plot to a boundaring 2-torus. ω_1^+ about 1/15 ω^-, $\omega_2^+ = \omega^-$.

Quasi-periodic solutions: 6300 < Re < 6750

Quasi-periodic solutions with two nonrational related frequencies ω_1, ω_2

$$x = x(Re,\omega_1 t,\omega_2 t,x_0). \tag{9c}$$

We discovered at least one torus-to-torus bifurcation in this interval. No third basic frequency has, so far, been observed.

Quasi-periodic to periodic transition: Re = 6750

ω_1^- is unknown, ω^+ is expected to be equal to ω_2^-.

Periodic to chaotic to periodic transition: 6750 < Re < 8000

Periodic, $\omega(6750)$, to broad-banded, $\omega(I)$, to periodic, $\omega(8000)$, transition.

$\omega(6750)$, $\omega(8000)$, nondependent, are almost equal to the frequencies just after and just before the hysteresis, respectively.

The time for converging to a stable solution was, as expected, strictly dependent on the distance to a critical value.

5. The Projection Method

We use a simple spatial discretization of second order to obtain a first formulation in ODE's. The discretization used by *Sorensen & Ta Phuoc Loc (1989)* is not available here because it contains implicit ADI-steps. We rewrite equations (4-5) in the vectorform,

$$\dot{\Gamma} = F_1(Re,\Gamma,\omega,u,v,\psi), \tag{10a}$$

$$\dot{\omega} = F_2(Re,\Gamma,\omega,u,v,\psi). \tag{10b}$$

Denoting X as (Γ,ω), utilizing the linearity between (Γ,ω) and u,v,ψ through a discretization of the definition equations (6) and the Poisson equation (7) we simply have

$$\dot{X} = F(Re,X) \ . \tag{11}$$

X is a vector of 40000 coordinates, equal to two times the number of nodes in the numerical grid.

To introduce a Galerkin approximation into the system of ODE's defined by (11), we need an orthogonal set of basevectors in spatial coordinates. Eigenfunctions of the linearized, adjungated eigenvalue problem are not available here, neither are eigenvectors of equation (11) (the system is too large), so we use the Proper Orthogonal Decomposition technique as suggested by *Sirovich & Sirovich (1989)*. The technique has been generalized in order to operate with known solutions in the bifurcation theory. Assume for example n_1 stationary solutions $S_1,\ldots,S_{n_1}$ and n_2 snapshots $X_1,\ldots,X_{n_2}$ of the full numerical system. A base system for the stationary solutions is then defined by,

$$\varphi_{S_i} = \sum_{j=1}^{n_1} V_{S_i}{}^{j}\,\hat{S}_j, \quad \langle\varphi_{S_i},\varphi_{S_j}\rangle = \lambda_{S_i}\,\delta_{ij}, \qquad i = 1,\ldots,n_1, \tag{12}$$

$$K_S = A_S{}^{T} A_S, \quad A_S = \{\hat{S}_1,\ldots,\hat{S}_{n_1}\}, \quad \hat{S}_i = S_i - X_0, \quad X_0 = \frac{1}{n_2}\sum_{j=1}^{n_2} X_j, \tag{13}$$

where $\{V_{S_i}, \lambda_{S_i}\}$ are the eigenpairs, organized according to decreasing eigenvalue, of the covariance matrix K_S. K_S is symmetric and positive definite.
Next we project the residuals of the snapshots into the space orthogonal to $\{\varphi_{S_i}\}$,

$$\hat{X}_i = (X_i - X_0) - \sum_{j=1}^{n_1} \langle (X_i - X_0), \varphi_{S_j}\rangle \varphi_{S_j} / \lambda_{S_j}, \quad i = 1,\ldots,n_2, \tag{14}$$

$$\hat{X}_0 = X_0 - \sum_{j=1}^{n_1} \langle X_0, \varphi_{S_j}\rangle \varphi_{S_j} / \lambda_{S_j}, \tag{14a}$$

and finally doing similar operations as in equations (12) - (13), we find

$$K = A^T A, \quad A = \{\hat{X}_1,\ldots,\hat{X}_{n_2}\}, \tag{15}$$

$$\varphi_i = \sum_{j=1}^{n_2} V_i{}^{j}\,\hat{X}_j, \quad \langle\varphi_i,\varphi_j\rangle = \lambda_i\,\delta_{ij}, \quad i = 1,\ldots,n_2. \tag{16}$$

As above the set $\{V_i, \lambda_i\}$ denotes the eigenpairs of K organized according to decreasing eigenvalue. Notice that any of the stationary solutions S_i, $i = 1,\ldots,n_1$, and $\hat{X}_0$ belong to the system (X_0; φ_{S_i}, $i = 1,\ldots,n_1$) due to the construction. Then defining for example $X_S = S_1$ ($Re = Re_S$), we simply write the m'th most describing Galerkin approximation as,

$$X(t) = X_S + \sum_{i=-(n_1-1)}^{m} a_i(t)\,\varphi_i, \tag{17}$$

where $\varphi_{1-i} = \varphi_{S_i}$, $\lambda_{1-i} = \lambda_{S_i}$, $i = 1,\ldots,n_1$, $0 \le n_1$, $m \le n_2$ and $a = 0$ for $Re = Re_S$.

Substituting (17) into (11) and dotting with the orthogonal basevectors, φ_i, i = -(n_1-1),...,m, the equations of the reduced system become,

$$\dot{a}_i(t) = \left\langle F(Re, X_S + \sum_{k=-(n_1-1)}^{m} a_k(t)\ \varphi_k), \varphi_i \right\rangle / \lambda_i, \qquad i = -(n_1-1),\ldots,m, \tag{18}$$

or collected in the vectorfunction, a, finally,

$$\dot{a} = b_0 + \frac{1}{Re} b_1 + A_0\, a + \frac{1}{Re} A_1\, a + B(a,a), \tag{19}$$

$$a(Re = Re_S) = 0, \quad \text{stationary solution,} \tag{19a}$$

where b_0, b_1 are constant vectors, A_0, A_1 are linear vectorfunctions and B is a bilinear vectorfunction. The Galerkin approximation describes, by the ratio,

$$\sum_{i=1}^{m} \lambda_i / \sum_{i=1}^{n_2} \lambda_i, \tag{20}$$

the part of the total variation of the n_2 snapshots $X_1,\ldots,X_{n_2}$ orthogonal to $\varphi_{-(n_1-1)},\ldots,\varphi_0$, due to the subspace spanned by $\varphi_1,\ldots,\varphi_m$. Next, (see figure 3), we present the determination of the critical Reynolds number for different series of snapshots and dependent on the dimension m, *Christensen et al. (1992).*

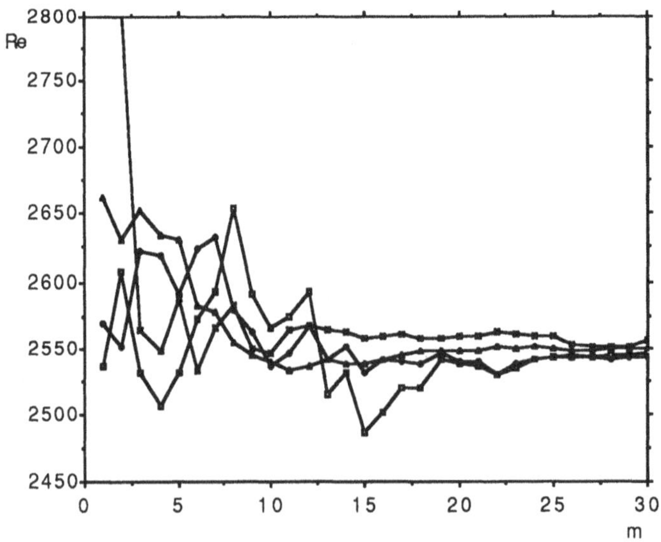

Figure 3. Determination of critical Reynolds number by different models.

6. Conclusion

It is well known that the axisymmetric boundary conditions do not guarantee axisymmetric streaming. Indeed, for a certain Reynolds number it will undergo a symmetribreaking bifurcation and becomes three dimensional. Possibly even after the first Hopf-bifurcation. This, however, will not deter us from continuing our research. Any results will help us in making the theory be comprehensive for three dimensional fluid flow problems.

With the present projection method it has been possible to reduce dramatically the dimension of the non-trivial driven cavity problem, and to identify the first transition from a stationary to an oscillatory motion. This is done as a supercritical Hopf-bifurcation at a critical Reynolds number and at a time period, which is in excellent agreement compared to the full numerical system (see *Christensen et al. (1992)*). The projection method is by no means confined to a local description of a simple phenomenon in dynamical sense. In fact, *Sirovich & Rodriguez (1987)* were able to follow the Ginzburg-Landau equation through a chaotic region with only six modes. However, considering the complexity of the given problem and the present results, one must expect a need for a somewhat larger number of modes. A projection method to describe the next transitions is therefore in preparation.

Acknowledgement

One of the authors (EAC) acknowledges financial support from the Danish Research Academy (Grant S910171), the Danish Technical Research Council (Grant 16-4967-2 OS), the Louis Dreyer Myhrwold's Fund and the Fisker & Nielsens Fund.

References

Aubry,N. (1991), On the hidden beauty of the Proper orthogonal Decomposition, Theoret.Comput. Fluid Dynamics, **2**, 339-352.

Bass,J. (1962), Les Fonctions Pseudo-Aleatoires, (Memorial Des Sciences Mathematiques), Gauthier-Villars, Paris.

Brons,M. (1990), A model for compressor flow instability with mode shifting, Mathl.Comput.Modelling, **14**, 746-749.

Christensen,E.A., Sorensen,J.N., Brons,M. and Christiansen,P.L. (1992), Low Dimensional Representations of Early Transition in Rotating Fluid Flow, (In preparation).

Daube,O. and Sorensen,J.N. (1989), Simulation numerique de l'ecoulement periodique axisymetrique dans une cavite cylindrique, C.R. Aca. Sci. Paris, 308, **2**, 443-469.

Escudier,M.P. (1984), Observations of the flow produced in a cylindrical container by a rotating endwall, Experiments in Fluids, **2**, 189-196.

Hassard,B.D., Kazarinoff,N.D., and Wan,Y-H. (1981), Theory and Applications of Hopf Bifurcation, Cambridge University Press, Cambridge.

Hopf,E. (1948), A mathematical example displaying features of turbulence, Commun. Pure Appl. Math., **1**, 303.

Joseph,D.D. (1981), Hydrodynamic stability and bifurcation, in Hydrodynamic Instabilities and the Transition to Turbulence, (Swinney,H.L. and Gollub,J.P. eds.), Springer-Verlag, TAP, **45**, 27-96.

Landau,L.D. (1944), On the problem of turbulence, C.R. Acad. Sci. USSR, **44**, 311.

Lopez,J.M. and Perry,A.D. (1992), Axisymmetric vortex breakdown. Part 3. Onset of periodic flow and chaotic advection, J. of Fluid Mech., **234**, 449-471.

Lorenz,E. (1963), Deterministic nonperiodic flow, J. Atmos. Sci., **20**, 130.

McCaughan,F.E. (1990), Bifurcation analysis of axial flow compressor stability, SIAM J.Appl.Math., **50**, 1232-1253.

Petersen,C.K. (1988), PATH - User's Guide, Dept. of Applied Math. and Nonlinear Studies, University of Leeds.

Reynolds,O. (1883), An experimental investigation of the circumstances which determine whether the motion of water shall be direct or sinuous, and of the law of resistance in parallel channels, Phil. Trans. R. Soc., **174**, 935.

Ruelle,D. and Takens,F. (1971), On the nature of turbulence, Commun. Math. Phys., **20**,167.

Sirovich,L. and Rodrigues,J. (1987), Coherent structeres and chaos: a model problem, Physics Letters A 120.

Sirovich,L. and Sirovich,C.H. (1989), Low Dimensional Description of Complicated Phenomena, in Nicolaenco,B. Foias,C. and Temam,R. (eds.). The Connection between Infinite Dimensional and Finite Dimensional Dynamical Systems, Comtemporary Mathematics, **99**, 277-305.

Sorensen,J.N. and Christensen,E.A. (1992), Numerical study of transition of rotating fluid in closed cylinder, (In preparation).

Sorensen,J.N. and Daube,O. (1989), Direct simulation of flow structures initiated by a rotating cover in a cylindrical vessel, Springer-Verlag, Adv. in Turb., 383-390.

Sorensen,J.N. and Ta Phuoc Loc (1989), High-Order Axisymmetric Navier-Stokes Code Description and Evaluation of Boundary Conditions, Int. J. for Num. Meth. in Fluids, **9**, 1517-1537.

SOME GEOMETRICAL ASPECTS OF INHOMOGENEOUS ELASTICITY

Marcelo EPSTEIN* and Gérard A. MAUGIN**

* University of Calgary, Department of Mechanical Engineering
2500 University Drive N.W., Calgary, Alberta, Canada T2N 1N4

**Université Pierre et Marie Curie, Laboratoire de Modélisation en Mécanique,
CNRS-URA 229, Tour 66, 4 place Jussieu, Boite 162
75252 Paris Cédex 05, France.

1. INTRODUCTION

The balance of *pseudomomentum* (canonical momentum) is a fully *material* equation [1]-[2] (i.e., with components on the abstract material manifold $\mathcal{M}^3$ and *not* in physical space $\mathbb{E}^3$) which allows one to compute directly either *inhomogeneity forces* (also called configurational forces) which are the main ingredients in devising fracture criteria in engineering [3] or, in another context, perturbing forces on soliton solutions in nonlinear elastic systems with dispersion [4]. Here we raise the question as to whether this balance law can be rewritten in a completely *covariant* manner on the material manifold. The answer (geometrical structure of $\mathcal{M}^3$) depends on the degree of fineness of the elastic description and the degree of singularity ascribed to the corresponding elastic field, i.e., the type of *continuous* distribution of structural defects admitted by the material. We illustrate this for nonlinear anisotropic *elastostatics* via the cases of *simple* and *second-grade* elastic materials which are likely to admit (translational) *dislocations* and/or *disclinations*, respectively, as elastic singularities. The results are obtained by studying the transformation properties of the elastic response on the material manifold (on the first and second tangent spaces). This follows previous works [5]-[8].

2. SIMPLE INHOMOGENEOUS ELASTIC MATERIALS

For a simple inhomogeneous anisotropic nonlinear elastic body B which is uniform in the sense of Noll [9] (i.e., made of identical material points $X \in \mathcal{M}^3$) in a global reference configuration $\hat{K}$, the deformation is described by the mapping $\mathbf{x} = \mathcal{X}(\mathbf{X})$ and we have the following kinematic, equilibrium (in the absence of extended forces) and constitutive equations (T=transpose)

$$\mathbb{F} = \frac{\partial \mathcal{X}}{\partial \mathbf{X}} \equiv \nabla_R \mathcal{X} \ , \ J_F = \det \mathbb{F} > 0 \ , \ \mathbf{C} := \mathbb{F}^T \mathbb{F} \ , \tag{2.1}$$

$$div_R \ \mathbf{T} = \mathbf{0} \ . \tag{2.2}$$

This contribution was written while the second author, who delivered the lecture at the Paris meeting, was a fellow of the *Institute for Advanced Study* of Berlin (Wissenschaftskolleg zu Berlin) for the year 1991/1992.

M. Flato et al. (eds.), Physics on Manifolds, 331–335.

$$\mathbf{T} = \partial\hat{W}/\partial\mathbb{F} \, , \, W = \hat{W}(\mathbb{F},\mathbf{X}) \, , \tag{2.3}$$

where $\mathbb{F}$ is the deformation gradient, $\mathbb{T}$ is the first Piola-Kirchoff stress and $\hat{W}$ is the elastic energy per unit volume of $\hat{K}$; the explicit dependence of $\hat{W}$ on $\mathbf{X}$ indicates the presence of *material inhomogeneities*. Following [5]-[6], let $\mathbf{K}^{-1}$ be a map which brings a small neighborhood of X into a stress-free state. Then the question of material homogeneity boils down to whether the $\mathbb{K}$ can be integrated in a *global* configuration ? Let $J_K = det\ \mathbb{K}$. The idea [5]-[6] is to rewrite $\hat{W}(\mathbb{F},\mathbf{X})$ as

$$\hat{W}(\mathbb{F},\mathbf{X}) = J_K^{-1}\check{W}(\mathbb{F}\mathbb{K}(\mathbf{X})) = \bar{W}(\mathbb{F},\mathbb{K},\mathbf{X}) \, . \tag{2.4}$$

On account of (2.2) we can then show that the following holds true :

$$div_R\mathbf{b} + \mathbf{f}^{inh} = \mathbf{0} \, , \, \mathbf{b}\, \mathbb{C} = \mathbb{C}\, \mathbf{b}^T \, , \tag{2.5}$$

wherein

$$\mathbf{b} = -\frac{\partial\bar{W}}{\partial\mathbb{K}}\mathbb{K}^T = \hat{W}\,\mathbb{1}_R - \mathbb{F}^T\mathbb{T} \, , \, \mathbf{f}^{inh} = -(\partial\hat{W}/\partial\mathbf{X})_{expl} \tag{2.6}$$

are the so-called *Eshelby*(material) stress tensor and the inhomogeneity force. The suffix notation "expℓ" indicates the explicit partial derivative. Clearly, an equation such as $(2.5)_1$ can be generated in a variational formulation considering variations $\delta_x\mathbf{X}$ (i.e., at fixed actual position $\mathbf{x}$), or else, it is easily proved that $(2.5)_1$ is none other than eqn. (2.2) multiplied to the left by $\mathbb{F}^T$ (the transformation uses simple kinematic identities and expressions (2.3)). However, the formulation invoking $\mathbb{K}$-maps is more fruitful for our present purpose. In particular, for a *uniform* body $(\partial\bar{W}/\partial\mathbf{X})_{expl} \equiv \mathbf{0}$ and eqn. $(2.5)_1$ delivers the equation

$$div_R\mathbf{b} + (\mathbf{b}\mathbb{K}^{-1})^T.\nabla_R\mathbb{K} = \mathbf{0} \; or \; b_{I.,J}^{.J} + b_J^{.K}(\mathbb{K}^{-1})^{\alpha}_{.K}\, K^{.J}_{.\alpha,I} = 0 \, . \tag{2.7}$$

Can we rewrite this using a *covariant derivative* based on $\mathbb{K}$? Indeed, $\mathbb{K}$ creates a *distant parallelism* and, in fact, provides a moving crystallographic frame $\mathbf{E}_\alpha = K^I_{.\alpha}\, \partial/\partial X^I$ and a (generally) *nonmetric* connection $\mathbf{\Gamma}$ and a *torsion* $\mathcal{T}$ such that [10] $(I,J,K = 1,2,3)$

$$\Gamma^I_{KJ} \equiv (\mathbb{K}^{-1})^{\alpha}_{.K,J}\, K^{.I}_{.\alpha} \, , \, \mathcal{T}^I_{JK} = 2\,\Gamma^I_{[JK]} \, . \tag{2.8}$$

On setting $\tilde{\mathbf{b}} = J_K\mathbf{b}$, we immediately check that $\tilde{\mathbf{b}}$ and $\mathcal{T}$ jointly satisfy the *covariant* equation [5]-[6]

$$\overset{\Gamma}{\nabla}\tilde{\mathbf{b}} = f(\tilde{\mathbf{b}},\mathcal{T}) \; or \; \tilde{b}^{.J}_{I..J} = \tilde{b}^{.J}_K\mathcal{T}^K_{JI} + \tilde{b}^{.J}_I\mathcal{T}^K_{JK} \, . \tag{2.9}$$

This answers the question raised for *simple* elastic materials. Furthermore, the *material torsion* $\mathcal{T}$ may be directly associated with a continuous distribution of (translational) *dislocations* by

$$\mathcal{T}^I_{JK} = 2\,\varepsilon_{JKM}\alpha^{IM}\,, \tag{2.10}$$

where $\boldsymbol{\varepsilon}$ is the alternator and $\boldsymbol{\alpha}$ is the *dislocation density* tensor usually defined from the Burgers vector (see classic works by Bilby, Kondo, Kröner [11]).

3. SECOND-GRADE INHOMOGENEOUS ELASTIC MATERIALS

In this case the elastic energy depends also on $\mathbb{B} \equiv \nabla_R \mathbb{F}$, and eqns. (2.2) and (2.3) are replaced by [2]

$$div_R \mathbf{T}^{eff} = \mathbf{0}\,,\ \mathbf{T}^{eff} = \mathbf{T} - div_R \mathbb{M}\,, \tag{3.1}$$

$$\mathbf{T} = \partial\hat{W}/\partial\mathbb{F}\,,\ \mathbb{M} = \partial\hat{W}/\partial\mathbb{B}\,,\ W = \hat{W}(\mathbb{F},\mathbb{B},\mathbf{X})\,, \tag{3.2}$$

where $\mathbb{M}$ is the *hyperstress* [12]. The same as above, via Noether's theorem it is possible to show that there is associated with $(3.1)_1$ a conservation law (for the parameters X^K) given by [2]

$$div_R \mathbf{b}^{eff} + \mathbf{f}^{inh} = \mathbf{0}\,, \tag{3.3}$$

wherein

$$\left.\begin{array}{llll} \mathbf{b}^{eff} \equiv \mathbf{b} - div_R \mathcal{B} & , & \mathbf{f}^{inh} \equiv -(\partial\hat{W}/\partial\mathbf{X})_{expl}\,, \\ \mathbf{b} = \hat{W}\mathbb{1}_R - \mathbb{F}^T\mathbf{T} - 2\mathbb{B}^T : \mathbb{M} & . & \mathcal{B} = \mathbb{F}^T\mathbb{M}\,. \end{array}\right\} \tag{3.4}$$

But, again, considering directly changes $\mathbf{Y} = \mathbf{Y}(\mathbf{X})$ on $\mathcal{M}^3$ is more rewarding for our purpose. Therefore, generalize (2.4) we write

$$\begin{aligned} \hat{W}(\mathbb{F},\mathbb{B},X) &= J_K^{-1}\bar{W}(\mathbb{F}\mathbb{K}, (\mathbb{B}\mathbb{K})^T + \mathbb{F}\mathbf{Q}) \\ &= \vec{W}(\mathbb{F}, \mathbb{B}, \mathbb{K}, \mathbf{Q}, \mathbf{X}) \end{aligned} \tag{3.5}$$

where $\mathbb{K} \in \mathcal{K}$ and $\mathbf{Q} \in \mathcal{Q}$ for a symmetry group $\mathcal{H} = \{\mathcal{K}, \mathcal{Q}\}$ on $\mathcal{M}^3$. Here $\mathbf{Q}$ is a map on the second tangent space at X. It was proved [8] that $\mathcal{H}$ is indeed a *group* with the composition law (with an obvious notation)

$$\begin{aligned} \{\mathbb{K}_3, \mathbf{Q}_3\} &= \{\mathbb{K}_1, \mathbf{Q}_1\} \circ \{\mathbb{K}_2, \mathbf{Q}_2\} \\ &= \left\{\mathbb{K}_1\mathbb{K}_2, (\mathbf{Q}_1\mathbb{K}_2)^T.\mathbb{K}_2 + \mathbb{K}_1\mathbf{Q}_2\right\}\,. \end{aligned} \tag{3.6}$$

This must be compared to the second of (3.5). Then

$$\mathbf{b} = -\left(\frac{\partial\bar{W}}{\partial\mathbb{K}}\mathbb{K}^T - \frac{\partial\bar{W}}{\partial\mathbf{Q}} : \mathbf{Q}^T\right),\ \mathcal{B} = -\frac{\partial\bar{W}}{\partial\mathbf{Q}} : \mathbb{K}\otimes\mathbb{K} \tag{3.7}$$

and (2.3) is shown to hold identically for $(\partial \bar{W}/\partial \mathbf{X})_{expl} = 0$. The same as in the foregoing section, by computing $(\partial \hat{W}/\partial \mathbf{X})_{expl}$ through the dependence of $\bar{W}$ upon $\mathbb{K}$ and $\mathbf{Q}$ for a *uniform* material and using the expressions of $\mathbf{b}$ and $\mathcal{B}$ we show that eqn. (3.3) takes on the form of the following fundamental identity :

$$div_R \mathbf{b}^{eff} + \left\{(\mathbf{b} - \mathcal{B}.\bar{\mathbf{Q}}).\mathbb{K}^{-1}\right\}^T.\nabla_R \mathbb{K} + (\nabla_R \mathbf{Q}).\mathcal{B} = 0 \tag{3.8}$$

where $\bar{\mathbf{Q}} = \mathbf{Q} : (\mathbb{K}^{-1} \otimes \mathbb{K}^{-1})$.

4. GEOMETRIZATION

The purpose of the "geometrization" of eqn. (3.8) is the more concise rewriting of this equation using covariant derivatives with respect to a *connection with curvature* and *torsion* which would reflect the possible presence of (translational) *dislocations* (through the torsion) and *disclinations* -dislocations of the rotationl type- (through the curvature). This geometrization is of the same technical and conceptual difficulty as the one met in general relativistic gravitational theories including spin effects [13] and obviously is much more tedious than the one proposed in eqn. (2.9) for *simple* elastic materials. Ideally, we expect that (3.8) be rewritten in the form

$$\left(div_\Gamma \bar{\mathbf{b}}^{eff}\right)_I = \mathcal{T}^{K}_{JI} \bar{b}^{.J}_{K} + \mathcal{T}^{P}_{PJ} \bar{b}^{.J}_{I} + \bar{\mathcal{B}}^{.KL}_{J} \mathcal{R}^{..K}_{IKL} , \tag{4.1}$$

where $\bar{\mathbf{b}}^{eff}$, $\bar{\mathbf{b}}$ and $\bar{\mathcal{B}}$ are weighed tensors and *both* material torsion $\mathcal{T}$ and *curvature* $\mathcal{R}$ are based on the *asymmetric* Γ *connection* defined in components by (cf. Eisenhart [14], eqn. (18.3))

$$\Gamma^{I}_{KJ} = -(\mathbb{K}^{-1})^{\alpha}_{.K} K^{I}_{.\alpha,J} + Q^{I}_{.\alpha\beta}(\mathbb{K}^{-1})^{\alpha}_{.K}(\mathbb{K}^{-1})^{\beta}_{.J} , \tag{4.2}$$

from which the Γ−covariant derivatives of a covariant and contravariant material vectors are defined by, e.g.,

$$\left.\begin{aligned} \left(\overset{\Gamma}{\nabla} \mathbf{W}\right)_{K;L} &= W_{K,L} - \Gamma^{I}_{KL} W_I , \\ 2\overset{\Gamma}{\nabla}_{[J} \overset{\Gamma}{\nabla}_{M]} U^K &= \mathcal{R}^{..K}_{IJM} U^I - \mathcal{T}^{P}_{JM} \overset{\Gamma}{\nabla}_P U^K . \end{aligned}\right\} \tag{4.3}$$

There exists thus a rather strong *similarity* between the Einstein-Cartan theory of gravitation with spin (and hyperstress) and the geometrical identity presumably satisfied by Eshelby's tensors of stress ($\mathbf{b}$) and hyperstress ($\mathcal{B}$) on a material manifold $\mathcal{M}^3$ endowed with dislocations of a general type. For the time being, however, there appears to be *no* unique solution to the material covariant writing (4.1). Three remarks are in order. First, the above geometrization seems to be realizable only in the framework of a *parth-dependent parallelism* on the material manifold (hence, non-Euclidean connections). Second, *disclinations* or "rotation dislocations" enter the picture. The later are characterized by the *rotation* of a trihedron attached at each point of the dislocation line [15]. The geometric characterization of continuous distributions of disclinations by the *curvature* of the material manifold was advocated by Anthony [16]. However, the present approach to defective materials also suggests the association of hyperstresses (in particular *couple* stresses [12]) with curvature, and thus disclinations. Finally, we notice that the last term in (4.1) is exactly of the form of the so-called *Mathisson "force"* in gravitation theory [17]. Here it is a material force which accounts for the *disclination density* in a continuum elastic description. We can only emphasize this inspiring similarity between gravitation-spin theory and inhomogeneous elasticity, an excellent opportunity to pay a tribute to Yvonne Choquet-Bruhat.

REFERENCES

[1] G.A. Maugin, *C. R. Acad. Sci. Paris*, II-311, 763-768 (1990).

[2] G.A. Maugin and C. Trimarco, *C. R. Acad. Sci. Paris*, II-313, 851-856 (1991) ; *Acta Mechanica*, 94, 1-28 (1992).

[3] G.A. Maugin, *The Thermomechanics of Plasticity and Fracture*, Cambridge University Press, U.K. (1992), Chapter 7.

[4] G.A. Maugin, Pseudomomentum in solitonic Elastic System, *J. Mech. Phys. Solids*, (in the press, 1992).

[5] M. Epstein and G.A. Maugin, *C. R. Acad. Sci. Paris*, II-310, 675-678 (1990).

[6] M. Epstein and G.A. Maugin, *Acta Mechanica*, 83, 127-133 (1990).

[7] M. Elzanowski and M. Epstein, *J. Elasticity*, (1991).

[8] M. Elzanowski and M. Epstein, On the Symmetry Group of Second-grade Materials, *Preprint*, Calgary (1990).

[9] W. Noll, *Arch. Rat. Mech. Anal.*, 27, 1-32 (1967).

[10] Y. Choquet-Bruhat, *Géométrie différentielle et systèmes extérieurs*, Dunod, Paris (1968).

[11] F.R.N. Nabarro, *Theory of Crystal Dislocations*, Clarendon Press, Oxford (1967).

[12] R.A. Toupin, *Arch. rat. Mech. Anal.*, 17, 85-112 (1964)

[13] F.W. Hehl, P. von der Heyde, G.D. Kerlich and J.M. Nester, *Rev. Mod. Phys.*, (1976).

[14] L.P. Eisenhart, *Non-Riemannian Geometry*, American Mathematical Society, New York (1928) - Sixth reprint (1968).

[15] J. Friedel, in : *Dislocations in Solids*, ed. F.R.N. Nabarro, vol. I, pp. 3-32, North-Holland, Amsterdam (1979).

[16] K.H. Anthony, *Arch. Rat. Mech. anal.*, 39, 43-88 (1970).

[17] M. Mathisson, *Acta Phys. Polon.*, 6, 163-200 (1937).

INTEGRATING THE KADOMTSEV-PETVIASHVILI EQUATION IN THE 1+3 DIMENSIONS VIA THE GENERALISED MONGE-AMPÈRE EQUATION: AN EXAMPLE OF CONDITIONED PAINLEVÉ TEST *

Tommaso BRUGARINO and *Antonio* M. GRECO

Dipartimento di Matematica ed Applicazioni Università di Palermo

Via Archirafi 34, 90123 PALERMO, *Italy*

Summary

We consider the Kadomtsev-Petviashvili equation in the 1 + 3 dimensions (1+3D KP):

$$[u_t + uu_x + u_{xxx}]_x + au_{yy} + 2bu_{yz} + cu_{zz} = 0,$$

where a, b and c are supposed to depend on t, x, y and z. It can be seen as the simplest not trivial generalization of the Kadomtsev-Petviashvili equation to the 1+3 dimensional case. We prove, in contrast with the 1+2D case, that the Painlevé test is not satisfied, no matter how the coefficients are chosen, due to the fact that the highest order resonance is not identically verified. This suggest that the 1+3D KP probably fails the complete integrability property. On the other hand, by imposing the aforesaid resonance it is the same as requiring the function giving the singularity manifold to solve a generalized Monge-Ampère equation. We are in the case of the so called *conditioned Painlevé test*. The 1+3D KP enjoys the *conditional integrability property*, and only a subclass of its solutions can be obtained.

* Work supported by the G.N.F.M. of C.N.R. and by M.U.R.S.T. (40% e 60%) of Italy.

M. Flato et al. (eds.), Physics on Manifolds, 337–345.

1. Introduction

In a recent paper [1] we considered the Kadomtsev-Petviashvili (KP) equation, which is concerned with dispersive non linear propagation in 1+2 dimensions and rules the weak transversal perturbations of the progressive waves arising as solutions of the Kortweg-de Vries equation. We accounted any possible dependence of the phenomenological coefficients on the independent variables and we found that the complete integrability property imposes some conditions to the coefficients, which permitted us to reduce the equation to the canonical form, i.e.: $[u_t + u\ u_x + u_{xxx}]_x + u_{yy} = 0$.

The purpose of this paper is to consider the simplest possible extension of the KP equation to the case of three space dimensions, motivated by the fact that the transversality of the perturbations appears to be a very restrictive requirement from a physical point of view and also by a direct mathematical interest. This can be achieved, by introducing a third space variable and supposing that the perturbation amplitude satisfies the following equation:

$$[u_t + u\ u_x + u_{xxx}]_x + au_{yy} + 2bu_{yz} + cu_{zz} = 0 \tag{1}$$

We prove, in contrast with the 1+2D case, that the Painlevé test is not verified, due to the fact that the highest order resonance is not identically satisfied.

This suggest that the 1+3D KP probably fails the complete integrability property, according to the fact that the solitonic behaviour probably is not permitted in the genuine three-dimensional spatial case and to the fact that the complete integrability property for non linear evolution equations seems to be strictly connected with the possibility of *supermposing solitonic solutions* to obtain solutions corresponding to *suitably arbitrary* given Cauchy data.

On the other hand, by imposing the aforesaid highest resonance it is the same as requiring that the function giving the singularity manifold solves a generalized Monge-Ampère equation. We are in the case of the so called *conditioned Painlevé test* [2,3] and the 1+3D KP owns the *conditional integrability property.*

In Sec.2 we perform the Painlevé analysis on the Eq. (1), and we see that the conditioned Painlevé test can be passed only if the coefficients a, b and c are constants. After this we show that the only possibility to fulfil identically all the resonances, without reducing the equation to the $1+2$D case, requires that the expansion function solves a generalized (three-dimensional) Monge-Ampère equation. Two possible choices for the coefficients appear to be actually significative, in fact the only two, apart trivial transformations of independent variables absorbing the constant coefficients.

Both choices will be discussed in the Sec.3, and, in both the cases, we will exhibit some explicit solution.

2. The equation and the conditioned Painlevé property

As said in introduction the aims of this paper is to test the integrability of the KP equation in the case of three space dimensions. So that, we start by considering the equation

$$(u_t + uu_x + u_{xxx})_x + au_{yy} + 2bu_{yz} + cu_{zz} \tag{1}$$

where up to now the coefficients a, b, and c are supposed to depend by the all independent variables, and where, to avoid inessential complications we considered the term involving the derivitaves with respect to x and t in its canonical form [1].

First of all, we inspect the equation (1), by using the Painlevé analysis [4], having in mind to verify if for a suitable choose of the coefficients a, b and c, the equation passes the Painlevé test.

Therefore, we look for a solution given by the following expansion

$$u = \phi^{-n} \sum_{j=0}^{\infty} u_j \phi^j \tag{2}$$

with:

(i) $\phi = \phi(t, x, y, z)$ analytic,

(ii) n integer,

(iii) for a suitable m, $u_j(t, x, y, z) \equiv 0$ when $j \geq m$,

(iv) u_j all analytic for $j < m$ in a neighboorhod of $\phi = 0$, some ones determined by the recursion formula, the other ones completely arbitrary, because of the so-called *resonances*.

If all the resonances result identically verified, as a consequence of the previous established conditions, the considered equation passes the Painlevé test. If some of them is not satisfied, we can try to impose it, by obtaining a further condition on the functions to be determined, which, of course, must result compatible with the other ones. If this is the case, we say that the considered equation passes the so-called *conditioned Painlevé test*. In the first case the equation is *completely integrable*, while no longer in the second case, in which we can find only some, may be not trivial, subclass of solutions (see [5] and quoted references).

For the Eq. (1), the analysis of the leading order term shows that $n = 2$ and gives $u_0 = -12\phi_x^2$.

After that, by inserting in Eq. (1) the expansion (2), with $n = 2$, we obtain a formal expansion in j-th powers of ϕ which, by taking into account the above determination of u_0, leads to the following recursion formula:

$$\mathcal{E}_j \equiv (j+1)(j-4)(j-5)(j-6)\phi_x^4 u_j - \mathcal{F}_j(t, x, y, z, u_k;\ k < j) = 0 \qquad (3)$$

where the dependence of $\mathcal{F}_j$ on the t, x, y, z comes from that of the coefficients a, b, c and their derivatives whit respect to independent variables.

The recursion formula (3) says that the resonances can accour for $j = -1, j = 4, j = 5, j = 6$, the first one corresponding to the arbitrariness of the fonction ϕ, the remaining ones to be verified, as far as its compatibility is concerned.

By proceeding in the standard way, we obtain for

$j = 1$:

$$\mathcal{E}_1 \equiv u_1 - 12\phi_{xx} = 0,$$

$j = 2$:

$$\mathcal{E}_2 \equiv \phi_t\phi_x + u_2\phi_x^2 + 4\phi_{xxx}\phi_x - 3\phi_{xx}^2 + \mathcal{A} = 0,$$

$j = 3$:

$$\mathcal{E}_3 \equiv \phi_t\phi_{xx} - \phi_{tx}\phi_x + u_3\phi_x^3 - \phi_{xxxx}\phi_x + 4\phi_{xxx}\phi_{xx} - 3\phi_{xx}^3/\phi_x + 2\mathcal{B} + \mathcal{A}\phi_{xx}/\phi_x - \phi_x\mathcal{C} = 0,$$

where we made the positions:

$$\mathcal{A} = a\phi_y^2 + 2b\phi_y\phi_z + c\phi_z^2,$$

$$\mathcal{B} = a_x\phi_y^2 + 2b_x\phi_y\phi_z + c_x\phi_z^2,$$

$$\mathcal{C} = a\phi_{yy} + 2b\phi_{yz} + c\phi_{zz}.$$

By inscrting at the orders $j = 4$ and $j = 5$ the values of u_1, u_2 and u_3 previously obtained, we easily see that the two corresponding resonances can be identically satisfied only by requiring that the coefficients a, b and c reduce to constants. If this is the case, at the order $j = 6$ the resonance condition which we find factorizes as follows:

$$(b^2 - ac)\mathcal{E} - 0$$

where

$$\mathcal{E} \equiv \phi_x^2(\phi_{yy}\phi_{zz} - \phi_{yz}^2) + \phi_y^2(\phi_{zz}\phi_{xx} - \phi_{zx}^2) + \phi_z^2(\phi_{xx}\phi_{yy} - \phi_{xy}^2) +$$

$$2\phi_x\phi_y(\phi_{xz}\phi_{yz} - \phi_{xy}\phi_{zz}) + 2\phi_y\phi_z(\phi_{yx}\phi_{zx} - \phi_{yz}\phi_{xx}) + 2\phi_z\phi_x(\phi_{zy}\phi_{xy} - \phi_{zx}\phi_{yy})$$

At this point we have two possibilities. The first one consisting in the requirement tha the first factor $b^2 - ac$ vanishs, the second in the condition for ϕ to be a solution of the $\mathcal{E} = 0$, which is a three-dimensional extension of the well known Monge-Ampère equation. As is easily to see, the vanishing of the first factor permits us to reduce immediately the considered equation to the standard $1 + 2$D case.

By supposing from now on that $b^2 - ac \neq 0$, it remains the more interesting second possibility, which is a true case of conditioned Painlevé property. We will study this case in the next section, by assuming without loss of generality, to avoid unuseful calculations, that the coefficients a, b and c reduce respectively to $1, 0$ and 1 or $0, 1/2$ and 0, correspoding to the two cases $b^2 - ac > 0$ or, $b^2 - ac < 0$. Of course the difference between the two above choices is only apparent if complex variables are permitted, but it become actually effective, if one wont to preserve the reality of the all involved variables, dependent and independent.

We conclude this section by observing that the above condition (iii) required to pass the Painlevé test, i.e. for a suitable m, $u_j \equiv 0$ when $j \geq m$, can be satisfied, as is easily to see, for $j \geq 3$ by assuming $u_3 = 0$, as we shall done from now on. This given, the function ϕ must be a solution of the two equations $\mathcal{E}_3 = 0$ and $\mathcal{E} = 0$.

Essentially in an equivalent way, if we retain u_2 as unknowm, and we require $u_3 = 0$, we obtain for the two unknowns ϕ and u_2 the three equations $\mathcal{E}_2 = 0, \mathcal{E}_3 = 0$ and $\mathcal{E} = 0$, which, in terms of u_2 write:

$$[u_{2t} + u_2 u_{2x} + u_{2xxx}]_x + a u_{2yy} + 2b u_{2yz} + c u_{2zz} = 0 \qquad (4).$$

This last can be seen as an autobäklund transformation, provided that ϕ solves both $\mathcal{E}_2 = 0$ and $\mathcal{E}_3 = 0$,

3. Some explicit solutions

We consider now the two cases annonced at the end of the previous section, having in mind to give some explicit solutions, may be not trivial, but surely not of solitonic type.

The solution u will be given in both cases by the sum

$$u = \frac{u_0}{\phi^2} + \frac{u_1}{\phi} + u_2. \qquad (5)$$

As matter of fact, the only solution of genuine solitonic type which we were able to construct, in both cases, is a one-wave type solution generated by the following function ϕ

$$\phi = 1 + h e^{ax+by+cz+Vt}$$

whit h, a, b, c and V arbitrary constants. It is a straghtforward calculation to show tha this function verifies all requested conditions, but essentially, it reduces the starting equation to a Kortweg-de Vries equation.

3.1 The first case (KP$_I$)

The equation write:

$$[u_t + uu_x + u_{xxx}]_x + u_{yz} = 0,$$

and the solution can be expressed by means of the formula (5), with $u_0 = -12\phi_x^2$, $u_1 = 12\phi_{xx}$, and where u_2 is given in terms of ϕ and its derivatives by the equation $\mathcal{E}_2 = 0$, which in this case write:

$$\mathcal{E}_{2I} \equiv \phi_t\phi_x + u_2\phi_x^2 + 4\phi_{xxx}\phi_x - 3\phi_{xx}^2 + \phi_y\phi_z = 0,$$

A function ϕ will generate in this way a solution of the KP$_I$ if it simultaneously solves the two equation $\mathcal{E} = 0$ and $\mathcal{E}_3 = 0$. This last reduces in this case to

$$\mathcal{E}_{3I} \equiv \phi_t\phi_{xx} - \phi_{tx}\phi_x - \phi_{xxxx}\phi_x + 4\phi_{xxx}\phi_{xx} - 3\phi_{xx}^3/\phi_x + \phi_y\phi_z\phi_{xx}/\phi_x - \phi_{yz}\phi_x = 0.$$

At first, by using an appropriate Legendre transformation, we select a restrict class of solutions of the Monge-Ampère equation expressed in terms of arbitrary functions of the two functionally independent invariants y/x and z/x and of t. By requiring later to these functions to be solutions of the equation $\mathcal{E}_{3I} = 0$, it was possible to construct two different kind of simple solutions.

The first one explicitely write:

$$\phi_1 = a\frac{x}{y} + b\frac{x}{z} + W(t),$$

where W is an arbitrary function of t and a and b are arbitrary constants. The solution generated by the fonction ϕ_1 by means of the sum (5) can be explicetely written as follows:

$$u = -\frac{yz}{\mathcal{P}}W_t - \frac{abx^2y^2z^2}{\mathcal{P}^2\mathcal{Q}^2}W^2 - 2\frac{abx^3yz}{\mathcal{P}^2\mathcal{Q}}W - \frac{12\mathcal{P}^2 + abx^4}{\mathcal{Q}^2}$$

where we put $\mathcal{P} \equiv az + by$ and $\mathcal{Q} \equiv \mathcal{P}x + Wzy$.

The second type of generating function is a stationary solution of both $\mathcal{E} = 0$ and $\mathcal{E}_{3I} = 0$ which explicetely write:

$$\phi_2 = a\frac{y}{x} + a^{-1}\frac{z}{x} + \frac{\sqrt{3}}{6}x + b,$$

with a and b arbitrary constants. Although it appear to be quite simple, the solution which it generates by means of the sum (5), become very involved. By chosing $a = 1$ and $b = 0$, the solution u can be expressed in the following simple form:

$$u = -\frac{12}{x^2}\left(\left(\frac{H^-}{H^+}\right)^2 + \frac{H^+}{H^-} - \frac{2\sqrt{3}}{3}\frac{P}{H^+H^-}\right)$$

where we have put $P = y + z$ and $H^\pm = \frac{\sqrt{3}}{6}x \pm \frac{P}{x}$.

3.2 The second case (KP$_{II}$)

The equation write:

$$[u_t + uu_x + u_{xxx}]_x + u_{yy} + u_{zz} = 0.$$

The solution can be again expressed by the formula (5) with u_2 given by the equation

$$\mathcal{E}_{2II} \equiv \phi_t\phi_x + u_2\phi_x^2 + 4\phi_{xxx}\phi_x - 3\phi_{xx}^2 + \phi_y^2 + \phi_z^2 = 0,$$

and ϕ solutions of both $\mathcal{E} = 0$ and

$$\mathcal{E}_{3II} \equiv \phi_t\phi_{xx} - \phi_{tx}\phi_x - \phi_{xxxx}\phi_x + 4\phi_{xxx}\phi_{xx} - 3\phi_{xx}^3/\phi_x +$$

$$(\phi_y^2 + \phi_x^2)\phi_{xx}/\phi_x - (\phi_{yy} + \phi_{zz})\phi_x = 0.$$

In this case the solutions of the Monge-Ampère equation above selected are not avalaible because they do not solve the equation $\mathcal{E}_{3II} = 0$.

A simple solution of both equations $\mathcal{E} = 0$ and $\mathcal{E}_{3II} = 0$ can be written in the following form:

$$\phi = W(t) + (ax + Vt)(by + cz)$$

where W is an arbitrary function of t and a, b, c and V are arbitrary constants. By making the positions $f = ax + vt$ and $g = by + cz$, the explicit form of the solution u becomes

$$u = -12\frac{a^2g^2}{(W + fg)^2} - \frac{ag(vg + W_t) + (b^2 + c^2)f^2}{a^2g^2}$$

4. Concluding remarks

We have proved that the 1+3D KP do not passes the Painlevé test, but by imposing to the function giving the singularity manifold to be a solution of a generalized Monge-Ampère equation, we constructed some explicit solutions. These solutions shall be used as input to construct new solutions via autobäcklund transformations, as we observed at the end of section 2. Moreover, we conjecture that the showed partial integrability leads to select a class of solutions which can be obtained by a suitable symmetry group via the standard reduction methods [6]. We are exploiting these aspects in a work in progress.

References

[1] T.Brugarino and A.M.Greco, J. Math. Phys. **32**, 69 (1991).

[2] J.Weiss, J. Math. Phys. **25**, 2226 (1984).

[3] R.Conte Phys. Lett. A. **140**, 383 (1989).

[4] J.Weiss, M.Tabor and G.Carnevale, J. Math. Phys. **24**, 522 (1983).

[5] G.M.Webb and G.P.Zank, Phys. Lett. A. **150**, 14 (1990)

[6] P. Olver, Applications of Lie Groups to Differential Equations, (Springer, New York, 1986).

SPINNING MASS ENDOWED WITH ELECTRIC CHARGE AND MAGNETIC DIPOLE MOMENT

V.S. Manko* **and N.R. Sibgatullin****

**Física Teórica, Universidad del País Vasco, Apartado 644, 48080 Bilbao, Spain*

***Department of Hydrodynamics, Moscow State University, Moscow 119899, Russia*

Abstract. Exact asymptotically flat four-parameter solutions of the Einstein-Maxwell equations able to describe the exterior gravitational field of a rotating star possessing an electric charge and magnetic dipole moment are presented in explicit form.

Despite the developement during the last two decades of the powerful superposition techniques for generating new exact solutions of the Einstein-Maxwell equations, the efforts to apply them for solving an astrophysically very important problem of describing the exterior gravitational field of a rotating compact massive source endowed with electric charge and magnetic dipole moment have been unsuccessful up to now. The known results are due to the papers of Kramer [1], Ernst and Wild [2], and Díaz [3], in the first of which the magnetised asymptotically flat generalisation of the Kerr solution [4] was constructed with the aid of the Backlund transformations, while in two others the magnetised Kerr and Kerr-Newman [5] metrics were obtained with the use of the Harrison symmetry transformation. However, the Kramer solution does not admit the static vacuum limit because of a special ansatz used in the generating technique, whereas those of Ernst and Wild, and of Díaz are not asymptotically flat since their magnetostatic limit is the Bonnor-Melvin magnetic universe [6].

In the present paper we use the method [7] developed by one of the authors for the construction of solutions of the Ernst equations [8], to solving of which reduces the stationary axisymmetric electrovacuum problem, by setting the behaviour of the Ernst potentials at the symmetry axis to obtain two examples of the asymptotically flat solutions of the Einstein-Maxwell equations possessing four independent parameters (mass, angular momentum, electric charge and magnetic dipole moment) and representing correctly the exterior field of a rotating, charged, magnetised source.

As was shown in [7], the complex potentials ε and Φ satisfying the Ernst equations [8] can be obtained from the integrals

$$\varepsilon = \frac{1}{\pi}\int_{-1}^{1}\frac{\mu(\sigma)e(\xi)d\sigma}{\sqrt{1-\sigma^2}}; \quad \Phi = \frac{1}{\pi}\int_{-1}^{1}\frac{\mu(\sigma)f(\xi)d\sigma}{\sqrt{1-\sigma^2}} \tag{1}$$

where the unknown function $\mu(\sigma)$ must be found from the integral equation

M. Flato et al. (eds.), Physics on Manifolds, 347–351.

$$\fint_{-1}^{1} \frac{[e(\xi)+\tilde{e}(\eta)+2f(\xi)\tilde{f}(\eta)]\mu(\sigma)d\sigma}{(\xi-\eta)\sqrt{1-\sigma^2}} = 0 \tag{2}$$

and the normalising condition

$$\int_{-1}^{1} \frac{\mu(\sigma)d\sigma}{\sqrt{1-\sigma^2}} = \pi . \tag{3}$$

Here $\xi \equiv z+i\rho\sigma$, $\eta \equiv z+i\rho\tau$ ($\sigma, \tau \in [-1, 1]$); $e(\xi)$ and $f(\xi)$ are the local holomorphic continuations of the functions $e(z) \equiv \varepsilon(\rho=0, z)$, $f(z) \equiv \Phi(\rho=0, z)$ into the complex plane $z+i\rho$ [ρ, z are the Weyl canonical coordinates which enter into (1)-(3) as parameters]; $\tilde{e}(\eta) \equiv [e(\eta^*)]^*$, $\tilde{f}(\eta) \equiv [f(\eta^*)]^*$ and $\fint$ denotes the principal value of the respective integral.

The function $\mu(\sigma)$ for arbitrary rational functions $e(z)$ and $f(z)$ should be searched in the form

$$\mu(\sigma) = A_0 + \sum_{k=1}^{m_1} A_1^k (\xi-\xi_1)^{-k} +...+ \sum_{k=1}^{m_n} A_n^k (\xi-\xi_n)^{-k} \tag{4}$$

where the coefficients A_0, A_i^k have to be then found from eqs. (2) and (3), the ξ_i being the roots of the equation

$$e(\xi) + \tilde{e}(\xi) + 2f(\xi)\tilde{f}(\xi) = 0 \tag{5}$$

with the multiplicity m_i.

Let us consider now two concrete examples of the application of this general scheme.

1. *Asymptotically flat magnetised Kerr-Newman solution.*

To obtain the required generalisation of the Kerr-Newman solution, one should choose the functions $e(z)$ and $f(z)$ in the following form

$$e(z) = \frac{z-m-ia}{z+m-ia}; \quad f(z) = \frac{q}{z+m-ia} + \frac{ib}{(z+m-ia)^2}, \tag{6}$$

Then from (6) and (5) it follows that the unknown function $\mu(\sigma)$ should be searched in the form

$$\mu(\sigma) = A_0 + \sum_{n=1}^{4} A_n(\xi-\alpha_n)^{-1} , \tag{7}$$

where α_n are the roots of the algebraic equation of the fourth order

$$[(\xi+m)^2+a^2](\xi^2-m^2+a^2+q^2) + b^2-2aqb = 0, \tag{8}$$

the complex α_n occuring only in the complex conjugate pairs.

The coefficients A_0, A_n depending on ρ, z, m, a, q and b can be found by substituting (7) into eqs. (2) and (3). Then, after integrating and setting to zero the coefficients at the independent powers of η, one comes to a closed system of five linear algebraic equations for the determination of A_0, A_n

$$A_0 + \sum_{n=1}^{4} \frac{A_n}{r_n} = 1 ; \quad A_0 - \sum_{n=1}^{4} \frac{A_n}{m+\alpha_n-ia} = 0 ; \quad \sum_{n=1}^{4} \frac{A_n}{(m+\alpha_n-ia)^2} = 0 ; \qquad (9)$$

$$\sum_{n=1}^{4} \frac{(\alpha_n-ia)A_n}{(m+\alpha_n-ia)r_n} = 0 ; \quad \sum_{n=1}^{4} \frac{[q(m+\alpha_n-ia)+ib]A_n}{(m+\alpha_n-ia)^2(m+\alpha_n+ia)r_n} = 0 ; \quad r_n \equiv [\rho^2 + (z-\alpha_n)^2]^{1/2},$$

which can be readily solved to give

$$A_0 = \frac{A}{A+B} ; \quad A_n = \frac{(m+\alpha_n-ia)^2(m+\alpha_n+ia)r_n N_n}{A+B} ; \quad n = \overline{1,4} ; \qquad (10)$$

$$A \equiv -(\alpha_1-\alpha_2)(m+\alpha_1+ia)a_{(2,3,4)}r_1r_2+(\alpha_1-\alpha_3)(m+\alpha_1+ia)a_{(3,2,4)}r_1r_3$$
$$-(\alpha_1-\alpha_4)(m+\alpha_1+ia)a_{(4,2,3)}r_1r_4-(\alpha_2-\alpha_3)(m+\alpha_2+ia)a_{(3,1,4)}r_2r_3$$
$$+(\alpha_2-\alpha_4)(m+\alpha_2+ia)a_{(4,1,3)}r_2r_4-(\alpha_3-\alpha_4)(m+\alpha_3+ia)a_{(4,1,2)}r_3r_4 ;$$

$$B \equiv -(\alpha_2-\alpha_3)(\alpha_2-\alpha_4)(\alpha_3-\alpha_4)b_{(1,2,3,4)}r_1+(\alpha_1-\alpha_3)(\alpha_1-\alpha_4)(\alpha_3-\alpha_4)b_{(2,1,3,4)}r_2$$
$$-(\alpha_1-\alpha_2)(\alpha_1-\alpha_4)(\alpha_2-\alpha_4)b_{(3,1,2,4)}r_3+(\alpha_1-\alpha_2)(\alpha_1-\alpha_3)(\alpha_2-\alpha_3)b_{(4,1,2,3)}r_4 ;$$

$$N_1 \equiv -a_{(2,3,4)}r_2+a_{(3,2,4)}r_3-a_{(4,2,3)}r_4 ; \quad N_2 \equiv a_{(1,3,4)}r_1-a_{(3,1,4)}r_3+a_{(4,1,3)}r_4 ;$$

$$N_3 \equiv -a_{(1,2,4)}r_1+a_{(2,1,4)}r_2-a_{(4,1,2)}r_4 ; \quad N_4 \equiv a_{(1,2,3)}r_1-a_{(2,1,3)}r_2+a_{(3,1,2)}r_3 ;$$

$$a_{(n,p,s)} \equiv (m+\alpha_n+ia)(\alpha_p-\alpha_s)\{q(m+\alpha_p-ia)(m+\alpha_s-ia)(m+\alpha_p+\alpha_s)$$
$$+ib[(m+\alpha_p)^2+(m+\alpha_s)^2+\alpha_p\alpha_s-m^2+a^2-ia(2m+\alpha_p+\alpha_s)]\} ;$$

$$b_{(n,p,s,t)} \equiv m(m+\alpha_n+ia)\{q(m+\alpha_p-ia)(m+\alpha_s-ia)(m+\alpha_t-ia)$$
$$+ib[(m+\alpha_p)(2m+\alpha_s+\alpha_t)+(m+\alpha_s)(m+\alpha_t)-a^2]\} .$$

Hence, now it is possible to construct the potentials ε and Φ corresponding to e(z) and f(z) from (6). Substituting (7) into (1) and accounting for (9) one obtains after the integration the final expressions for ε and Φ in terms of A_n

$$\varepsilon = 1 - \sum_{n=1}^{4} \frac{2mA_n}{(m+\alpha_n-ia)r_n}; \quad \Phi = \sum_{n=1}^{4} \frac{[q(m+\alpha_n-ia)+ib]A_n}{(m+\alpha_n-ia)^2 r_n}, \tag{11}$$

The formulae (11), (10) fully determine the desired asymptotically flat magnetised generalisation of the Kerr-Newman solution. It possesses all the required stationary and static limits to describe correctly the exterior gravitational field of a charged, magnetised, rotating mass: the ordinary Kerr-Newman solution, whose limiting cases are well-known, is recovered from (11), (10) by setting the magnetic parameter $b = 0$; in the magnetostatic case ($q = a = 0$) it reduces to the solution for a massive magnetic dipole; in the absence of rotation and electromagnetic field ($a = q = b = 0$) one comes to the Schwarzschild solution [it should be mentioned that for some special cases the roots of eq. (8) may become multiple that essentially requires the use of the l'Hospital rule to perform the limiting transformations in (11)].

2. Although the dependence of ε and Φ on r_n in solution (11) is very simple, the expressions for the Ernst potentials contain the coefficients which need the explicit form of α_n, the latter being of course rather complicated in the general case. Below we shall derive another asymptotically flat four-parameter solution of the Ernst equations describing the field of a rotating, charged, magnetised source by choosing the initial e(z) and f(z) in such a way that eq. (5) could admit simple algebraic roots, so that the final expressions for ε and Φ will contain only the parameters m, α, β and ν representing, respectively, mass, angular momentum, electric charge and magnetic dipole moment of the source. The new choice of e(z) and f(z) is the following

$$e(z) = 1 - \frac{2m}{z} - \frac{4m^2(\beta^2+i\alpha)}{z^2} - \frac{16m^4\nu^2}{z^4}; \quad f(z) = \frac{2m\beta}{z} + \frac{4im^2\nu}{z^2}, \tag{12}$$

where the terms in e(z) containing β and ν are introduced to obtain expliciply the roots of eq. (5). Then from (5) one finds that $\xi_{1,2,3} = 0$, $\xi_4 = 2m$, and therefore the function $\mu(\sigma)$ should be looked for in the form

$$\mu(\sigma) = A_0 + A_1\xi^{-1} + A_2\xi^{-2} + A_3\xi^{-3} + A_4(\xi-2m)^{-1}. \tag{13}$$

Similar to the previous case, the A_0 and A_n should be found from (2) and (3), and the final expressions for ε and Φ then will follow from (1). Omitting some lengthy calculations we shall give the resulting form of ε and Φ which can be written most elegantly in the prolate spheroidal coordinates (x, y) related to ρ and z by the formulae $\rho = m(x^2-1)^{1/2}(1-y^2)^{1/2}$, $z = mxy$ (in the expressions for e and F below we have also performed the shift $z \to z+m$ along the symmetry axis to provide the reduction of ε to the Schwarzschild solution in the absence of rotation and electromagnetic field)

$$\varepsilon = A_-/A_+; \quad \Phi = 2(x+y)C/A_+; \tag{14}$$

$$A_\mp \equiv x\mp1-(x\pm1)(1\mp y)(1\pm y)^3\{a^2+b^4+2b^2c^2[2y(x\mp1)^2(9y\mp4)+2(x\mp1)(1\pm y)(\mp3+9y\pm4y^2)$$

$+(1\pm y)^2(5\pm 4y+y^2)]\mp 2b^2c^4(xy+1)(x\mp 1)(x\pm 1)^2(1\mp y)^2(1\pm y)^3+c^4[2(x\mp 1)^4(-3\pm 16y\mp 144y^3$
$+163y^4)+8y(x\mp 1)^3(1\pm y)(13\mp 83y+66y^2\pm 36y^3)+4(x\mp 1)^2(1\pm y)^2(9\mp 98y+56y^2\pm 104y^3+25y^4)$
$+4(x\mp 1)(1\pm y)^3(\mp 21+13y\pm 43y^2+25y^3\pm 4y^4)+(1\pm y)^6(3\pm y)^2]\mp 2c^6(x\mp 1)(x\pm 1)^2(1\mp y)^2(1\pm y)^3$
$\times[\mp 2(x\mp 1)^3(1\mp 4y+4y^2\mp y^3)\pm 4y(x\mp 1)^2(1\pm y)(1\pm 9y+2y^2)+y(x\mp 1)(1\pm y)^2(5\pm y)(7\pm y)+(1\pm y)^3(3$
$\pm y)^2]-c^8(x\mp 1)^3(x\pm 1)^5(1\mp y)^5(1\pm y)^7\}\mp 2(1\pm y)\{b^2(xy+1)+c^2[2(x\mp 1)^3(\pm 1-4y\mp 4y^2+9y^3)+4y$
$\times(x\mp 1)^2(1\pm y)(\mp 8+9y\pm 2y^2)+(x\mp 1)(1\pm y)^2(\mp 12+11y\pm 12y^2+y^3)+(1\pm y)^5]\}\mp 2i(1\pm y)\{a(2xy\mp x$
$+y^2\mp y+1)\mp 2ac^2(x\pm 1)(1\mp y)(1\pm y)^2[2xy(xy+1)-x^2-y^2+2]\pm bc(1\mp y)(2x+y\mp 1)(2xy+y^2+1)\pm 2bc^3$
$\times(xy+1)(x\pm 1)(1\mp y)(1\pm y)^2[4(x\mp 1)^2(1\mp 2y+3y^2)\pm 2(x\mp 1)(1\pm y)(1\pm 3y+4y^2)+(1\pm y)^3(3\pm y)]$
$+c^4(x\mp 1)(x\pm 1)^3(1\mp y)^3(1\pm y)^5[a(6xy\mp x+y^2\mp y+5)+bc(6xy+y^2+5)(6xy\mp 2x+y^2\mp 2y+5)]\}$;

$C\equiv b+ic(2xy+y^2+1)+(1-y^2)\{iab+b^3+c(a-ib^2)(2xy+y^2+1)+bc^2[(x-1)^2(-3+19y^2)+2(x-1)(1+y)$
$\times(-3+15y+4y^2)+(1+y)^2(3+y)^2]+ic^3[2y(x-1)^3(1+15y^2)+3(x-1)^2(1+y)(1+y+19y^2+11y^3)$
$+2(x-1)(1+y)^2(3+5y)(1+4y+y^2)+(1+y)^4(3+y)^2]\}-(x^2-1)(1-y^2)^3\{iabc^2+b^3c^2-c^3(a-ib^2)(6xy$
$+y^2+5)+bc^4[(x-1)^2(-3+19y^2)+2(x-1)(1+y)(-3+15y+4y^2)+(1+y)^2(3+y)^2]-ic^5[2y(x-1)^3(53y^2$
$-5)-(x-1)^2(1+y)(9+21y-253y^2-65y^3)+2(x-1)(1+y)^2(-9+95y+51y^2+7y^3)+(1+y)^3(5+y)(3+y)^2]$
$+c^6(x^2-1)^2(1-y^2)^3[b-ic(6xy+y^2+5)]\}$;

$$a\equiv \alpha/(x+y)^2\ ;\quad b\equiv \beta/(x+y)\ ;\quad c\equiv \nu/(x+y)^3\ .$$

The solution obtained also possesses all the required stationary and static limits to represent correctly the field of a charged, magnetised, rotating compact mass, and it is different from the magnetised Kerr-Newman metric defined by (11), (10) since the multipole structure of two solutions determined by their behaviour at the symmetry axis is different.

REFERENCES

[1] D.Kramer, Class. Quantum Gravity **1**, L45 (1984).

[2] F.J.Ernst, and W.J.Wild, J. Math. Phys. **17**, 182 (1976).

[3] A.G.Díaz, J. Math. Phys. **26**, 155 (1985).

[4] R.P.Kerr, Phys. Rev. Lett. **11**, 237 (1963).

[5] E.T.Newman, E.Couch, K.Chinnapared, A.Exton, A.Prakash, and R.Torrense, J. Math. Phys. **6**, 918 (1965).

[6] W.B.Bonnor, Proc. Phys. Soc. **67A**, 225 (1954).

[7] N.R.Sibgatullin, *Oscillations and Waves in Strong Gravitational and Electromagnetic Fields* (Nauka, Moscow) 1984; English translation: Springer-Verlag, 1991.

[8] F.J.Ernst, Phys. Rev. **168**, 1415 (1968).

EQUATIONS DE VLASOV EN THÉORIE DISCRÈTE

G. PICHON

Equation de Boltzmann en Relativité

(V, g) variété riemanienne, $T(V)$ fibré tangent (coordonnées x^α, p^α).

X champ de vecteurs géodésique sur $T(V)$

$$X = p^\alpha \frac{\partial}{\partial x^\alpha} - \Gamma^\alpha_{\lambda\mu} p^\lambda p^\mu \frac{\partial}{\partial p^\alpha}$$

$\mathcal{D}(V)$, $\mathcal{D}(T(V))$: fonctions C^∞ à support compact

$\mathcal{D}'(V)$, $\mathcal{D}'(T(V))$: espaces duaux

$\langle \quad \rangle$, $(\quad)$: crochets de dualité

Un choc en $x \in V$ est une application de

$$T^2_x(V) \longrightarrow T^2_x(V) : (x, p, q) \longrightarrow (x, p', q')$$
$$p' = p'(x, p, q, \alpha) \quad q' = q'(x, p, q, \alpha)$$

α paramètre, $\alpha \in G$.

On désigne par H un opérateur linéaire : si $\psi \in \mathcal{D}(T(V))$, $(H\psi)(x, p, q)$ définic par

$$(H\psi)(x, p, q) = \sum_{\alpha \in G} \varepsilon(x, p, q, \alpha) \, [\psi(x, p') + \psi(x, q') - \psi(x, p) - \psi(x, q)]$$

ε fonction dépendant de la théorie envisagée.

Théorie continue et chocs élastiques $\alpha \in S^2$.

Théorie discrète : sommation discrète.

L'équation de Boltzmann pour $F(x, p)$

$$(XF, \psi) = (F \oslash F, H(\psi))$$

$F \oslash F = F \otimes F * \delta(x - y)$.

M. Flato et al. (eds.), Physics on Manifolds, 353–356.

Théorie relativiste discrète

En chaque point $x \in V$, p ne peut prendre qu'un nombre fini de valeurs $p_i(x)$ $(i = 1, \ldots, r)$ chacune ayant pour densité $a_i(x)$. On définit l'élément $a_i\mu(p_i) \in \mathcal{D}'(T(V))$ par

$$(a_i\mu(p_i), \psi) = \langle a_i(x), \psi(x, p_i(x))\rangle, \quad \psi \in \mathcal{D}(T(V))$$

et on étudie la distribution f

$$F = \sum_{i=1}^{r} a_i\, \mu(p_i).$$

Pour définir H on se donne r^4 fonctions $\varepsilon_{ijk\ell}$ sur V vérifiant

$$\varepsilon_{ijk\ell} = \varepsilon_{jik\ell} = \varepsilon_{k\ell ij} \quad (i, j, \ldots = 1, \ldots, r)$$

et on pose pour $\psi \in \mathcal{D}(\mathcal{T}(\mathcal{V})$, $\psi_k(x) = \psi(x, p_k(x))$, puis

$$(H\psi)(x, p_i(x), p_j(x)) = \sum_{k<\ell} \varepsilon_{ijk\ell}(\psi_k(x) + \psi_\ell(x) - \psi_i(x) - \psi_j(x)).$$

L'équation

$$(X(F), \psi) = (F \oslash F, H(\psi))$$

s'écrit

$$X\Big(\sum_{i=1}^{r} a_i\mu(p_i)\Big) = \sum_{\substack{i<j \\ k<\ell \\ m=1}}^{r} (a_ia_j - a_ka_\ell)\varepsilon_{ijk\ell}(\delta_k^m + \delta_\ell^m - \delta_i^m - \delta_j^m)\,\mu(p_m)$$

(δ_k^m symbole de Kronecker). Equations vérifiées par les a_i et les p_i

$$\begin{cases} \nabla_\alpha(a_m p^\alpha) = \displaystyle\sum_{\substack{i<j \\ k<\ell}} (a_ia_j - a_ka_\ell)\,\varepsilon_{ijk\ell}(\delta_m^k + \delta_m^\ell - \delta_m^i - \delta_m^j) \\ p_i^\alpha \nabla_\alpha p_i^\beta = 0 \qquad (i, m, \ldots = 1, \ldots, r) \quad (\alpha, \beta, \ldots = 1, \ldots, 4) \end{cases}$$

Tenseur d'impulsion énergie associé à F :

$$T_F = \sum_{i=1}^{r} a_i\, p_i \otimes p_i.$$

Si les chocs sont élatiques i.e.

$$p_i + p_j = p_k + p_\ell$$

T_F est conservatif : $\nabla_\alpha T_F^{\alpha\beta} = 0$.

Plasma avec vitesses discrètes

$$V = \mathbb{R} \times \mathbb{R}^3 \quad (t, x)\ , \quad p_i = (p_i^0, b_i), p_i^0 = \sqrt{m^2 + \sum_{k=1}^{3} (b_i^k)^2}.$$

Chaque particule a pour charge e. E et H : champs électrique et magnétique moyens engendrés par les particules. Le champ de vecteur X

$$X = \frac{\partial}{\partial t} + \frac{b}{p^0} \nabla_x + e(E + \frac{b}{p^0} \times H) \nabla_b$$

(champ de Vlasov).

La distribution f est

$$f(t, x, p) = \sum_{i=1}^{r} a_i(t, x)\, \delta(b_i)$$

$\delta(b_i)$ mesure de Dirac en b_i sur $\mathbb{R}^3$.

Le système d'équations :

$$\begin{cases} \dfrac{\partial E}{\partial t} - \nabla \times H + \sum\limits_i e_i a_i b_i (p_i^0)^{-1} = 0 \\ \dfrac{\partial H}{\partial t} + \nabla \times E = 0 \\ \dfrac{\partial b_i}{\partial t} + (p_i^0)^{-1} b_i \cdot \nabla_x b_i + e_i (E + (p_i^0)^{-1} b_i \times H) = 0 \\ \dfrac{\partial a_i}{\partial t} + (p_i^0)^{-1} b_i \nabla_x a_i + a_i \mathrm{div}(b_i (p_i^0)^{-1}) = \mathcal{I}_i \cdot (p_i^0)^{-1} \end{cases}$$

avec $\mathcal{I}_i = \sum\limits_{j,m,\ell} (\varepsilon_{ijm\ell} a_m a_\ell - \varepsilon_{m\ell ij} a_i a_j)$.

Conditions initiales :

$$E(0, x) = E_0(x), H(0, x) = H_0, a_i(0, x) = a_{i,0}, b_i(0, x) = b_{i,O}.$$

Contraintes :

$$\mathrm{div} E_0 = \sum_i e_i a_{i0} \quad \mathrm{div} H_0 = 0.$$

THÉORÈME Y. CHOQUET. — *Le système (S) d'inconnues E, H, a_i, b_i est non strictement hyperbolique au sens de Leray-Ohya. Le problème de Cauchy a une solution unique si les données initiales sont dans des classes de Gevrey. Le domaine de dépendance est déterminé par le cône de lumière.*

Bibliographie

A. ALVES : Eur. J. Mech. B., fluids voL.9 n° 5, 1990, p. 457-467 (Consider the case of a discrete Boltzmann equation in a Newtonian field).

D. BANCEL et Y. CHOQUET BRUHAT : Existence, uniqueness and local stability for the Einstein -Maxwell - Boltzmann system. Comm. Maths-Phys. 33, 1973, p.83-96.

J. BROADWELL : F. Fluid Mecanics 19, 1964, p.367-370 and The Physics of Fluids, 7, 1964, p.1243-1247.

H. CABANNES : The discrete Biltzmann equation. Lecture Notes University of California Berkeley (1980).

Y. CHOQUET-BRUHAT et G. PICHON : Plasmas with discrete velocities. Colloque Euro-Mech. Coïmbra Portugal. Septembre 1990.

K. KAJIKANI : Local solution of Cauchy problem for non linear hyperbolic systems in Gevrey classes. Hokkaido Math. Journal XII n° 3, (1983), p. 434-460.

J. LERAY et Y. OHYA : Systèmes non linéaires hyperboliques non stricts. Math. Ann. 70 (1967) p.167-205.

G. PICHON : Distribution de Boltzmann discrète sur une variété riemannienne. CR. Acad. Sci. Paris t.309, série 1, p.407-408, (1989).

G. PICHON et M. HUYNH-SERVET : Distribution de Boltzmann discrète sur un fibré tangent. J. Math. Pures et Appl. 71, p. 21 à 32 (1992).

Convexity and Symmetrization in Classical and Relativistic Balance Laws Systems

Tommaso Ruggeri

Department of Mathematics and Research Center for Applied Mathematics - C.I.R.A.M.- University of Bologna Via Saragozza 8, 40123 Bologna, Italy.

Les équations de la Physique-Mathématique et, en particulier celles de la Mécanique des milieux continus expriment usuellement des lois non linéaires de bilan. De même, on sait bien que les systèmes d'équations différentielles dont ci-dessus ont validité tout-à-fait générale et pourtant le nombre des équations est plus petit que celui des champs inconnus. Au but de balancer le système, il faut assigner des relation entre les champs: les *équations constitutives.* Dans la moderne théorie qui régit les équations constitutives, celles-ci devront à leur fois satisfaire à des principes gènèraux. En particulier, les classes qu'on peût accepter devront être telles que le système complet soit compatible avec les principes de *objectivité* et d' *entropie.* On démande, de plus, en particulier pour les théories causales du type hyperbolique, que le densité d'entropie vérifie une condition naturelle de concavité, qui assure *a priori* la stabilité thermodynamique. On peut caractérizer d'une façon tout-à-fait générale, aussi dans le cas classique qu'en celui relativiste, les conséquences des ces principes sur la structure des équations constitutives; donc aussi sur celle des équations finales. Il faut cependant ici remarquer qu'il-y-a une différence essentielle sur l'hypothèse de concavité de l'entropie lorsqu'on passe du cas classique à celui relativiste. En effet, dans les cas classique la dite hypothèse assure que le système differentiel est symétrique et hyperbolique; donc la *bonne position* (locale) pour le problème de Cauchy; au contraire, dans le cas relativiste, la concavité aussi que la symétrisation dépendent du choix de la congruence du genre temps. Comme cette congruence n'est pas constante en tant qu'elle est influencée par le champ même, on peut vérifier qu'en générale la condition de stabilité thermodynamique (concavité) n'implique pas la symétrisation du système. On peut toutefois démontrer l'existence d'une congruence privilégiée du genre temps pour laquelle la dite symétrisation est possible; congruence, celle-ci, qu'on peut obtenir par une convenable trasformation du type de Legendre a partir du quadri-vecteur d'entropie. Pour l'observateur privilégié, dont ci-dessus, la concavité implique pourtant la symétrisation. En ajoutaut à cette condition l'hypothèse que les vitesses charactéritiques soient plus petites que la vitesse de la lumière, il est posssible démontrer que le système est symétrique hyperbolique par rapport à une congruence quelconque du genre temps. Les considerations générales dont ci-dessus sont ensuite appliquées à la moderne théorie de la thermodynamique *étendue* classique et relativiste.

M. Flato et al. (eds.), Physics on Manifolds, 357–365.

1 Balance Laws and Universal Principles

In many physical applications a $\Re^N$ vector field $\mathbf{u}(\mathbf{x}, t)$ satisfies a system of N balance laws of the form

$$\frac{\partial \mathbf{F}^o}{\partial t} + \frac{\partial \mathbf{F}^i}{\partial x^i} = \mathbf{f}. \tag{1.1}$$

The *constitutive laws* are related to the field $\mathbf{u}$ in a local manner, i.e.:

$$\mathbf{F}^o \equiv \mathbf{F}^o(\mathbf{u}); \qquad \mathbf{F}^i \equiv \mathbf{F}^i(\mathbf{u}); \qquad \mathbf{f} \equiv \mathbf{f}(\mathbf{u}). \tag{1.2}$$

The previous hypotheses are equivalent to request that our field equations form a first order quasi-linear system of PDE's. The restrictions arise from the following universal principles:

- *Objectivity Principle*:
  - the field equations (in the classic case) are covariant with respect to the Galilean transformations;
  - the non convective parts of (1.1) and (1.2) characterising the true constitutive quantities are independent form a generic observer.

- *Entropy Principle*:

  there exists an entropy h^o and an entropy flux $-h^i$ such that all the solutions of the system of balance laws must satisfy the scalar inequality

  $$\frac{\partial h^o}{\partial t} + \frac{\partial h^i}{\partial x^i} \leq 0; \qquad h^o = h^o(\mathbf{u}); \qquad h^i = h^i(\mathbf{u}).$$

- *Causality and Stability*:

  h^o is a convex function of the densities $\mathbf{F}^o$. This request corresponds to a *Thermodynamical stability condition.* We shall see in the next section that this condition is related to the *causality* of the differential system.

1.1 Galilean Invariance

In this section we concentrate our attention on the first statement of the Objectivity Principle in the classical case, i.e. the covariance of the field equations with respect to a Galilean transformation for a general system of balance laws. In this kind of problem the velocity $\mathbf{v}$ plays an important role and so it is convenient to divide the field $\mathbf{u}$ in a pair:

$$\mathbf{u} \equiv (\mathbf{w}, \mathbf{v}); \qquad \mathbf{w} \in \Re^{N-3}; \quad \mathbf{v} \in \Re^3;$$

where $\mathbf{w}$ are objective quantities. It is possible to prove the following results (Ruggeri 1989 [1]; see also Shugrin [2]):

Theorem 1.1 *There exists a linear operator $(N \times N)$ $\mathbf{X}(\mathbf{v})$ such that:*

$$\mathbf{F}^o(\mathbf{w},\mathbf{v}) = \mathbf{X}(\mathbf{v})\mathbf{F}^o(\mathbf{w},\mathbf{0}) \quad \mathbf{G}^i(\mathbf{w},\mathbf{v}) = \mathbf{X}(\mathbf{v})\mathbf{G}^i(\mathbf{w},\mathbf{0}) \quad \mathbf{f}^o(\mathbf{w},\mathbf{v}) = \mathbf{X}(\mathbf{v})\mathbf{f}^o(\mathbf{w},\mathbf{0})$$

$$\mathbf{G}^i \doteq \mathbf{F}^i - \mathbf{F}^o v^i$$

where

$$\mathbf{X}(\mathbf{v}) = e^{\mathbf{A}^r v_r} = \mathbf{I} + \mathbf{A}^r v_r + \frac{1}{2}\mathbf{A}^r\mathbf{A}^s v_r v_s + \dots$$

$$\mathbf{A}^r = \text{const.}, \qquad \mathbf{A}^r\mathbf{A}^s = \mathbf{A}^s\mathbf{A}^r. \qquad \forall r,s = 1,2,3.$$

Moreover if in the system

$$\frac{\partial \mathbf{F}^o}{\partial t} + \frac{\partial}{\partial x^i}\left(\mathbf{F}^o v^i + \mathbf{G}^i\right) = \mathbf{f}$$

the densities, the (non–convective) fluxes and productions have the form

$$\mathbf{F}^0 = \begin{bmatrix} F \\ F_{k_1} \\ F_{k_1 k_2} \\ \vdots \\ F_{k_1 k_2 \cdots k_p} \end{bmatrix}; \qquad \mathbf{G}^i = \begin{bmatrix} G^i \\ G^i_{k_1} \\ G^i_{k_1 k_2} \\ \vdots \\ G^i_{k_1 k_2 \cdots k_p} \end{bmatrix}; \qquad \mathbf{f} = \begin{bmatrix} f \\ f_{k_1} \\ f_{k_1 k_2} \\ \vdots \\ f_{k_1 k_2 \cdots k_p} \end{bmatrix}$$

($F_{k_1 k_2 \dots k_j}, G^i_{k_1 k_2 \dots k_j}, f_{k_1 k_2 \dots k_j}$ symmetric tensors), than the matrices $\mathbf{A}^r$ are sub triangular block matrices with blocks of zero in the main diagonal and they are nil–potent of degree $p+1$, i.e.

$$\mathbf{A}^{k_1}\mathbf{A}^{k_2}\dots\mathbf{A}^{k_{p+1}} \equiv 0 \quad \textit{for all} \quad k_1, k_2 \dots k_{p+1} = 1,2,3.$$

Therefore, the matrix $\mathbf{X}(\mathbf{v})$ is a polynomial matrix in $\mathbf{v}$ of order p:

$$\mathbf{X}(\mathbf{v}) = e^{\mathbf{A}^r v_r} = \mathbf{I} + \mathbf{A}^r v_r + \dots + \frac{1}{p!}\mathbf{A}^{k_1}\dots\mathbf{A}^{k_p} v_{k_1}\dots v_{k_p}.$$

In addition, $\mathbf{A}^r$ are completely determined and then the dependence on the velocity in densities, fluxes and productions is full determined by the Galilean Invariance.

1.2 Entropy Principle

The problem of exploiting this principle at this level of generality was considered by different authors (Godunov 1961 [3], Friedrichs & Lax 1971 [4] - Boillat 1974 [5], T. Ruggeri & A. Strumia [6]). In particular in [5,6] the following theorem was proved:

Theorem 1.2 *There exists a* **main field** $\mathbf{u}'$ *and* 4 **potentials** h'^o, h'^i *such that*

$$\mathbf{F}^o = \frac{\partial h'^o}{\partial \mathbf{u}'}; \qquad \mathbf{F}^i = \frac{\partial h'^i}{\partial \mathbf{u}'}, \tag{1.3}$$

$$\textit{where} \quad h^o = \mathbf{u}' \cdot \mathbf{F}^o - h'^o, \qquad h^i = \mathbf{u}' \cdot \mathbf{F}^i - h'^i; \qquad \mathbf{u}' \cdot \mathbf{f} \leq 0. \tag{1.4}$$

Inserting (1.3), the system (1.1) becomes

$$\mathbf{A}'^{o}\frac{\partial \mathbf{u}'}{\partial t} + \mathbf{A}'^{i}\frac{\partial \mathbf{u}'}{\partial x^i} = \mathbf{f}(\mathbf{u}')$$

where

$$\mathbf{A}'^{o} = \frac{\partial^2 h'^{o}}{\partial \mathbf{u}'\partial \mathbf{u}'}; \qquad \mathbf{A}'^{i} = \frac{\partial^2 h'^{i}}{\partial \mathbf{u}'\partial \mathbf{u}'}.$$

For this reason if $\mathbf{A}'^{o}$ is positive definite the system is Symmetric Hyperbolic.

1.3 Convexity and Symmetrizability

Therefore, in order to have the system in a symmetric form with nice consequences for the well position of the Cauchy problem, it is necessary that the following quadratic form is positive:

$$Q = \delta\mathbf{u}' \cdot \mathbf{A}'^{o} \cdot \delta\mathbf{u}' = \delta\mathbf{u}' \cdot \delta\mathbf{F}^{o} = \delta\mathbf{u}' \cdot \frac{\partial^2 h'^{o}}{\partial \mathbf{u}'\partial \mathbf{u}'} \cdot \delta\mathbf{u}' > 0 \quad \forall \delta\mathbf{u}'. \tag{1.5}$$

We call this mathematical condition the *Symmetrizability Condition* and taking into account the last expression in (1.5) we note that the symmetrizability corresponds to the convexity of h'^{o} with respect to the main field components $\mathbf{u}'$. Taking into account $(1.4)_1$ we observe that h'^{o} is the Legendre transformation of h^{o} and $\mathbf{u}'$ and $\mathbf{u}$ are conjugate variables:

$$\mathbf{u}' = \frac{\partial h^{o}}{\partial \mathbf{u}}; \qquad \mathbf{u} = \frac{\partial h'^{o}}{\partial \mathbf{u}'},$$

and then $Q = \delta^2 h^{o}$. So in the classical case the symmetrizability condition is equivalent to the physical assumption of convexity. In fact convexity of $-h^{o}$, i.e. the concavity of entropy density, is in the thermodynamical theories a natural request of thermodynamical stability. Then we conclude that a thermodynamical assumption has as consequence the symmetric form of the system and for well known theorems the Cauchy problem (locally in time) is well posed for smooth initial data.

1.4 Compatibility between Entropy Principle and Galilean Invariance

The two previous theorems gives different constraints and therefore it is necessary to check their compatibility. Taking into account that it is possible to verify the existence of the following relation [1]

$$\mathbf{u}' = \mathbf{X}^{-1}(\mathbf{v})\hat{\mathbf{u}}', \qquad (\hat{g} = g(\mathbf{w}, \mathbf{0}))$$

the Entropy Principle turns out to be compatible with the Galilean Invariance and becomes a constraint only for the non convective densities, the fluxes and productions in terms of the non convective field $\mathbf{w}$, i.e. only for the *true constitutive functions*

$$\hat{\mathbf{F}}^{o} \equiv \hat{\mathbf{F}}^{o}(\mathbf{w}); \quad \hat{\mathbf{F}}^{i} \equiv \hat{\mathbf{F}}^{i}(\mathbf{w}); \quad \hat{\mathbf{f}} \equiv \hat{\mathbf{f}}(\mathbf{w}).$$

1.5 Conclusions in the Classical Case

Therefore, these universal principles give in the classical case very strong limitations on the structure of the balance laws system. Resuming, we have:

- Galilean Invariance dictates a precise dependence on velocity in the field equations;
- Entropy Principle is a strong condition for the proper constitutive equations with the existence of potentials;
- Convexity of h^o (thermodynamical stability) is equivalent to the condition of having a (very special) symmetric hyperbolic system and the Cauchy problem is (locally) well posed.

2 Extended Thermodynamics

Using only the previous argumentation's it was possible to construct a continuum approach to non equilibrium thermodynamics of gases (Extended Thermodynamics [7]). The surprising result was that only using macroscopic arguments the so obtained field equations coincide, in the case of the ideal gases, with those deduced by means of the Grad 13-moments procedure from the Boltzmann equation.

The first three equations are the usual conservation laws while the other new equations govern the evolution of the heat flux and the shear stress and reduce to the Fourier and Navier Stokes equations when two relaxing times vanish. The coincidence in the case of ideal gases between our approach and the microscopic one gives a good test on the validity of the extended thermodynamics arguments with the advantage that it is possible to use this approach also for non ideal gases, solids, etc.- A recent interesting utilisation of extended thermodynamics is the construction of a model for a phonon gas in a rigid heat conductor [8]. In this way, it is possible to explain some experimental evidences not described using the usual approach of an ideal phonon gas.

3 Relativistic Case

In the relativistic case we can proceed in the same way as in the classical one. The only difference lies in the convexity problem. In fact, in the relativistic case it is necessary first of all to define the time-like congruence ξ_α and the entropy depends on this choice. The reason is that the entropy is the component of the entropy four-vector h^α in the direction ξ_α: $h = h^\alpha \xi_\alpha$. On the other hand, this congruence is in general field dependent (for example in the fluid case it is usually coincident with the fluid four-velocity). Therefore, it is not surprising that here there is no coincidence between the symmetrizability condition $Q > 0$ and the convexity one $\delta^2 h > 0$. In fact, there exists the following relation (Ruggeri 1990 [9]):

$$Q = \delta^2 h + h'^{\alpha} \delta^2 \xi_\alpha.$$

Moreover in the case of non constant congruences, in general $h = h^\alpha \xi_\alpha$ and $h' = h'^\alpha \xi_\alpha$ are not related by a Legendre transformation and the fields $\mathbf{u}'$ and $\mathbf{u}$ are not conjugate variables.

A natural question is about the existence of a privileged time-like congruence for which the physical condition of convexity guarantees the (mathematically suitable) symmetrizability condition. An answer is contained in the following theorem [9]:

Theorem 3.1 *If h'^α is a time-like vector oriented towards the future, then there exists the congruence $\overline{\xi}_\alpha$:*

$$\overline{\xi}_\alpha = \frac{h'_\alpha}{c\sqrt{h'^\beta h'_\beta}},$$

collinear with h'^α, such that if $\delta^2 \overline{h} > 0$, $(\overline{h} = h^\alpha \overline{\xi}_\alpha)$ with respect to the field $\overline{\mathbf{u}} = \mathbf{F}^\alpha \overline{\xi}_\alpha$, we have $\overline{Q} = \delta \mathbf{u}' \cdot \delta \mathbf{F}^\alpha \overline{\xi}_\alpha > 0$.

Moreover $\overline{h}$ and $\overline{h}'$ are Legendre transformation and $\mathbf{u}'$ and $\overline{\mathbf{u}}$ are conjugate field variables.

Therefore with respect to this privileged observer the physical assumption of convexity gives the same consequences of the classical one. But if one changes the time-like congruence we do not know a priori if the system remains symmetric. Nevertheless, it is possible to prove that, by adding another physical request i.e. that the signals propagate with velocities smaller than the light one, then the symmetrization remains true not only for the privileged observer but for all the time-like congruences. In fact, there exists the theorem [9]:

Theorem 3.2 *If $\overline{h} = h^\alpha \overline{\xi}_\alpha$ is a convex function with respect to the field $\overline{\mathbf{u}} = \mathbf{F}^\alpha \overline{\xi}_\alpha$ and the characteristic hypersurfaces $\phi(x^\mu) = 0$ of the system are space - or light like so that*

$$\phi^\alpha \phi_\alpha \leq 0 \quad \text{with} \quad \phi_\alpha = \partial_\alpha \phi,$$

then

$$Q = \delta \mathbf{u}' \cdot \delta \mathbf{F}^\alpha \xi_\alpha > 0$$

holds for all time-like ξ_α oriented towards the future, i.e. the system is symmetric hyperbolic with respect to all choices of the time direction.

Therefore, we conclude that the symmetrizability for all time-like congruences is guaranteed by the following physical requirements:

- Convexity of the privileged entropy $-\overline{h}$ with respect to the privileged time-like congruence;

- Characteristic Hypersurfaces space or light like.

Now, it remains the interesting problem of identifying in the physical cases the privileged time-like congruence and the privileged entropy density. We give as examples the cases of a relativistic fluid with and without dissipation. In this last one the Extended Thermodynamics approach is exploited in analogy with the classical case presented in the previous section.

3.1 Example of Non Dissipative Fluid

In this well known case we have the usual 5 fields and the conservation laws of particle number and energy momentum:

$$\partial_\alpha V^\alpha = 0, \qquad \partial_\alpha T^{\alpha\beta} = 0.$$

Now the entropy four-vector $-h^\alpha$ and the dual one are given by [6]

$$h^\alpha = -nSu^\alpha; \quad h'^\alpha = \frac{p}{T}u^\alpha$$

(n, S, u^α, p, T have the meaning of number of particles, density of entropy in the rest frame, four-velocity, pressure and absolute temperature). Then using the results of theorem 3.1 we obtain

$$\overline{\xi}_\alpha = \frac{u_\alpha}{c^2}; \quad \overline{h} = -nS,$$

i.e. the privileged congruence coincides (except for the normalisation factor) with the four-velocity and the privileged entropy with the usual one.

Ruggeri and Strumia [6] have founded, exploiting the convexity condition with respect to the privileged time-like direction, the well known stability conditions:

$$c_p > 0, \qquad \left(\frac{\partial p}{\partial \rho}\right)_S > 0.$$

Therefore, we have the guarantee that our system is symmetric with respect to the fluid flow direction. But if one adds the condition that the characteristic velocities are bounded compared to c:

$$\left(\frac{\partial p}{\partial \rho}\right)_S \leq c^2$$

then, according to the theorem 3.2, the system is symmetric hyperbolic for all the other time-like congruences under both the previous conditions, i.e.:

$$c_p > 0, \qquad 0 < \left(\frac{\partial p}{\partial \rho}\right)_S \leq c^2.$$

In [6] this result was verified directly.

3.2 Relativistic Extended Thermodynamics

In the case of an ideal gas, the system of equations of Extended Thermodynamics is formed by 14 balance laws

$$\partial_\alpha V^\alpha = 0, \quad \partial_\alpha T^{\alpha\beta} = 0, \quad \partial_\alpha A^{\alpha\beta\mu} = I^{\beta\mu}$$

for 14 fields

$$n, T, u^\alpha, \pi, q^\alpha, t^{<\alpha\beta>}$$

(Liu - Müller - Ruggeri - Ann. Phys. 1986 [10]). Here new quantities appear as the non equilibrium pressure π, the heat four-vector flux q^α and the stress tensor $t^{<\alpha\beta>}$.

Now, the privileged time-like congruence is different a priori with respect to the normalised four-velocity. In fact, in [9] we find:

$$c^2\overline{\xi}^\alpha = u^\alpha + \frac{\Omega}{p\gamma} q^\alpha,$$

where Ω is expressed in terms of complicate integrals:

$$\Omega = \frac{1}{k} \frac{I_{4,1}J_{4,0} - I_{4,0}J_{4,1}}{I_{4,0}I_{4,2} - (I_{4,1})^2},$$

$$J_{m,n}(\alpha,\gamma) = \int_0^\infty \frac{\sinh^m \rho \cosh^n \rho}{\exp(\alpha/k + \gamma \cosh \rho) \mp 1} d\rho, \quad I_{m,n} = \frac{\partial}{\partial \alpha} J_{m,n}.$$

However, it is very interesting and surprising that for the non degenerate gases the following relations hold:

$$I_{m,n} = -\frac{1}{k} J_{m,n} \quad \rightarrow \quad \Omega = 0 \quad \rightarrow c^2\overline{\xi}^\alpha = u^\alpha.$$

So, the privileged congruence coincides with the usual one even at non-equilibrium. Therefore, we expect a contribution of the heat flux in the expression of $\overline{\xi}_\alpha$ only in the degenerate case, where $\Omega \neq 0$.

References

[1] **T. Ruggeri**, *Galilean Invariance and Entropy Principle for Systems of Balance Laws.* Cont. Mech. Thermodyn. **1** (1989).

[2] **S. M. Shugrin**, *Galilean Systems of Differential Equations.* Differential Equations **16** (1980) n. 12, 1402 (1981).

[3] **S. K. Godunov**, *An interesting class of quasilinear systems.* Sov. Math. **2** (1961).

[4] **K.O. Friedrichs & P.D. Lax**, *Systems of conservation equations with a convex extension.* Proc. Nat. Acad. Sci. USA., **68** (1971).

[5] **G. Boillat**, *Sur l' éxistence et la recherche d'équations de conservation supplémentaires pour les systémes hyperboliques.* C.R.Acad.Sci., Paris, **278** A (1974).

[6] **T. Ruggeri & A. Strumia**, *Main field and convex covariant density for quasi-linear hyperbolic systems. Relativistic fluid dynamics.* Ann. Inst. H.Poincaré, **34** (1981).

[7] **I. Müller & T. Ruggeri**, *Extended Thermodynamics* Springer- Verlag - Tracts on Natural Philosophy. In press 1992.

[8] **T. Ruggeri, A.Muracchini, L. Seccia**, *Shock Waves and Second Sound in a Rigid Heat Conductor: A Critical Temperature for NaF and Bi.* Phys. Rev. Lett. **64** (No. 22), 2640 (1990). *Continuum Approach to Phonon Gas and Shape Changes of Second Sound via Shock Waves Theory.* Submitted to Physical Review B.

[9] **T. Ruggeri**,, *Convexity and Symmetrization in Relativistic Theories. Privileged time-like Congruence and Entropy.* Continuum Mech. Thermodyn. **2**, 163-177 (1990).

[10] **I-Shih Liu, I. Müller, T. Ruggeri**, *Relativistic Thermodynamics of Gases.* Annals of Physics, **169** (1986).

Mathematical Physics Studies

Publications:

1. F.A.E. Pirani, D.C. Robinson and W.F. Shadwick: *Local Jet Bundle Formulation of Bäcklund Transformations.* 1979 ISBN 90-277-1036-8
2. W.O. Amrein: *Non-Relativistic Quantum Dynamics.* 1981 ISBN 90-277-1324-3
3. M. Cahen, M. de Wilde, L. Lemaire and L. Vanhecke (eds.): *Differential Geometry and Mathematical Physics.* Lectures given at the Meetings of the Belgian Contact Group on Differential Geometry held at Liège, May 2–3, 1980 and at Leuven, February 6–8, 1981. 1983 ISBN 90-277-1508-4 (pb)
4. A.O. Barut (ed.): *Quantum Theory, Groups, Fields and Particles.* 1983 ISBN 90-277-1552-1
5. G. Lindblad: *Non-Equilibrium Entropy and Irreversibility.* 1983 ISBN 90-277-1640-4
6. S. Sternberg (ed.): *Differential Geometric Methods in Mathematical Physics.* 1984 ISBN 90-277-1781-8
7. J.P. Jurzak: *Unbounded Non-Commutative Integration.* 1985 ISBN 90-277-1815-6
8. C. Fronsdal (ed.): *Essays on Supersymmetry.* 1986 ISBN 90-277-2207-2
9. V.N. Popov and V.S. Yarunin: *Collective Effects in Quantum Statistics of Radiation and Matter.* 1988 ISBN 90-277-2735-X
10. M. Cahen and M. Flato (eds.): *Quantum Theories and Geometry.* 1988 ISBN 90-277-2803-8
11. Bernard Prum and Jean Claude Fort: *Processes on a Lattice and Gibbs Measures.* 1991 ISBN 0-7923-1069-1
12. A. Boutet de Monvel, Petre Dita, Gheorghe Nenciu and Radu Purice (eds.): *Recent Developments in Quantum Mechanics.* 1991 ISBN 0-7923-1148-5
13. R. Gielerak, J. Lukierski and Z. Popowicz (eds.): *Groups and Related Topics.* Proceedings of the First Max Born Symposium. 1992 ISBN 0-7923-1924-9
14. *To be published*
15. M. Flato, R. Kerner and A. Lichnerowicz (eds.): *Physics on Manifolds.* 1993 ISBN 0-7923-2500-1

Kluwer Academic Publishers – Dordrecht / Boston / London

The manufacturer's authorised representative in the EU is Springer Nature Customer Service Centre GmbH, Europaplatz 3, 69115 Heidelberg, Germany. If you have any concerns regarding our products, please contact ProductSafety@springernature.com

Printed and bound by CPI Group (UK) Ltd, Croydon, CR0 4YY

27/07/2026

02177365-0003